THE NEW WORLD OF QUANTUM CHEMISTRY

ACADÉMIE INTERNATIONALE
DES SCIENCES MOLÉCULAIRES QUANTIQUES

INTERNATIONAL ACADEMY
OF QUANTUM MOLECULAR SCIENCE

THE NEW WORLD OF QUANTUM CHEMISTRY

PROCEEDINGS OF THE SECOND INTERNATIONAL CONGRESS
OF QUANTUM CHEMISTRY HELD AT NEW ORLEANS, U.S.A.,
APRIL 19–24, 1976

Edited by

BERNARD PULLMAN
Institut de Biologie Physico-Chimique, Fondation Edmond de Rothschild, Paris

and

ROBERT PARR
Chemistry Department, University of North Carolina, Chapel Hill, U.S.A.

D. REIDEL PUBLISHING COMPANY
DORDRECHT-HOLLAND / BOSTON-U.S.A.

Library of Congress Cataloging in Publication Data

International Congress of Quantum Chemistry, 2d, New
Orleans, 1976.
The new world of quantum chemistry.

At head of title: International Academy of Quantum
Molecular Science.
Includes bibliographies.
1. Quantum chemistry—Congresses. I. Pullman,
Bernard, 1919– II. Parr, Robert G., 1921–
III. International Academy of Quantum Molecular Science.
IV. Title.
QD462.AlI57 1976 541'.28 76–43051
ISBN 90–277–0762–6

Published by D. Reidel Publishing Company,
P.O. Box 17, Dordrecht, Holland

Sold and distributed in the U.S.A., Canada and Mexico
by D. Reidel Publishing Company, Inc.
Lincoln Building, 160 Old Derby Street, Hingham,
Mass. 02043, U.S.A.

Printed in The Netherlands

TABLE OF CONTENTS

OPENING REMARKS

Professor Bernard PULLMAN

President of the International Academy
of Quantum Molecular Science

Ladies and Gentlemen, my dear colleagues,

It is a great honor and pleasure for me to inaugurate officially the 2nd International Congress of Quantum Chemistry. Like the 1st one which was held 3 years ago in Menton, France, this Congress is organized under the auspices of the International Academy of Quantum Molecular Science. It translates into action one of the major aims of the Academy which is to stimulate contacts, collaboration and friendship between quantum-molecular scientists at the International level, whatever be their fields of interest. At this particular period when the scientific exchanges become rather difficult, because of the strong increase in the costs of transportation and an unfortunate concommitant decrease in the funds available from public sources for scientific research in general, this gathering of quantum molecular scientists of widely distributed interests seems an especially wellcome venture. The development of the applications of quantum-mechanical concepts, theories and methodologies to broader and new fields of exploration is a continuous and striking phenomenon. Considered originally as limited essentially to the field of chemistry, the quantum theories are now flowing freely through biology, pharmacology, solid state physics etc. From studies of molecules in free space they evoluate more and more towards the study of molecules in their environments. The programm of the present meeting testifies to this extension and broadening. We do not doubt that the encounters which will occur here, the exchanges of ideas and informations will stimulate new developments and new collaborations.

It is my pleasant duty to thank my colleagues of the Scientific Organizing Commitee of the Academy for their efforts and collaboration in the establishment of the scientific program. Our next thanks are due to the Executive Commitee of the Congress and in particular to its chairman, Prof. Robert Parr, for the extremely efficient action which they have conducted for the sake of this organization and to the success of which many of us owe the possibility of being here. I believe that it may be more judicious to reserve for the closing session of the Congress the expression of our appreciation to the Local Commitee and especially to its chairman, Prof. Politzer, for what looks like a most entertaining and rich program of social and cultural events.

Because of the material difficulties that I mentioned earlier, we are less numerous here than we expected to be. Those of us who are present feel therefore particularly priviledged. We hope that this situation will have the possible advantage of enabling deeper discussions and closer associations. I wish you all a profitable and pleasant stay.

Institut de Biologie Physico-Chimique
Laboratoire de Biochimie Théorique
associé au C.N.R.S.
13, rue P. et M. Curie - 75005 Paris -

ACKNOWLEDGMENTS

Generous grants in support of this Congress were received from the National Science Foundation, the Petroleum Research Fund of the American Chemical Society, and the Energy Research and Development Administration. The United States Air Force kindly arranged transportation for two invited speakers from Europe, through its Window on Science Program. Thanks also are due to several of the invited speakers for arranging their travel expenses from other sources, and to the Hewlett-Packard Corporation for sponsoring our first coffee break.

I thank my fellow Executive Committee members, Professors John Pople, Klaus Ruedenberg and Peter Politzer, for much help in preparation for the Congress. Professor Politzer's contribution was enormous, for making the wonderful local arrangements. In this he was assisted by Professors Basil Anex and Robert Flurry, to whom we also are grateful.

The Congress Secretary, Mrs. Shirley Ritter, labored hard toward the success of the Congress, for over a year. A very special thanks to her.

ROBERT G. PARR

Chairman of the Executive Committee

SYMPOSIUM I. FOUNDATIONS OF QUANTUM CHEMISTRY

Chairman : R. Ruedenberg

Department of Chemistry,
Iowa State University
Ames, Iowa, U.S.A.

PRESENT STATUS OF THE CORRELATION PROBLEM

R. McWeeny

Department of Chemistry, The University,
Sheffield, U.K.

1. Introduction

Almost exactly 10 years ago, at the Slater Symposium on Sanibel Island, I gave a review lecture [1] on "The Nature of Electron Correlation in Molecules". Since that time the output of papers and review articles on the so-called "correlation problem" has continued undiminished, but new perspectives have emerged and work in this area has, to some extent, been overshadowed by advances in other fields. To quote from a recent review [2] "Nowadays about 90% of the quantum-mechanical calculations on molecules are performed by the self-consistent-field (SCF) method, using more or less extended sets of basis functions, without any consideration of the possible effects of correlation". There are, no doubt, two main reasons for this trend: (i) the ready availability of efficient SCF programmes, and (ii) the totally unexpected success of the resultant Hartree-Fock (or near-Hartree-Fock) wavefunctions as a means of predicting molecular properties at a quantitative level. Nevertheless, even if it might now be considered a minority interest, important work on electron correlation is in progress and until it has reached fruition, probably within the next decade, "chemical accuracy" for systems of moderate size will not be achieved. It is therefore again opportune to attempt a reappraisal and an exposition of current trends.

2. What is correlation?

Even the concept of electron correlation presents difficulties: it may be approached in terms of the energy difference between an independent-particle model (in which, by definition, correlation

B. Pullman and R. Parr (eds.), The New World of Quantum Chemistry, 3–31.

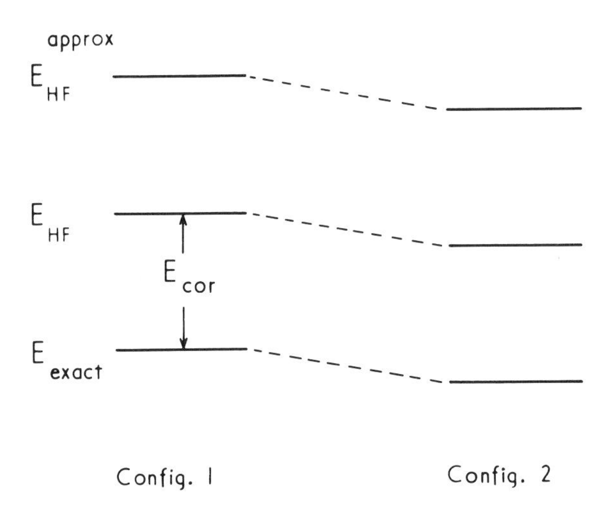

Fig. 1. Correlation energies

$E_{cor} = E_{exact} - E_{HF}$ may be much smaller than $E_{exact} - E_{HF}^{approx}$, but is usually much larger than the experimental energy difference between two molecular conformations (e.g. a barrier height).

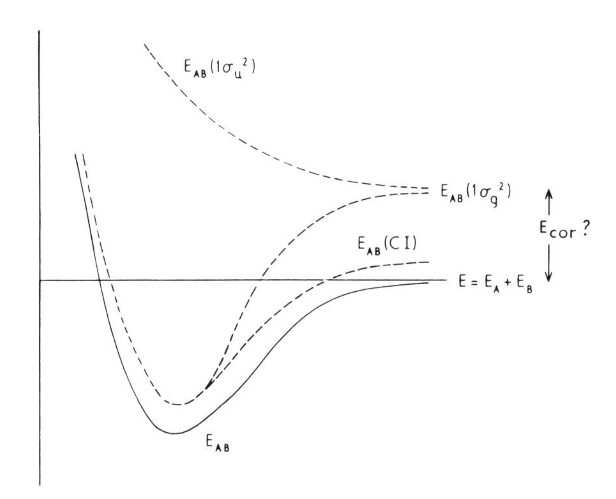

Fig. 2. Behaviour of E_{HF} in bond-breaking situation

Single-configuration functions for $1\sigma_g{}^2$ and $1\sigma_u{}^2$ give wrong asymptotic behaviour. CI leads to a more realistic result.

is absent) and the actual system; or it may be approached "dynamically" in terms of the probability of finding two electrons simultaneously at two nearby points in space - a quantity which will be diminished by their mutual repulsion (i.e. their lack of independence). Let us examine each approach in turn:

(i) Energy definition

As long ago as 1934, Wigner proposed as a measure of "correlation" the correlation energy defined as

$$E_{cor} = E_{Exact} - E_{IPM} \tag{2.1}$$

where E_{IPM} was the energy of the Hartree-Fock model. E_{Exact} is normally accepted as "exact for the system described by a non-relativistic, fixed-nucleus Hamiltonian"(a limitation we shall assume throughout).

For a molecule in its equilibrium geometry, with a non-degenerate ground state, well-described by a single Hartree-Fock (HF) determinant of doubly occupied orbitals, this may seem a satisfactory numerical measure of correlation. But, except for very small systems, lack of completeness of the one-electron basis implies that we can obtain only E_{IPM}^{approx}; and the energy difference then includes a component which could be eliminated simply by improving the orbitals (and hence the independent-particle contributions to the energy). The situation is depicted in Fig. 1, which also emphasizes the fortuitious nature of successful predictions (e.g. of barrier heights) made with neglect of correlation energy. Even with a complete basis, giving exact HF functions, success depends upon a rather accurate constancy of the correlation energy; while with approximate HF functions it would depend also on "equal accuracy" of the basis set limits achieved in different molecular calculations. There is a fair amount of "experimental" evidence for approximate constancy of correlation and basis set errors in many situations; but empirical rules are notoriously unreliable.

For a molecule in non-equilibrium geometry, more particularly in a bond-breaking geometry, such rules break down completely - as indeed does the basis of the definition (2.1). To see what happens we need only consider a homonuclear diatomic with two valence electrons in a bonding MO; this is an IPM description and the MO, being an eigenfunction of a 1-electron model Hamiltonian (with full molecular symmetry), normally turns out to be of σ_g symmetry, extending equally over both centres. It the molecule is now stretched, the σ_g MO and its antibonding (σ_u) partner come closer together in energy; the σ_g^2 determinant then gives a very poor energy (Fig. 2), whose correction requires the admission of a heavy σ_u^2 contamination. Is E_{Cor} at large distances really a correlation error? How can it be, when the asymptotic situation corresponds to two remote uncorrelated valence electrons (one on each atom)? Clearly, the σ_g^2 energy is wrong (and by an amount much larger than is commonly ascribed to

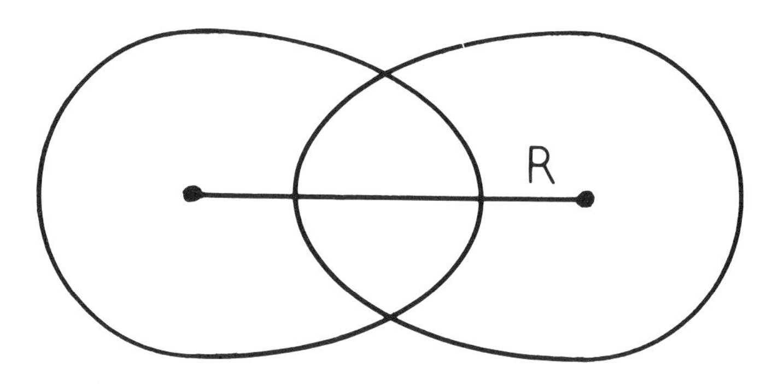

Fig. 3. Localized orbitals in H_2

The MO + CI function is equivalent to a 1-configuration VB function with strongly overlapping localized orbitals, giving correct atomic states (no correlation) on dissociation ($R \to \infty$).

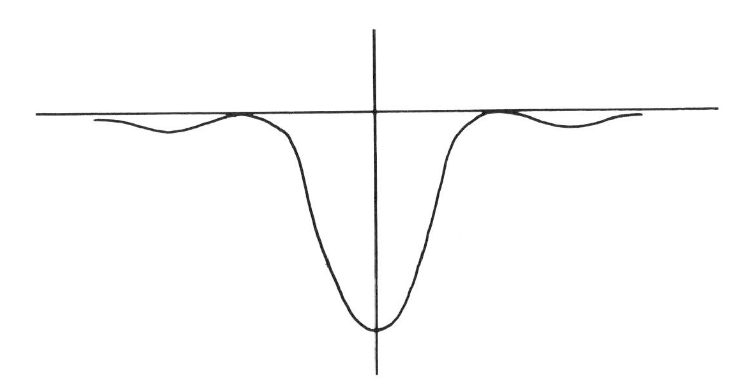

Fig. 4. Form of the Fermi hole for a free-electron gas

correlation effects) because the wavefunction is not sufficiently flexible to describe the asymptotic situation. Long before the bond is broken, the single antisymmetrised product of IPM orbitals is inadequate and CI is essential; the IPM reference state in the definition (1) is therefore unclear when near-degeneracies arise. As Coulson and Fischer first showed [3] the mixture of determinants is equivalent to a one-configuration function provided the σ_g and σ_u MO's are replaced by localized non-orthogonal combinations (Fig. 3), giving a wavefunction of valence-bond type; but these orbitals are no longer eigenfunctions of a single model Hamiltonian; nor do they belong to irreducible representations of the molecular symmetry groups - we have an illustration of the "symmetry dilemma" [4] and of the "instability" [5] of the one-determinant (restricted) Hartree-Fock function above a critical internuclear distance. An alternative procedure is that of unrestricted Hartree-Fock (UHF) theory [6] in which the IPM counts α- and β-spin electrons as distinct, yielding localized orbitals and a correct dissociation limit; but this involves a further symmetry violation insofar as the wavefunction no longer describes a true spectroscopic state but rather a mixture of singlet, triplet, ••• (or doublet, quartet, •••) components. What is clear is that the most suitable IPM reference level in (2.1), from which to measure a correlation energy, is not easy to define. Perhaps we should admit, in such situations, a <u>multi</u>-determinant "root" or reference function allowing what has been called "first order CI"; but then the definition of correlation energy becomes somewhat blurred.

(ii) <u>Dynamical definition</u>

We next recall the definition of correlation in terms of the probability of approach of two particles. Electrons which repel each other are <u>less likely to be found simultaneously close together than in the absence of repulsion</u>; this intuitive concept can be given quantitative form in terms of the "correlation hole" [1,7,8]. If $P_1(r_1)$ is the probability/unit volume of finding an electron at point r_1 and $P_2(r_1,r_2)$ is that of finding any two electrons simultaneously at points r_1 and r_2, then

$$\frac{P_2(r_1,r_2)}{P_1(r_2)} = \text{probability/unit volume of an electron at } r_1 \text{ when one is known to be at } r_2$$

For close approach, this <u>conditional</u> probability is expected to be less than $P_1(r_1)$, when the particles repel, and we use the function

$$F_{r_2}(r_1) = \frac{P_2(r_1,r_2)}{P_1(r_2)} - P_1(r_1) \qquad (2.2)$$

to describe the <u>correlation hole</u> (negative for small values of r_{12}) surrounding a reference electron at point r_2. Another way of putting this is to say that in the absence of correlation we should have $P_2(r_1,r_2) = P_1(r_1) \times P_1(r_2)$ (independent events); in which case F_{r_2} would vanish everywhere. On introducing a correlation factor, through the equation

$$P_2(r_1,r_2) = P_1(r_1)P_1(r_2)[1 + f(r_1,r_2)], \quad (2.3)$$

the hole function takes the form

$$F_{r_2}(r_1) = P_1(r_1)f(r_1,r_2) \quad (2.4)$$

Similar functions may be defined for different spin configurations (e.g. both "up", "up" at r_1"down" at r_2, etc.). For a one-determinant wavefunction the hole vanishes for unlike spins, but there is a "Fermi hole" (first calculated for a free-electron gas by Wigner and Seitz [7] which arises from antisymmetry and has the form shown in Fig. 4. The dynamical effect of coulomb repulsion, however, applies irrespective of spin and is thus completely ignored at the Hartree-Fock level. The Fermi hole has been discussed extensively by Slater [9] and is at the basis of the "X_α-method [10]; and some actual calculations have been made for atoms [11,12]. The "coulomb hole" has received much less attention (since meaningful results can be obtained only from very extensive CI calculations) but a few preliminary studies have been made [13,14]. The correlation factor $f(r_1,r_2)$ is illustrated for a small molecule in Figs. 5 and 6, for various positions of the reference electron (r_2). The hole is apparently not well localised and its behaviour near $r_{12} = 0$ does not show the expected sharp cusp due to the coulomb singularity. Such holes may be calculated at any level of approximation; the hole for H_2 in Heitler-London approximation is indicated in Fig. 7 for the reference electron (at r_2) near the left hand nucleus. In this example some correlation is evidently admitted, as the second electron (at r_1) is pushed towards the other nucleus, but the detailed form of the hole is an artefact of the calculation, the cusp being "tied" to the nuclei by the nature of the orbital basis.

The difficulty of allowing for correlation effects should now be apparent. Limited CI provides a means of recognising "large scale" or "long range" correlation effects and, as a consequence, of ensuring that the electrons correctly separate on pulling the molecule apart into fragments. But only <u>infinite order</u> CI (with admission of continuum functions or their equivalents) could build up the correct cusp associated with <u>short</u>-range correlation; the situation is analogous to that of expanding a delta function in terms of its Fourier components. The cusp behaviour is important, owing to the singularity in the repulsion energy, and it is for this reason that wavefunctions containing r_{12} explicitly score such impressive successes [16-20].

Before turning to methods of calculation we look briefly at the cusp conditions. The explicit appearance of interelectronic variables is suggested by development of the exact (spin free) wavefunctions around the point $R = \frac{1}{2}(r_1 + r_2)$ - the Schrödinger

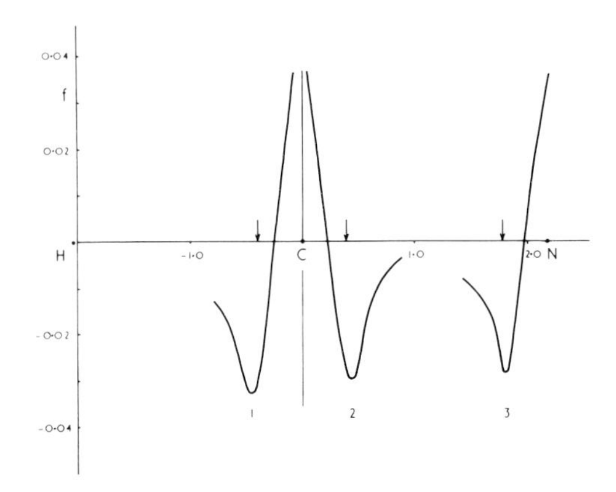

Fig. 5. Coulomb hole in HCN

The three curves indicate the hole around a reference electron in each of the three arrowed positions; the function plotted is $f(r_1,r_2)$ for fixed r_2. (Taken from Ref. [14]).

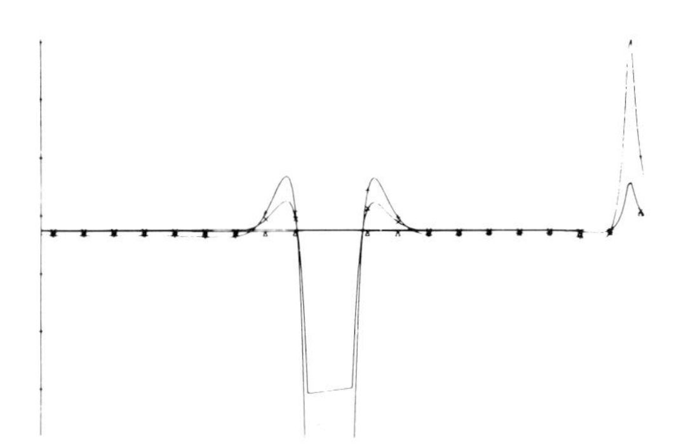

Fig. 6. Coulomb hole in HCN

The hole is indicated by a "difference plot" (+ points) of $P_2 - P_2^{HF}$, the reference electron being near the carbon (cf. Fig. 5). The Δ and x points refer to like and unlike spin components. (From Ref. [14]).

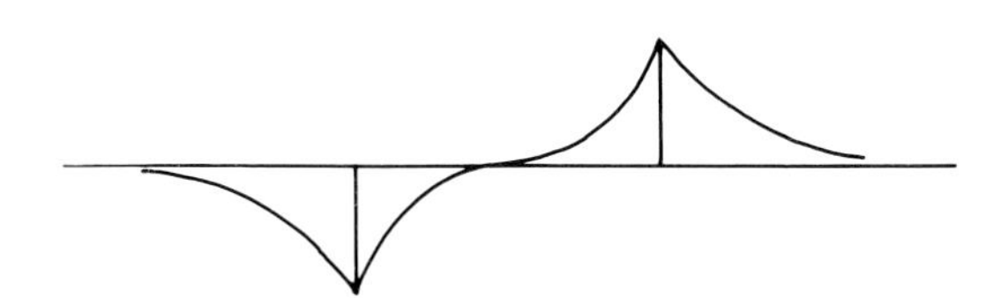

Fig. 7. Form of the coulomb hole in H_2 (Heitler-London approximation)

The reference electron is near the left-hand centre.

equation is then satisfied, for $r_{12}\to 0$ by [21]

$$\Phi(R,R,r_3,\ .\ .\ .r_N)[1+\tfrac{1}{2}r_{12}] + r_{12}.c(R,r_3,\ .\ .\ .r_N) + .. \quad (2.5)$$

where c is a vector function of the coordinates $R,r_3,\ .\ .\ .r_N$, vanishing as the electron density approaches uniformity. This form of the cusp in the wavefunction was known to Hartree who, with Ingman [20], explored a variety of r_{12}-type correlation factors in calculations on the helium atom. Their conclusion was that the improvement in energy was hardly dependent on the behaviour of the correlation functions for larger r_{12} values so long as they gave the correct form $1 + \tfrac{1}{2}r_{12}\ .\ .\ .$, for $r_{12}\to 0$. In other words, to get really good energies the cusp must be correct but longer-range behaviour is unimportant; there is thus an "insensitivity of the correlation energy" which does not seem to have been widely recognised. In molecules, where the electron density is more extended, the long-range correlation (adequately computed by CI methods) is clearly more significant, but the short-range effect is still important - and extremely difficult to handle. We must now turn to practical methods: these are of essentially two types - those which depend entirely on CI in one form or another, and those which introduce r_{12} explicitly.

3. The cluster development

To obtain a unified approach to current methods of calculation it is useful to start from a "cluster development" (see, for example, Refs. [21-24]) of the exact wavefunction. We start from a leading determinant Φ_0 (e.g. a Hartree-Fock function) built from occupied spin-orbitals

$$\{\psi_i\} = \{\psi_1,\psi_2,\ .\ .\ .\ \psi_i,\ .\ .\ .\ \psi_N\}$$

and consider an infinite CI expansion based on a complementary set of "virtual orbitals"* (all orbitals assumed orthonormal)

$$\{\psi_m\} = \{\psi_{N+1},\psi_{N+2},\ .\ .\ .\ \psi_m,\ .\ .\ .\}$$

The simplest form of the development is just

$$\Psi = C_0\Phi_0 + \sum_{i,m} C_i^m \Phi_i^m + \sum_{\substack{i<j\\ m<n}} C_{ij}^{mn}\Phi_{ij}^{mn} + \ .\ .\ . \quad (3.1)$$

where i,j,k,ℓ refer to occupied orbitals, m,n,p,q to virtual, and the terms are in the order single "excitations" $(i\to m)$, double

---

* In what follows we shall loosely use the term "orbitals" even when referring to spin-orbitals. By "occupied" we shall mean occupied in the leading determinant Φ_0.

"excitations" ($ij \rightarrow mn$), etc. If we use Ω_0 for the basic Hartree product, and A for the N-electron antisymmetriser ($= N!^{-1} \sum (-1)^p P$), we may recast (3.1) as (using x_i for space-spin variables)

$$\Psi = \sqrt{N!}\,\mathbf{A}\{\phi_0\Omega_0 + \sum_i \phi_i(x_i)\Omega_i(x_1,..x_N) + 1/\sqrt{2}\sum_{i<j}\phi_{ij}(x_i,x_j)\Omega_{ij}(x_1,..x_N) + ..\} \quad (3.2)$$

where Ω_{ij}, for instance, is the product Ω_0 with the ψ_i and ψ_j factors deleted, while

$$\phi_0 = C_0,$$

$$\phi_i(x_i) = \sum_m C_i^m \psi_m(x_i), \quad (3.3)$$

$$\phi_{ij}(x_i,x_j) = \sum_{m<n} C_{ij}^{mn}(1/\sqrt{2})\det|\psi_m(x_i)\psi_n(x_j)|, \text{ etc.}$$

The functions ϕ_i, ϕ_{ij}, ϕ_{ijk}, . . . are 1-cluster, 2-cluster, 3-cluster, . . . functions for increasing numbers of electrons, while Ω_i, Ω_{ij}, . . . are the residual orbital products with factors ψ_i, $\psi_i\psi_j$, . . . excluded. In particular, ϕ_i may be viewed as a correction to the function ψ_i, and ϕ_{ij} as a correction to the 2-electron product $\psi_i\psi_j$ in Φ_0; since ϕ_{ij} is a general function it is capable of describing a "correlated pair".

It is easy to see, however, that 2-clusters may arise in the expansion of a function which does not describe correlated particles. If we take a single determinant, Φ_0, and then let $\psi_i \rightarrow \psi_i + \phi_i$ ($i = 1, ...N$) we shall get a cluster expansion of the new determinant (Ψ, say) of the form (3.2) but with

$$\phi_{ij}(x_i,x_j) = (1/\sqrt{2})\det|\phi_i(x_i)\phi_j(x_j)|$$

i.e. an antisymmetrized product of two 1-cluster functions. This 2-cluster is said to be an "unlinked product" of 1-clusters; in terms of virtual orbitals, the 2-cluster coefficients in (3.3) would be expressible as products of 1-cluster coefficients,

$$C_{ij}^{mn} = C_i^m C_j^n$$

and no specifically 2-body effects would be admitted. Similarly, the expansion of a pair-correlated product would contain 4-clusters; but these would be unlinked products of the 2-clusters and would not introduce real 4-body correlations. We are thus obtaining an interpretation of correlation at the "operational" level of a particular type of wavefunction expansion. In nearly all current work it is assumed that "linked" or "irreducible" clusters for more than two particles can be neglected. This means, for example, that although 4-cluster functions may be important they are for the most part describing antisymmetrized products of 2-cluster functions which may both be significant at a point in configuration space where $x_i \to x_j$ and $x_k \to x_\ell$ (i.e. where correlation is important within each of two separated pairs of electrons - but not between the two pairs).

The systematic separation of linked and unlinked clusters is most neatly obtained by second-quantization methods (e.g. [24]). Briefly, if $|0\rangle$ is the reference ket in Fock space corresponding to determinant Φ_0, the vector corresponding to $\sum_{i,m} C_i^m \ {}_i^m$ would be generated by applying

$$v_i = \sum_m C_i^m a_m^\dagger a_i \tag{3.4a}$$

and the 2-cluster term applying

$$v_{ij} = \sum_{m<n} C_{ij}^{mn} a_m^\dagger a_n^\dagger a_j a_i \tag{3.4b}$$

If we define new coefficients and operators, distinguished by a roof $\hat{}$, we can obtain an alternative expansion in the form

$$|\Psi\rangle = e^T|0\rangle \tag{3.5a}$$

where
$$T = \sum_i \hat{v}_i + \sum_{i<j} \hat{v}_{ij} + \ldots = T_1 + T_2 + \ldots \tag{3.5b}$$

The terms in the expansion of the exponential operator then represent the clusters of each order in terms of irreducible and reducible parts. Thus we find

$$e^T = 1 + \sum_i \hat{v}_i + \sum_{i<j} (\hat{v}_{ij} + \hat{v}_i\hat{v}_j) + \sum_{i<j<k} (\hat{v}_{ijk} + \hat{v}_i\hat{v}_{jk} + \hat{v}_j\hat{v}_{ik} + \hat{v}_k\hat{v}_{ij} + \hat{v}_i\hat{v}_j\hat{v}_k) + \ldots \tag{3.6}$$

where, for example, $\hat{v}_{ij}$ produces an irreducible 2-cluster (representing a real correlation) while $\hat{v}_i\hat{v}_j$ gives simply a product of 1-cluster functions (i.e. of orbital corrections). In principle, by correctly choosing the orbital basis, the 1-cluster functions may be eliminated: the corresponding orbitals, which maximize the overlap $\langle 0|\Psi\rangle$, are usually called the Brueckner orbitals [24]. If we assume Brueckner orbitals*, and neglect irreducible clusters for three or more particles, the corresponding cluster development is

$$\Psi = \sqrt{N!}\, A\Big\{\phi_0\Omega_0 + \sum_{(ij)} 2^{-\frac{1}{2}}\phi_{ij}\,\Omega_{ij} + \sum_{(ij)(k\ell)} 2^{-\frac{1}{2}}\phi_{ij}\, 2^{-\frac{1}{2}}\phi_{k\ell}\,\Omega_{ijk\ell} + \ldots\Big\} \tag{3.7}$$

where (ij) $(k\ell)$ are pairs with no common index.

4. Correlated pair theories

There is a very extensive literature concerned with the

* For closed-shell systems, the Hartree-Fock orbitals are fair approximations.

variational implementation of (3.7); the original developments were associated largely with the names of Szass [25], Sinanoglu [26], Nesbet [27], Kutzelnigg and co-workers [28], and many important contributions have been made by others. Here we shall simply comment on the various approaches and levels of approximation, without entering into technical details.

First we note that (3.7) corresponds to use of the exponential (3.4) in the truncated form

$$e^{T} \simeq \exp\left(\sum_{i} \hat{v}_{i} + \sum_{(ij)} \hat{v}_{ij}\right) = e^{(T_1+T_2)} \tag{4.1}$$

and with neglect of the 1-cluster creators contained in T_1. We may distinguish several types of approximation:

(i) CI with all single and double excitations

A popular method of limited CI corresponds to inclusion of all single and double excitations and thus to truncation of the series expansion to

$$e^{T} \simeq 1 + \sum_{i} \hat{v}_{i} + \sum_{(ij)} \hat{v}_{ij} \tag{4.2}$$

which works on $|0\rangle$ to produce a result equivalent to the terms shown in (3.1).

(ii) Independent electron pair approximations

If we focus attention on a particular pair ij (i.e. consider one pair at a time), and choose the orbitals so as to eliminate single-excitation terms, we can take $T = \hat{v}_{ij}$ and $e^{T} \simeq 1 + T$ to produce the variation function

$$|\Psi_{ij}\rangle = |0\rangle + \sum_{(mn)} \hat{C}_{ij}^{mn} |\Phi_{ij}^{mn}\rangle \tag{4.3}$$

On optimizing the coefficients for every pair independently and then summing to obtain a resultant $\sum_{(ij)} \hat{v}_{ij}$ in (3.4) and (3.5) we obtain an IEPA or "independent-electron-pair-approximation". This approach, in which Φ_0 is usually taken to be a single Hartree-Fock determinant, was first systematically developed by Sinanoglu [26] under the name MET (many electron theory), and by Nesbet [27] who pointed out a connection with the work of Bethe and Goldstone [29,30]. The wavefunction is then taken to be of the form (3.7) with all pairs included, and the energy contributions from higher-order unlinked clusters can, in principle, be admitted if desired. The energy E_{ij} associated with the function (4.3) may be used to define a pair correlation energy $\varepsilon_{ij} = E_{ij} - E_{HF}$, and a first-order estimate of the correlation energy is then

$$E_{cor} \simeq \sum_{(ij)} \varepsilon_{ij} \tag{4.4}$$

This "decoupling" of the pairs and their correlation energy contributions is not a completely rigorous procedure; results look encouraging until one obtains more than 100% of the correlation energy! The reason, of course, is that the energy estimate is not a variational upper bound to the exact energy. Nesbet [31] has extended this type of approach to compute systematically the "net increments" in quantities such as E_{cor} due to inclusion of clusters of successively higher order.

(iii) Coupled electron pair approximations

In going beyond IEPA, one meets great technical difficulties associated with the simultaneous optimization of parameters referring to all the different pairs. These have been overcome to some extent by Meyer [32], who writes (3.7) in the form (cf.(3.1))

$$\Psi = C_0\Phi_0 + \sum_{P,(mn)} C_P^{mn}\,\Phi_P^{mn} + \sum_M C_M\Phi_M \tag{4.5}$$

where P is a given pair and M is a "multiple pair excitation", and then sets up a system of secular equations to determine the C_P^{mn} simultaneously for all pairs. The resultant coupled electron pair approximation (CEPA) has been used with great success in obtaining highly accurate wavefunctions for small molecules [33]; but the computational cost is heavy.

(iv) Geminal product approximations

When the electron pairs can in good approximation be strongly localized into spatially "separate" regions (e.g. electron-pair bonds, lone pairs, etc. in simple saturated molecules), an antisymmetrised product of the resultant "geminals" provides a particular type of CEPA. The corresponding exponential operator is

$$e^T \simeq e^{(T_A+T_B+\ldots)} = e^{T_A}\,e^{T_B}\;\ldots \tag{4.6}$$

where
$$T_R = \sum_i \hat{v}_{i_R} + \sum_{(i_Rj_R)} \hat{v}_{i_Rj_R} \tag{4.7}$$

and i_R, j_R label spin-orbitals "belonging to" pair R. In order that the geminals, A and B produced by e^{T_A} and e^{T_B} should remain strongly-orthogonal it is necessary that the virtual spin-orbitals m, n, ... should also be divided into exclusive sets (m_A, n_A, ...) (m_B, n_B, ...); and this constraint is the main weakness of the method. Such functions [34] may be generalized [35] to include quasi-independent groups containing any numbers of electrons; the correlation effects <u>within</u> each group may be admitted to infinite order, but <u>inter</u>-group correlation is neglected. The groups are also <u>coupled</u> in the sense that each group function refers to the electrons of that group in the field of all other groups. There is some hope that such procedures, which involve iteration to "self-consistency" will be extended to the CEPA method, in order to further optimize the individual pair functions. It was pointed out by McWeeny and Steiner [36] that exactly the same "group function" formalism applies, irrespective of whether the groups (in this case the pairs and the remaining "sea") are localized or non-localized.

5. Diagrammatic techniques

The explicit determination of a wavefunction requires the calculation of the coefficients in (3.3), which appear in the exponential operator (3.4) or in some truncated form of it. Unfortunately, the number of coefficients is normally much too large for the application of variation techniques, even though impressive progress has been made in the solution of high order secular equations [37], and for this reason there has been a growing interest in perturbation methods; the introduction of diagrammatic analysis is simply a powerful way of handling and summing various types of terms in a perturbation expansion. A detailed exposition would be out of place in this review, and many good reviews already exist but, in view of their growing importance and relative unfamiliarity to many quantum chemists, it seems worth recalling the main ideas. A derivation similar to that presented here has been given by Cizek and Paldus (1975) (see Ref. [40]).

It is usual to introduce a "model" Hamiltonian, H_0, of which the leading term, Φ_0 is an exact* eigen-function. Thus H_0 is a sum of one-electron operators incorporating some kind of effective field of Hartree-Fock type. We then write

$$H = H_0 + \lambda h \qquad (\lambda \to 1) \tag{5.1}$$

and require the solution of

$$H|\Psi\rangle = (H_0 + \lambda h)|\Psi\rangle = E|\Psi\rangle \tag{5.2}$$

Instead of using the exponential form (3.5) we may write

$$|\Psi\rangle = W(\lambda)|0\rangle \tag{5.3}$$

where the "wave operator" $W(\lambda)$ may be developed as

$$W(\lambda) = W_0 + \lambda W_1 + \lambda^2 W_2 + \ldots \tag{5.4}$$

The most convenient normalization is $W_0 = 1$, and since $W_n|0\rangle$ may be assumed orthogonal to $|0\rangle$ this corresponds to

$$\langle\Psi|0\rangle = 1 \tag{5.5}$$

---

* In practice, "exact" only for a projected Hamiltonian defined within the subspace determined by a given choice of basis set.

Substitution of (5.4), and a similar energy expansion, into (5.2) leads to the nth order equation

$$(E_0 - H_0)W_n|0> = h\, W_{n-1}|0> - \sum_{j=1}^{n} E_j\, W_{n-j}|0> \qquad (5.5)$$

which is a recurrence relation for generating the Rayleigh-Schrodinger (RS) series. The nth order energy is (multiplying by $<0|$)

$$E_n = <0|h\, W_{n-1}|0> \qquad (5.7)$$

and we need a systematic way of building up the RS series.

If (E_0-H_0) had an inverse Q we could multiply (5.6) throughout by Q to obtain

$$W_n|0> = Qh\, W_{n-1}|0> - \sum_{j=1}^{n} E_j\, Q\, W_{n-j}|0> \qquad (5.8)$$

The inverse does not exist, for the spectral resolution would contain a term $(E_0-E_n)^{-1}|\Phi_n><\Phi_n|$ which would "blow up" when working on any ket with a non-zero 0-component: but if we introduce the projection operator $P = [1 - |0><0|]$, onto the subspace complementary to $|0>$, this difficulty is avoided and we may use

$$Q = P(E_0 - H_0)^{-1} P = \frac{P}{(E_0 - H_0)} \qquad (5.9)$$

as a pseudo-inverse (i.e. an inverse in the subspace orthogonal to $|0>$). We then obtain by recursion, using (5.8),

$$W_1|0> = Qh|0>\,, \quad W_2|0> = (Qh)(Qh)|0> - E_1(Qh)|0>, \qquad (5.10)$$

etc., and observe that the general term $W_n|0>$ will require the evaluation of $(Qh)^n|0>$.

It is at this point that diagram techniques become useful; for we shall rapidly lose track of the many terms which arise from $(Qh)^n$ unless we can classify and collect them. Again, second quantization methods are useful (though not indispensable) in deriving the results. We write the two parts of the Hamiltonian as

$$H_0 = E_0 + \sum_p \varepsilon_p N(a_p^\dagger a_p)$$
$$h = \tfrac{1}{2} \sum_{p,q,r,s} <pq|g|rs>\, N(a_p^\dagger a_q^\dagger a_s a_r) \qquad (5.11)$$

where p, q, r, s, run over all one-electron states (holes and particles) and the <u>normal products</u> contain the operators in their "most destructive" order relative to the reference state $|0>$; this means that operators of type $a_i^\dagger$ (which destroy $|0>$ by trying to produce electrons in states which are already occupied), and of type a_m (which destroy $|0>$ by trying to annihilate electrons in

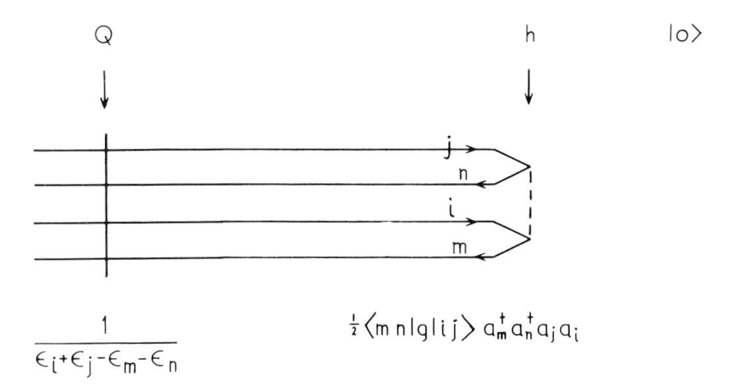

Fig. 8. Association of diagrams with operators

Reading from right to left, each term of h produces from $|0\rangle$ a "doubly excited" state indicated by the hole (i,j) and particle (m,n) lines, and multiplies by a matrix element. The operator Q gives an energy denominator corresponding to the hole and particle lines cut by the vertical line.

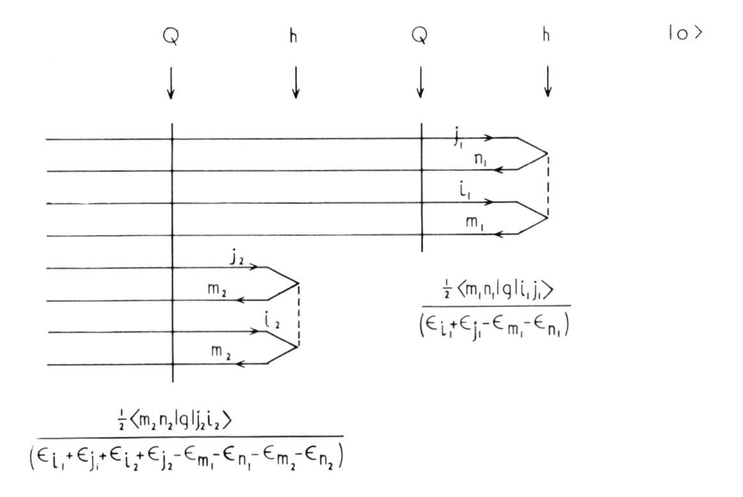

Fig. 9. Diagram for a four-fold excitation

The lines emerging on the left indicate a resultant excitation $i_1\ j_1\ i_2\ j_2 \rightarrow m_1\ n_1\ m_2\ n_2$; this multiplied by the matrix elements (from the h's) and energy denominators (from the Q's) as shown.

states already empty), must be put as near to $|0>$ as possible. The expectation value of any such product is then zero, and $E_1 = 0$.

In forming $W_1|0> = Qh|0>$, with h defined in (5.11), we note that a non-zero result only arises if r, s = i, j, ... (occupied in $|0>$) and p, q = m, n,.. (unoccupied in $|0>$); $a_j a_i$ then create holes, while $a_m^\dagger a_n^\dagger$ create particles in the virtual orbitals. Thus

$$W_1|0> = Qh|0> = \sum_{m,n,i,j} \left[\frac{P}{(E_0-H_0)}\right]\left[\tfrac{1}{2}< mn|g|ij > a_m^\dagger a_n^\dagger a_j a_i\right] |0>$$

and is a sum of "double-excitation" terms, contributing to the $\sum C_{ij}^{mn}\ \Phi_{ij}^{mn}$ in the cluster development (3.1). The factors in such an expression, coming from the Q and the h, may be indicated pictorially as in Fig. 8, which should be read starting from the $|0>$ on the extreme right. The typical matrix element in h is represented by the broken line while its ends, or "vertices", are joined by lines <u>entering</u> (i,j) and lines <u>leaving</u> (m,n); these labelled lines, called "hole" and "particle" lines, respectively, select particular creation-annihilation pairs, $a_m^\dagger a_i$ and $a_n^\dagger a_j$ (responsible for these two excitations), and thus indicate the particular matrix element involved and the particular excitation produced "at the level Q". The Q operator, at this level, operating on the doubly-excited eigenket of H_0, merely gives a factor $(\varepsilon_i + \varepsilon_j - \varepsilon_m - \varepsilon_n)^{-1}$. The "outgoing lines" on the left indicate the nature of the state produced; and summation over all such diagrams (labellings) yields the resultant $W_1|0>$.

Less trivial results arise in second order. Thus, with $Qh\ Qh|0>$ we could associate diagrams of the type shown in Fig. 9 where the two matrix elements* correspond to choice of labels i_1 j_1 m_1 n_1 and i_2 j_2 m_2 n_2, respectively, each represented by a part of the diagram similar to that in Fig. 8. The outgoing lines on the left indicate a contribution to the quadruple excitation term $C_{i_1j_1i_2j_2}^{m_1n_1m_2n_2}\ \Phi_{i_1j_1i_2j_2}^{m_1n_1m_2n_2}$ in the cluster development, and the energy denominator factors arising from the Q's at level 1 and level 2 are $(\varepsilon_{i_1} + \varepsilon_{j_1} - \varepsilon_{m_1} - \varepsilon_{n_1})^{-1}$ and $(\varepsilon_{i_1} + \varepsilon_{j_1} + \varepsilon_{i_2} + \varepsilon_{j_2} - \varepsilon_{m_1} - \varepsilon_{n_1} - \varepsilon_{m_2} - \varepsilon_{n_2})^{-1}$ respectively.

More interesting diagrams also arise from $Qh\ Qh|0>$ with a different selection of hole and particle labels. Thus, if the first and second matrix elements (right to left) are of the type shown in Fig. 10a it is clear that the a_{m_2} operator at level 2 can annihilate the particle produced at level 1 by $a_{m_1}^\dagger$ on taking $m_2 = m_1$. The full result of the normal product analysis is simply that <u>non-zero results arise only from diagrams in which similarly</u>

* The second is displaced downwards simply to "open up" the diagram and make it clearer.

labelled "ends" are joined together, any free ends coming out on the left. Thus on choosing $n_2 = n_1$, $m_2 = m_1$, and $i_2 = i_1$, another non-zero contribution to $Qh\ Qh|0 >$ is represented by the diagram in Fig. 10b. This contribution is clearly a one-cluster contribution namely $|\Phi_{j1}^{p2} >$ multiplied by

$$\frac{1}{\varepsilon_{j_1} - \varepsilon_{p_2}} \tfrac{1}{2}< p_2 i_1|g|n_1 m_1 > \frac{1}{\varepsilon_{i_1} + \varepsilon_{j_1} - \varepsilon_{m_1} - \varepsilon_{n_1}} \tfrac{1}{2}< n_1 m_1|g|j_1 i_1 > \qquad (5.12)$$

In the diagram we have omitted the Q lines (one at each level), as is customary, and have indicated a typical "loop" and an "internal hole line": the sign with which this contribution appears is determined as $(-1)^{\ell+h}$ where ℓ is the number of loops and h that of internal hole lines, giving +1 in the example shown.

The literature contains a large and bewildering variety of types of diagram, and of abbreviated forms, but the principles are essentially the same. The important thing is the way the lines are connected and many diagrams may be "topologically equivalent", leading to identical contributions. Thus if the right hand matrix element in Fig. 10b is "turned over" without breaking any lines we get the diagram in Fig. 11a, which gives the same result as that in Fig. 10b., namely (5.12). Such diagrams need not be considered if, for a diagram with n matrix elements, we supply a factor 2^n to the original result (e.g. (5.12)) - thus exactly cancelling the n factors of ½. Also "exchange" diagrams such as that in Fig. 11b, produced by exchanging the entering or the leaving lines at any matrix element, may be disregarded if $< rs|g|tu >$ is replaced by $[< rs|g|tu > - < rs|g|ut >]$. This second step is sometimes indicated by replacing the broken line - - - - in the original diagram (e.g. Fig. 10b) by a modified line such as ∿∿∿ (e.g. Fig. 11c). The two steps together are sometimes indicated by shrinking both vertices into a single • to give the so-called "Hugenholz diagrams" (Fig. 11d).

The importance of the diagrammatic methods arises firstly from the Goldstone theorem [30], which states that

(A) The nth order correction $|\Phi_n >$ arises from linked* diagrams only

(B) E_n arises from those which may be closed on the left by a single ⋮

and secondly from the way it brings into evidence the possibilities of summing to infinite order the contributions from certain types of diagram.

---

* In this usage "containing no subdiagrams, which could be removed without breaking any lines".

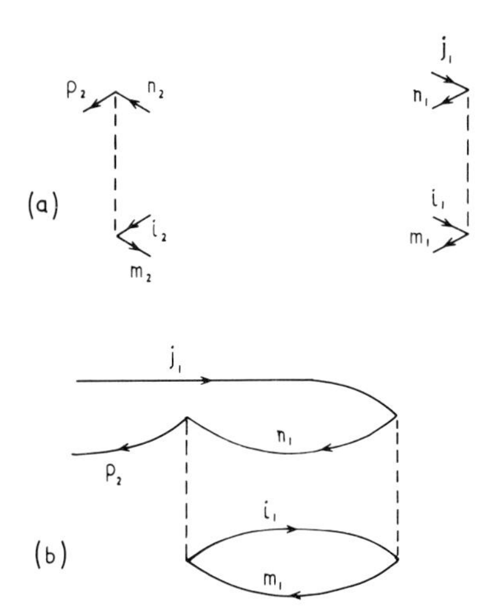

Fig. 10. An alternative second-order diagram

The two h-terms shown in (a) may be connected to give the diagram in (b).

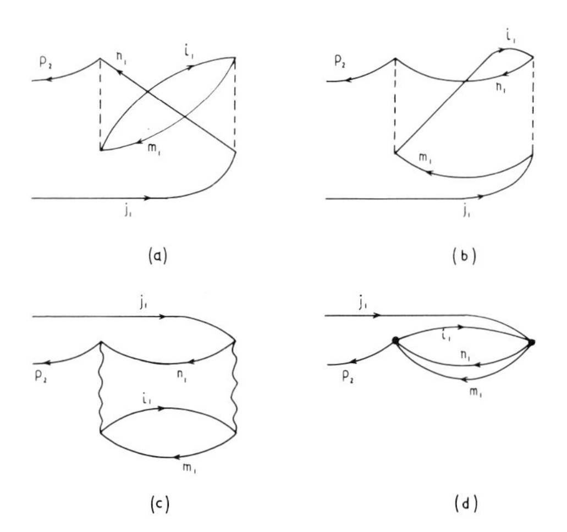

Fig. 11. Various second-order diagrams

(a) is topologically equivalent to the diagram in Fig. 10(b)

(b) is an "exchange" diagram

(c) is an abbreviation indicating inclusion of exchange

(d) is a further abbreviation (Hugenholz diagram)

The pioneering work in this field was done by Gell-Mann and Brueckner [38] for a free-electron gas, and by Kelly [39] for atoms. In both cases infinite-order summations were performed for certain classes of diagram, this being essential in the case of the free-electron gas in order even to get a correlation energy proportional (at given density) to the number of particles. By Kelly's methods it is possible to obtain well over 90% of the correlation energy for simple atoms. The extension of such work to molecules raises severe computational problems. The use of diagrams is only an aid; it does not alter the fact that we are doing a CI calculation, whose convergence is basis-set dependent and is normally rather slow; but it can suggest alternatives to the direct solution of enormous secular equations. Such an alternative was proposed by Cizek [40]; and it may be shown [41] that Cizek's equations are closely related to those of the coupled-pair many-electron theory (CPMET). Thus, on writing (3.7) in the form

$$\Psi = \exp(T_2)\Phi_0 \tag{5.13}$$

the expansion coefficients in T_2 (equations (3.4b) and (3.5b)) could be obtained by solving secular equations equivalent to

$$< \Phi_0|(H - E)|e^{T_2}\,\Phi_0 > = 0 \tag{5.14a}$$

$$< e^{T_2}\,\Phi_{ij}^{mn}|(H - E)|e^{T_2}\,\Phi_0 > = 0 \tag{5.14b}$$

The first equation gives $E-E_0$ in the form, remembering (5.5),

$$E_c = < \Phi_0|H|\sum_{i<j} T_{ij}\,\Phi_0 > \tag{5.15}$$

but to determine the coefficients in the T_{ij} it is necessary to solve (5.14b); with neglect of <u>linked</u> 4-cluster terms this can be reduced to

$$< \Phi_{ij}^{mn}|H|\Phi_0 > + \sum_{k<\ell} < \Phi_{ij}^{mn}|(H-E)|T_{k\ell}\,\Phi_0 >$$
$$+ \sum_{k<\ell} < \Phi_{ij}^{mn}|H|(T_{ij}T_{k\ell} + T_{ik}T_{j\ell} + T_{i\ell}T_{jk})\Phi_0 > = 0 \tag{5.16}$$

which is another form of Cizek's equations. An actual application [42] to the molecule BH_3 yielded $\Delta E_c = -0.048048$ Hartree, compared with a value -0.048050 obtained by direct CI with inclusion of unlinked 4-clusters; and the method is therefore very promising, though its implementation is on the borders of present-day computational techniques. When further approximations are made the equations may be reduced to more tractable form and are then closely related to those of Meyer's CEPA. These methods, along with direct CI itself (see for instance Ref. [43] for an account of recent advances in computational techniques), appear to take us to the limits of what can be done within a finite basis of orbitals.

It should be added, however, that the cluster expansion and diagrammatic methods in the form just described all start from the assumption of a one-determinant "root function". Whilst this may be a good starting point for a molecule in its equilibrium geometry, and assuming a non-degenerate ground state, it is certainly not so (cf. Section 2(i)) when the molecule is deformed and degeneracies or near-degeneracies arise. A multi-determinant root function is then essential, in order to recognise the long-range "correlation" which ensures correct dissociation. In this case, very severe complications arise and the whole diagrammatic analysis must be extended to include the so-called "folded diagrams" [44,45]. Work is in progress in this field (see, for example, Ref. [46]) but the computational implementation of such methods, which will be essential in, for example, reaction surface calculations, will probably be an order of magnitude more difficult than for the one-determinant root function.

6. Correlation factor methods

Finally, we turn to approximations in which the short range correlation does not depend on CI expansion but is dealt with analytically by introducing inter-electronic variables. Ever since the early work ot Hylleraas [15] and of James and Coolidge [16] it has been appreciated that inclusion of r_{12} in a 2-particle wave-function can lead to solutions of extraordinary accuracy; such work culminated in the helium atom calculations of Pekeris [17] and the hydrogen molecule calculations of Kolos and Wolniewicz [18]. Unfortunately, the generalization to many-electron systems raises difficulties (particularly in matrix element evaluation) which have so far proved insuperable.

The most promising approaches seek simplifications which take them outside the realms of the variation method in its orthodox form; but it must be remembered that this is also a feature of the many-body techniques (both diagrammatic perturbation theory and propagator, or Green's function, methods). We consider two such approaches.

(i) The method of Boys and Handy

Boys [47] proposed new methods of using a correlated wave-function of the form

$$\Psi = C\,\Phi_0 \tag{6.1}$$

where the factor C is a symmetric function containing the inter-electronic variables

$$C = \prod_{i>j} F(r_i, r_j) \tag{6.2}$$

while Φ_0 may be, for example, a single Slater determinant. The simplest correlation function would be $F = F(r_{12})$, a function of the interelectronic distance alone; but this would fail to describe the dependence of correlation on electron density and it is therefore desirable to use a function of both vectors r_1 and r_2 or, alternatively, of

$$R_{12} = \tfrac{1}{2}(r_1 + r_2), \quad r_{12} = r_2 - r_1 \tag{6.3}$$

Unfortunately, with the function (6.1), the variational energy expression becomes intractable (involving 6-electron integrals). Instead, therefore, Boys and Handy [48] considered the variational procedure based on

$$< \delta\Psi^{\dagger}|(H-E)|\Psi > = < \delta(C^{-1}\Phi_0)|(H-E)|\Psi > = 0 \tag{6.4}$$

where the left-hand function $\Psi^{\dagger}$ contains an inverse correlation factor. For arbitrary variations, this quantity still vanishes only when $H\Psi = E\Psi$ and the use of the "wrong" correlation factor in $\Psi^{\dagger}$ is, in that sense, immaterial. For approximate functions, on the other hand, Boys showed that the error in $< \Psi^{\dagger}|H|\Psi >$, regarded as an estimate of the energy, was determined by the product $(\mu\mu^{\dagger})$ of the errors in Ψ and $\Psi^{\dagger}$, respectively; and at the same time the functional

$$< \Psi^{\dagger}|H|\Psi > = < \Phi_0|C^{-1} H C|\Phi_0 > \tag{6.5}$$

which can be regarded as the expectation value of a (non-Hermitian) "transcorrelated Hamiltonian", is rather easier to handle than $< \Psi|H|\Psi >$. The use of suitable functional forms of $F(r_i,r_j)$ has in fact yielded, in applications to the atom Ne [49] and the molecule LiH [50], extremely accurate right-hand functions (Ψ) and consequently highly accurate estimates of the energy.

This method, whose full implementation required great computational ingenuity, is certainly worthy of further exploration. It has been considered further by Hall and Soloman [51] and by Armour [52]; but formidable difficulties stand in the way of its application to larger molecules.

(ii) Methods based on the correlation hole

There have been many attempts to "estimate" the correlation energy by non-rigorous methods, in order to "correct" the results of ab initio Hartree-Fock-type calculations. These range from a rough allowance of "so much per electron pair" (depending on its type and degree of localization), through semi-theoretical formulae like that of Wigner [53] and the formulation of "rules" based on accurate calculations [54], to more elaborate attempts to "graft on" some analytical form of correlation hole within the general framework of a Hartree-Fock calculation. This latter procedure, which is equivalent (see Ref. [1]) to replacing the interelectronic coulomb potential by a suitably screened potential (recognizing the cusp in the correlation hole for $r_{12}\rightarrow 0$), can be developed in a semi-rigorous way and seems to me to offer great promise. Above all, it is conceptually simple and can be implemented computationally with no great difficulty even for rather large molecules.

For simplicity we consider a closed shell ground state with a wavefunction of the form (6.1). We wish to know how the Hartree-Fock two-body density matrix $P_2(r_1,r_2)$ (only the diagonal element is needed) is modified by the presence of the correlation factor C. Perhaps the simplest approach is that of Colle and Salvetti [55], though a similar form can be deduced by starting from the cluster development (3.7) and using an elaboration of the arguments in Ref. [1]. Colle and Salvetti take

$$F(r_i,r_j) = 1 - \phi(r_i,r_j) \tag{6.6}$$

where $\phi(r_i,r_j)$ is of essentially short range, introducing the cusp, but is a function of both position vectors or equivalently, of the variables (6.3). In deriving $P_2(r_1,r_2)$ from (6.1) by integrating over (N-2) electronic variables, the factors $\phi(r_i,r_j)$ for $i,j \neq 1,2$ are of little consequence since over most of the integration range $r_i - r_j$ is large and $\phi(r_i,r_j)$ consequently zero. Thus* we obtain for the general (off-diagonal) element

$$P_2(r_1,r_2;r_1',r_2') = P_2^0(r_1,r_2;r_1',r_2')[1 - \phi(r_1,r_2) - \phi(r_1',r_2') + \phi(r_1,r_2)\phi(r_1',r_2')] \tag{6.7}$$

The real problem is to choose the functional form of ϕ so as to describe the variation in size and shape of the correlation hole, over the electron distribution, without violating the requirements of N-representability [56]: thus, for example, we require

$$(N-1)P_1(r_1;r_1') = \int P_2(r_1,r_2;r_1',r_2)\, dr_2 \tag{6.8}$$

Since, however, pair correlations do not modify the functional form of P_1^0, we may also ask that $P_1(r_1;r_1') = P_1^0(r_1;r_1')$; these two requirements then give

$$\int P_2^0(r_1,r_2;r_1',r_2)[\phi(r_1,r_2)\phi(r_1',r_2) - \phi(r_1,r_2) - \phi(r_1',r_2)]dr_2 = 0 \tag{6.9}$$

The energy corresponding to (6.7) is

$$E = E_0 + \tfrac{1}{2}\int P_2^0(r_1,r_2)\,\frac{[\phi^2(r_1,r_2) - 2\phi(r_1,r_2)]}{r_{12}}dr_1 dr_2 \tag{6.10}$$

---

* This argument may be made more respectable in terms of the mean value theorem.

(E_0 being the Hartree-Fock energy expression) and it follows that the correlation correction is equivalent to replacing $1/r_{12}$ by the screened interaction [1]

$$g'(r_1,r_2) = \frac{[1 - 2\phi(r_1,r_2) + \phi^2(r_1,r_2)]}{r_{12}} \tag{6.11}$$

It will be possible to compute correlation energies in this way provided we can choose a realistic ϕ which satisfies (6.9).

The form recommended by Salvetti is essentially*

$$\phi(r_1,r_2) = e^{-\beta^2 r^2}[1 - \Phi(R)\ (1 + \frac{r}{2a_0})] \tag{6.12}$$

where R and r (dropping subscripts) are the variables (6.3) and $\Phi(R)$ allows the form of the hole to depend on mean position of the two particles. The function (6.12) satisfies the basic cusp condition embodied in the leading term of (2.5), while the parameter β determines the size of the hole and may be expected to depend on electron density (and hence on the mean position R).

By writing $r_1 = R - \frac{1}{2}r$, $r_2 = R + \frac{1}{2}r$ and making a Taylor expansion of $P_2^{\,0}$, equation (6.9) yields a relationship between $\Phi(R)$ and $\beta(R)$ which may be fitted with remarkable accuracy by the function

$$\Phi \simeq \frac{\sqrt{\pi} a_0 \beta}{1 + \sqrt{\pi} a_0 \beta} \tag{6.13}$$

For a free-electron gas, the one remaining parameter β could depend only on the electron density, P; if we assume that it does so generally, then a dimensional argument suggests

$$\beta = q\ P^{1/3} \tag{6.14}$$

where q is a dimensionless constant. The form (6.12) then appears to have the intuitively expected dependence on electron density; where P is large, the hole is deep but tightly localized, while in the low density region it becomes shallow but somewhat more diffuse (the "size" being measured by the radius at which the first node appears).

Colle and Salvetti have made numerical studies of some atoms and small molecules, fixing q to get a good result for helium; the correlation energies for 8 systems are then obtained with over 95% accuracy. It therefore seems worth making a corresponding attack on the classic problem of the free-electron gas; for the results of Gell-Mall and Brueckner, Sawada, Pines, and others are in the form of series which are convergent only in the high-density and low-density regions - not for the intermediate densities more commonly encountered in molecules and solids.

---

* For dimensional consistency we have inserted the Bohr radius a_0.

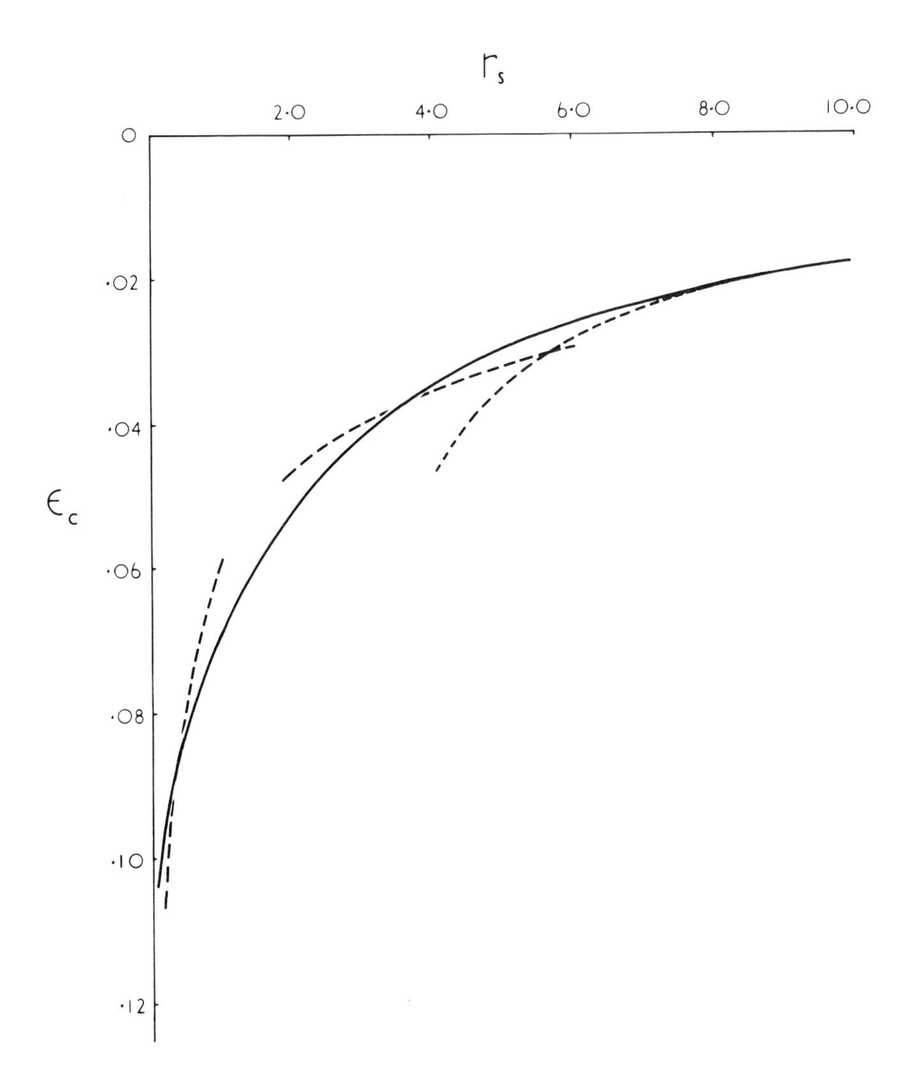

Fig. 12. Correlation energy for a uniform electron gas

The solid line is the function derived in the text; the broken lines indicate the high and low density series approximations (r_s small, and r_s large), and a medium density approximation suggested by Pines.

For a uniform gas, consisting of N electrons in a box of volume V, $P_1{}^0(r_1;r_2)$ takes the well known form

$$P_1{}^0(r_1;r_2) = \frac{N}{V} Q(2\pi k_F r_{12}) \qquad (6.15)$$

where k_F is the Fermi sphere radius defined by

$$\frac{4}{3}\pi k_F{}^3 V = \tfrac{1}{2}N \qquad (6.16)$$

and Q is the function

$$Q(x) = \frac{3(\sin x - x\cos x)}{x^3} \qquad (6.17)$$

The correlation energy follows from (6.10), where the Hartree-Fock $P_2{}^0$ is related to $P_1{}^0$ in the usual way. Substitution of (6.15) and (6.17) then yields an expression which, with ϕ defined by (6.12)-(6.14), may be evaluated by numerical integration for any given density.

We have performed such calculations using Salvetti's value of q (chosen by reference to a helium atom calculation). The results are shown in Fig. 12, along with those from the asymptotic series of Gell-Mann and Brueckner [38] and of Carr [57]. As is customary, we have introduced a dimensionless quantity r_s, the radius of a sphere (in units of a_0) which would hold just one electron at the given density ($P = N/V$): thus $a_0 r_s = (3/4\pi P)^{1/3}$ The present calculation reproduces the high and low density results (said to be valid for $r_s<0.5$ and $r_s>10.0$, respectively) almost perfectly and spans the whole range. In fact our calculated correlation energy per electron at density P may be fitted with high accuracy over the whole range of densities by the simple formula

$$\varepsilon_{cor} = \frac{E_{cor}}{N} = \frac{-1}{9.652 + 2.946/P^{1/3}a_0} \quad \frac{e^2}{4\pi\varepsilon_0 a_0} \qquad (6.18)$$

This appears to be a remarkable confirmation of the general validity of the correlation-hole approach, with the type of correlation factor suggested by Salvetti. Of course, for a uniform gas the hole must have spherical symmetry as assumed in the form (6.12); but even when the density varies rapidly and the hole is strongly distorted (cf. Figs. 5-6) the evaluation of integrals containing the scalar interaction ($r_{12}{}^{-1}$) automatically performs a spherical averaging; it therefore seems to be the case, from calculations made so far, that (6.12) yields a good representation of the spherically averaged correlation hole. Work is now in progress on the further applications of this approach.

7. Conclusion

It is clear that much progress has been made during the last 10 years, in spite of the somewhat diminished popularity of work on electron correlation. One is struck by the generality of the problems involved - not peculiar to electrons in molecules and certainly equally important in the solid state and in nuclei - and by the diversity of the techniques available for their solution. One also feels - in view of the high degree of sophistication of some of the techniques, the tremendous computational difficulties involved in their implementation, and their heartbreakingly slow convergence - that somehow we may be doing the wrong kind of mathematics. Perhaps we should be more adventurous once in a while, forgetting the rigours of the variation theorem, N-representability, and the like, and concentrating a bit more on the physics.

REFERENCES

[1] R. McWeeny, Int. J. Quant. Chem., 15, 351 (1967).

[2] P. V. Herigonte, Structure and Bonding, 12, 1 (1972).

[3] C. A. Coulson and I. Fischer, Phil. Mag. 40, 386 (1949).

[4] P. O. Lowdin, Rev. Mod. Phys., 35, 496 (1963).

[5] N. Moiseyev and J. Katriel, Chem. Phys. Let., 29, 69 (1974).

[6] R. K. Nesbet, Proc. Roy. Soc. (Lond.) A230, 312 (1955).

[7] E. Wigner and F. Seitz, Phys. Rev., 43, 804 (1973); see also F. Seitz "Modern Theory of Solids" (McGraw-Hill, New York, 1940) p.245.

[8] R. McWeeny, Rev. Mod. Phys., 32, 335 (1960).

[9] J. C. Slater, Phys. Rev. 81, 385 (1951).

[10] For a comprehensive review see J. C. Slater "Quantum Theory of Molecules and Solids, Vol. IV" (McGraw-Hill, New York, 1974).

[11] V. W. Maslen, Proc. Phys. Soc. (Lond) A69, 734 (1956).

[12] For related work see R. J. Boyd and C. A. Coulson, J. Phys. B (G.B.) 6, 782 (1973) and references therein.

[13] M. Suard and G. Joubet, Atti. Soc. Nat. (Modena) 100, 59 (1969); M. A. Besson and M. Suard, Int. J. Quant. Chem., 10, 151 (1976).

[14] C. N. M. Pounder "Correlation Effect in Molecules" Ph.D. Thesis, University of Newcastle upon Tyne (1975).

[15] E. A. Hylleraas, Z. Phys. 54, 347 (1929). For a review see

E.A. Hylleras in "Adv. Quant. Chem. 1, 1 (1964).

[16] H. James and A. S. Coolidge, J. Chem. Phys., 1, 825 (1933).

[17] C. L. Pekeris, Phys. Rev. 112, 1649 (1958); ibid. 115, 1216. (1959)

[18] W. Kolos and L. Wolniewicz, J. Chem. Phys., 49, 404 (1968).

[19] W. A. Bingel, Theor. Chim. Acta (Berl.) 8, 54 (1967).

[20] D. R. Hartree and A. L. Ingman, Mem. Manchester Lit. and Phil. Soc. 77, 69 (1933).

[21] W. Brenig, Nucl. Phys. 4, 363 (1957).

[22] F. Coester and H. Kummel, Nucl. Phys. 17, 477 (1960).

[23] O. Sinanoglu Rev. Mod. Phys. 35, 517 (1963).

[24] H. Primas in "Modern Quantum Chemistry. Istanbul Lectures" ed. O. Sinanoglu (Academic Press, New York, 1965).

[25] L. Szass, Z. Naturforsch 14a, 1014 (1959).

[26] O. Sinanoglu, Proc. Nat. Acad. Sci. (U.S.A.) 47, 1217 (1961); J. Chem. Phys. 36, 706, 3198 (1962); Adv. Chem. Phys. 6, 315 (1964).

[27] R. K. Nesbet, Phys. Rev. 109, 1632 (1958); Adv. in Chem. Phys. 9, 321 (1965).

[28] W. Kutzelnigg, Theor. Chim. Acta, 1, 327 (1963); Ahlrichs, W. Kutzelnigg and W. Bingel, Theor. Chim. Acta 5, 289 (1966); Ahlrichs and W. Kutzelnigg, J. Chem. Phys., 48, 1819 (1968).

[29] H. A. Bethe and J. Goldstone, Proc. Roy. Soc. (Lond.) A238, 551 (1957).

[30] J. Goldstone, Proc. Roy. Soc. (Lond.) A239, 267 (1957).

[31] R. K. Nesbet, Phys. Rev. 155, 57, 56 (1967); ibid 175, 2 (1968). For a review see R. K. Nesbet, Adv. Chem. Phys. 14, 1 (1969).

[32] W. Meyer, J. Chem. Phys. 58, 1017 (1973).

[33] W. Meyer, Proceedings of SRC Atlas Symposium No. 4 P.97 (Oxford, 1974).

[34] A. C. Hurley, J. E. Bernard-Jones and J. A. Pople, Proc. Roy. Soc. (Lond.) A220, 446 (1953).

[35] R. McWeeny, Proc. Roy. Soc., A253, 242 (1959); Rev. Mod. Phys. 32, 335 (1960) See also M. Klessinger and R. McWeeny J. Chem. Phys., 42, 3343 (1965).

[36] R. McWeeny and E. Steiner, Adv. in Quant. Chem., 2, 93 (1965) See also R. McWeeny and B. T. Sutcliffe "methods of Molecular Quantum Mechanics" (Academic Press, London, 1969) Sect. 7.5.

[37] For a recent review see G. W. Stewart in "Information Processing 74" (proceedings of IFSP Congress 74, Stockholm) (North Holland, Amsterdam, 1974). See also Ref. 43 .

[38] M. Gell-Mann and K. A. Brueckner, Phys. Rev. 106, 364 (1957)

[39] H. P. Kelly, Phys. Rev. 131, 684 (1963); ibid. 136, 396 (1964); ibid. 144, 39 (1966). For a review see H. P. Kelly, Int. J. Quant. Chem. 1s, 25 (1967).

[40] J. Cizek, J. Chem. Phys., 45, 4256 (1966); Adv. Chem. Phys. 14, 35 (1969). See also J. Paldus and J. Cizek, Adv. Quant. Chem. 9, 106 (1975).

[41] V. R. Saunders and M. A. Robb (1976), private communication. See also G. H. Van der Velde, Ph.D. Thesis (Groningen 1974).

[42] J. Paldus, J. Cizek and I. Shavitt, Phys. Rev., A5, 50 (1972).

[43] I. Shavitt in "Modern Theoretical Chemistry Vol. II. Electron Structure: Ab Initio Methods (M. F. Schaeffer III, ed.; Plenum Press, New York 1976).

[44] B. H. Brandow, Rev. Mod. Phys. 39, 771 (1967).

[45] P. G. H. Sanders, Adv. Chem. Phys. 14, 365 (1969).

[46] D. Mukherjee, R. K. Moitra and A. Mukhopadhyay, Mol. Phys. 30, 1861 (1975).

[47] S. F. Boys, Proc. Roy. Soc., A309, 195 (1969).

[48] S. F. Boys, and N. C. Handy, Proc. Roy. Soc., A310, 43 (1969).

[49] S. F. Boys and N. C. Handy, Proc. Roy. Soc., A310, 63 (1969).

[50] S. F. Boys and N. C. Handy, Proc. Roy. Soc., A311, 309 (1969).

[51] G. G. Hall and C. E. Soloman, Chem. Phys. Lett., 4, 352 (1969).

[52] E. A. G. Armour, Mol. Phys. 24, 163, 181 (1972).

[53] E. Wigner, Phys. Rev. 46, 1002 (1934).

[54] E. Clementi, J. Chem. Phys., 38, 996, 1001, 2248 (1963); ibid 39, 175 (1963); ibid 41, 295, 303 (1964). See also R. K. Nesbet, J. Chem. Phys. 40, 3619 (1964); ibid 43, 530 (1965).

[55] R. Colle and O. Salvetti, Theor. Chim. Acta (Berl.) 37, 329 (1975).

[56] A. J. Coleman, Rev. Mod. Phys. 35, 668 (1963).

[57] W. J. Carr, Phys. Rev. 122, 1437 (1961); see also W. J. Carr, R. A. Coldwell-Horsfall and A. E. Fein, Phys. Rev. 124, 747 (1961).

QUANTUM MECHANICAL FACETS OF CHEMICAL BONDS

R. DAUDEL

Centre de Mécanique Ondulatoire Appliquée, CNRS,
23, Rue du Maroc, 75019-Paris and Sorbonne.

To define the concept of bond is both a central problem of quantum chemistry and a difficult one. The concept of bond appeared little by little in the mind of chemists from empirical observations. From the wave-mechanical viewpoint it is not an observable. Therefore there is no precise operator associated with that concept. As a consequence there is not a unique approach to the idea of chemical bond. This is why we prefer to present various quantum mechanical facets of that important concept.

THE ENERGETIC FACET

K. Ruedenberg has analyzed carefully the nature of the chemical bond from the energetic point of view. He pointed out that the usual presentation is not convenient. Let us consider the hydrogen molecule-ion. It is often said that the bond energy of that molecule is due to the fact that when a proton approaches an hydrogen atom a region of low electric potential is created between the two nuclei because in that region the electron is attracted by two nuclei and not only by one as in the hydrogen atom. Ruedenberg said that such statement is not convaincing because in fact the regions in which the electric potential has its lowest value are the neighbourhood of each nucleus and not the bonding region. Therefore we have to explain why the electron like to visit the bonding region and why this phenomenon is a source of bond energy.

B. Pullman and R. Parr (eds.), The New World of Quantum Chemistry, 33–56.

Ruedenberg has shown that in fact the main origin of the bond energy is the appearance between the nuclei of a region in which the electric potential varies slowly. As a consequence the wave function in that region varies slowly. Therefore the contribution of that region to the kinetic energy is low as the kinetic operator depends on the second derivatives of the wave function. The electron will go slowly when visiting that region and therefore will remain a certain time there : this is why the contribution of that region of relatively low potential will be significant even if it is not the best from the point of view of the value of the potential. More precisely Ruedenberg(1) reaches the conclusion that : "It is in the behaviour of the kinetic energy that the molecule differs essentially from the free atom. At large distances, this leads to incipient bonding due to kinetic energy lowering. Near the equilibrium distance it leads to binding by lowering the kinetic resistance against the nuclear suction. This result in a closer attachment of the electronic cloud to the nuclei with a concomitant lowering of the potential energy and increase in the kinetic energy" This is why both the modulus of the potential energy and the modulus of the kinetic energy increase simultaneously as required by the virial theorem.

THE DENSITY FACET

The knowledge of a wave-function associated with a molecule permits to calculate the electronic density ρ (M) at any point M of the molecule and therefore to draw contour maps showing the lines of isodensity

However such maps do not clearly show the effect of binding on the electronic distribution in a molecule. Figure 1 shows for example how the electronic density changes during the formation of the hydrogen molecule. No striking features appear. This is why I introduced(2) the density difference function δ (M) at point M in a molecule which is simply the difference between the

---

(1) K. RUEDENBERG in Localization and Delocalization, Chalvet, Daudel, Diner, Malrieu ed, Reidel pub. (1975) p. 223.

(2) R. Daudel, C.R. Acad. Sci. <u>225</u>, 886 (1952).
M. Roux and R. Daudel, C.R. Acad. Sci. <u>240</u>, 90 (1955).

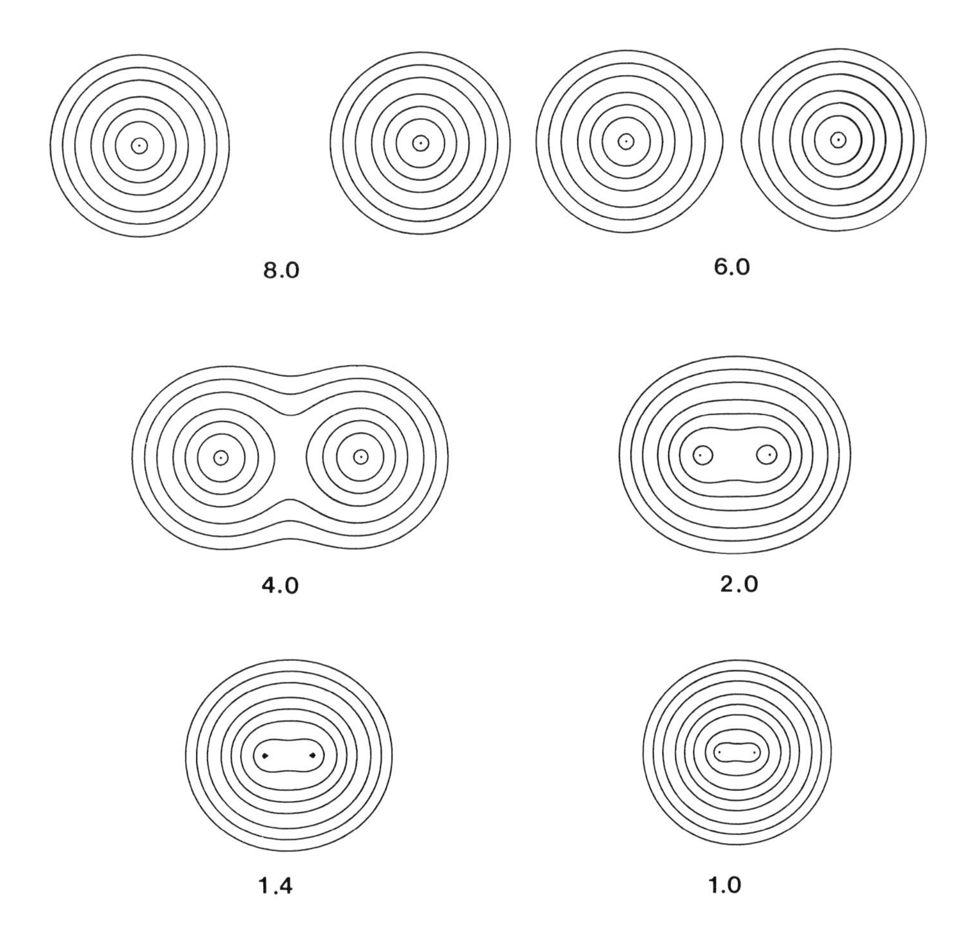

Figure 1 - Electronic density change during the formation of the hydrogen molecule. (reproduced by permission from R.F.W. BADER, An introduction to the electronic structure of atoms and molecules, Clarke, Irwin and Cy. Lim. TORONTO, 1970)

actual density ρ (M) and the virtual density ρ^V (M) which would result from the addition of the densities in the free atoms :

$$\delta (M) = \rho (M) - \rho^V (M)$$

Therefore at a point where δ is positive, the bonding has led to an increase of the electronic density. At a point where δ (M) is negative the bonding has led to a decrease of the electronic density. This is why the δ function is also called the bond density function. It shows the effect of the formation of the bonds on the electronic distribution.

Figure 2 shows for example the contour maps of the density difference function during the formation of the hydrogen bond. Dashed lines correspond to region in which δ (M) is negative. Solid contours correspond to positive values of the δ function. A striking electron transfer from the ends of the molecule to the central region appears for the equilibrium distance.

Figure 3 gives a more precise analysis of the variation of δ (M) along the line of the nuclei because the calculations have been made with the help of the very good function of James and Coolidge. The δ function is positive everywhere between the nuclei. This result agrees with Ruedenberg statement. The electron like to stay between the nuclei as a result of the lowering of the kinetic energy : the increase of density near the nucleus agrees with the idea of an increase of the nuclear suction resulting from that lowering. Many authors have calculated that function for various molecules[1]. Figure 4 shows typical results of such calculations. For a homonuclear molecule such as N_2 the effect of binding is

a) an increase of the electronic density in a central region of the molecule
b) an increase of the electronic density in the "lone pair" region. For a heteronuclear molecule like LiF we see :
a) an increase of the electronic density in all the fluorine region except for a small volume near the nucleus
b) a decrease of the density in the lithium region except for a small volume not far from the nucleus. The overall effect

---

(1) M. ROUX, M. CORNILLE and L. BURNELLE, J. Chem. Phys. 37, 1009 (1962).
B.J. RANSIL and J.J. SINAI, J. Chem. Phys. 46, 4050 (1967).
M.J. HAZELRIGG and P. POLITZER, J. Am. Chem. Soc. 73, 1009 (1969).
R.F.W. BADER, W.H. HENNEKER and P.E. CADE, J. Chem. Phys. 46, 3341 (1967).
P.E. CADE, Trans. Am. Crystallogr. Ass. 8, 1 (1972).

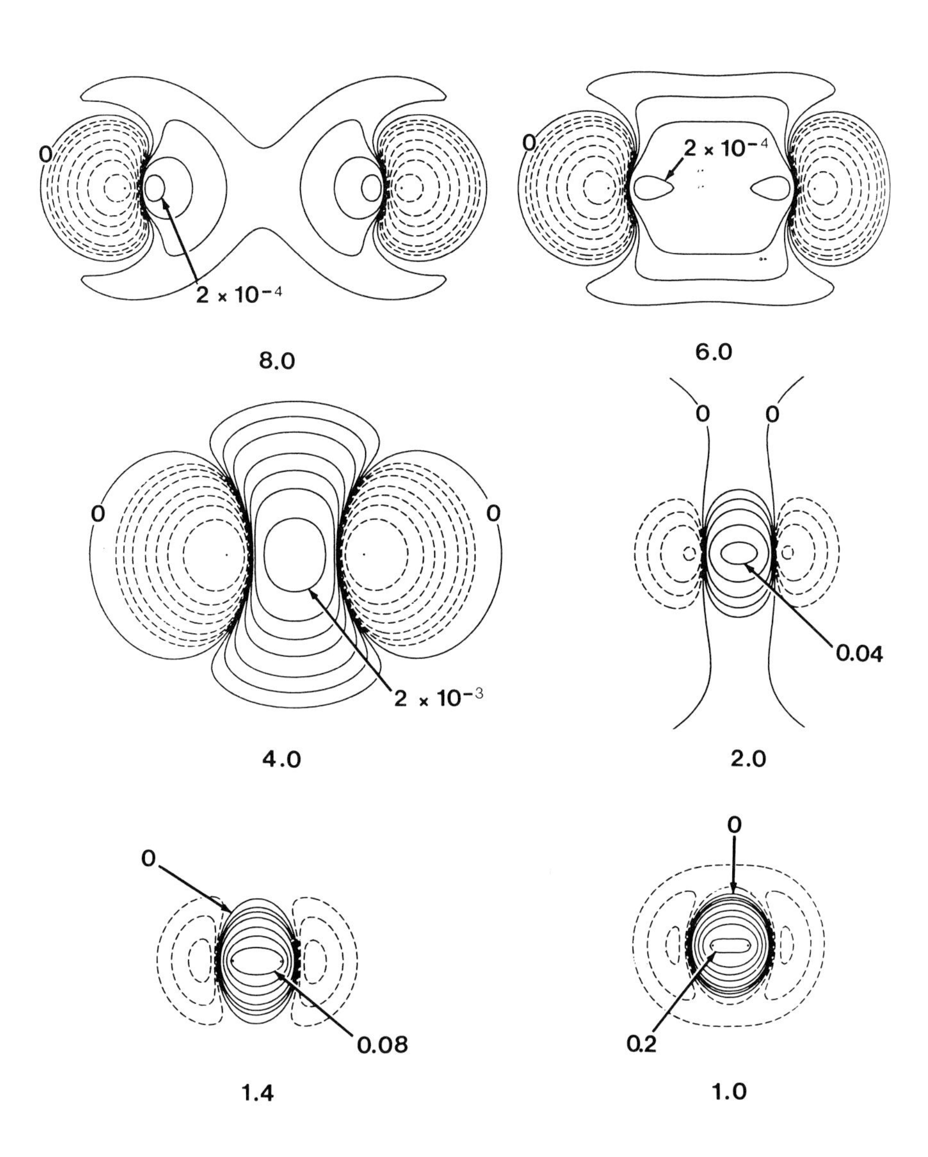

Figure 2 - Density difference function during the formation of the hydrogen molecule (after BADER, loc. cit.) Dashed lines corresponds to negative values, solide contours to positive values.

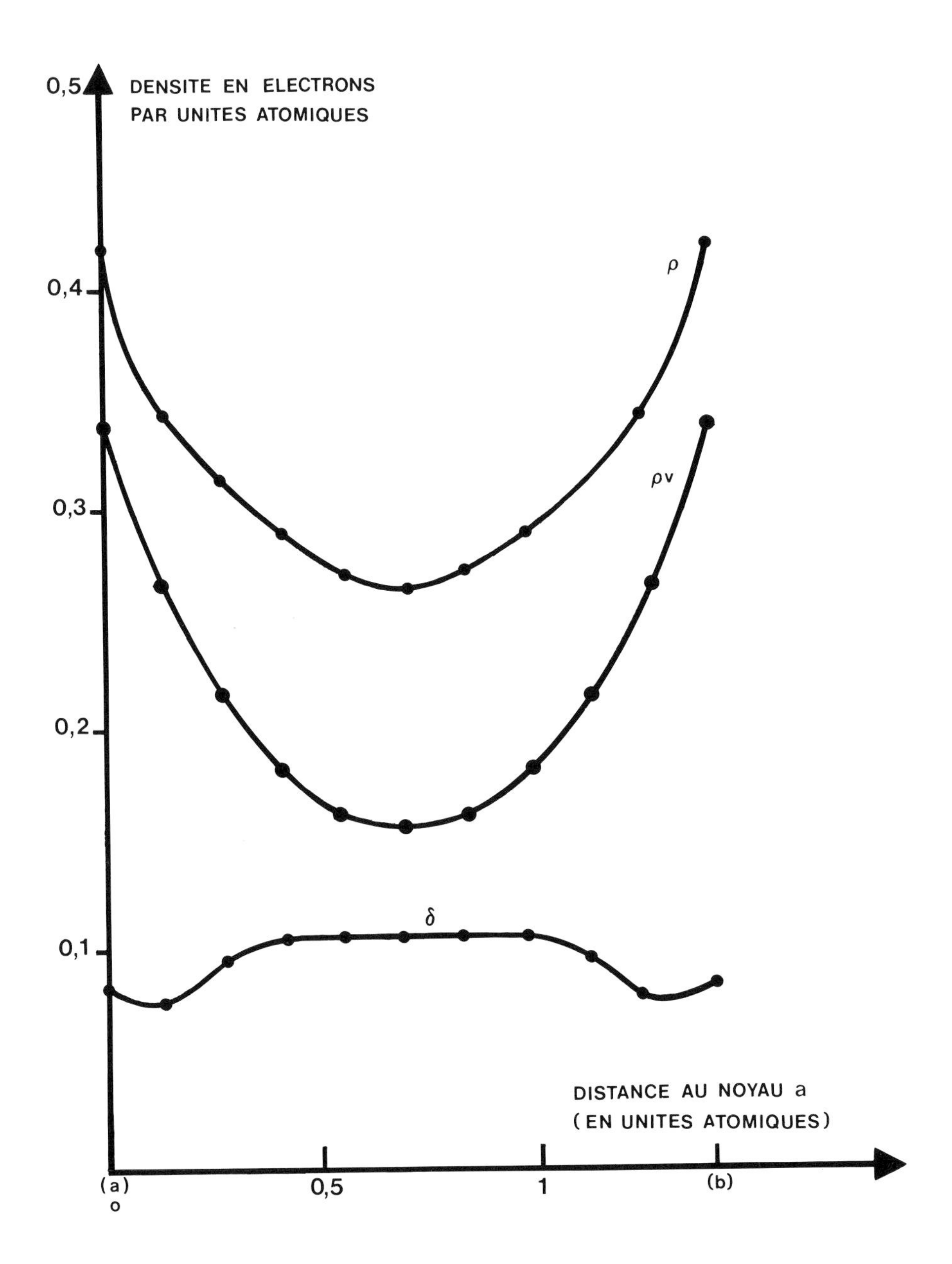

Figure 3 - Density difference function along the line of the nuclei in the hydrogen molecule.

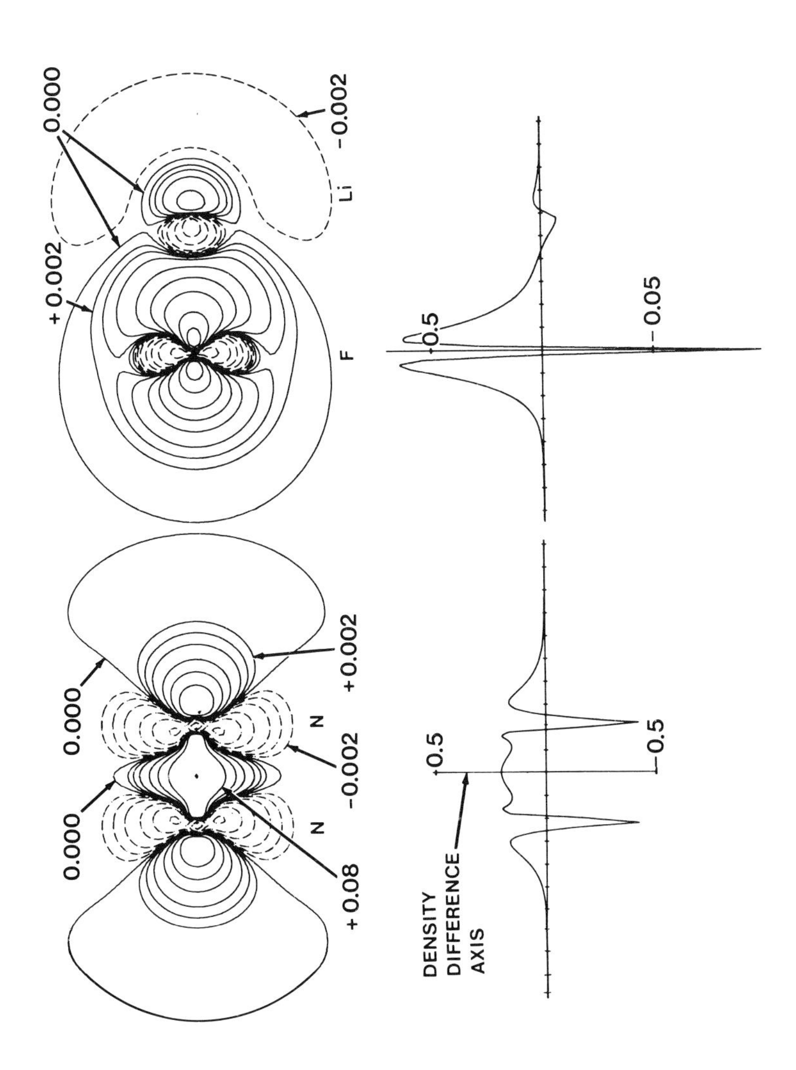

Figure 4 - Density difference function in N_2 and FLi (after BADER loc. cit.)

amounts to a transfer of charge from the lithium region to the fluorine one.

Figure 5 shows the contour map of the density difference function along the hydrogen bond of a formamide dimer(1). "It is seen that the first effect is a flying away of the electron from the intermolecular region and that it is a localized effect touching essentially the O, C, H, N atoms. On the contrary the total effect is a delocalized effect due to rearrangements over the whole molecular peripheries. On the whole the proton is denuded and the NH bond becomes more polar".

Progress has been recently made in the measurement of electron density in molecular crystals. By using both X ray diffraction at low temperature and neutron diffraction with the help of a good theory of the extinction of the X rays in the crystal it becomes possible to obtain an experimental determination of the δ (M) function. Figure 6 shows the result of such measurement in cyanuric acid(2). Figure 7 shows the contour map of the same function but calculated from a minimal basis set SCF wave function.(3) It is seen that the minimal basis set calculation underestimates the bond density function. An extended basis set is necessary to reproduce correctly the experimental data. It is better to include a certain amount of configuration interaction.

Near the nuclei the results of the measurement of the δ function by diffraction methods are not precise because of the movement of the nuclei. The effect of bonding on the electronic density near the nuclei can be achieved more precisely by measuring the rate of decay of some radioactive nuclei. This procedure has been proposed by Daudel(4) and Segré(5). The relation between the rate of decay of some radioactive nuclei and the electronic structure of a molecule has been found to be so sensitive that

---

(1) A. PULLMAN in Aspects de la Chimie Quantique Contemporaine, R. Daudel et A. Pullman ed. CNRS pub. (1971) p. 253.
(2) P. COPPENS, MTP International Review of Science Physical Chemistry Series II, Butterworths, London (1975).
(3) D.S. JONES, D. PAUTLER and P. COPPENS, Acta Crystallographica A28, 635 (1972).
(4) R. DAUDEL, Rev. Scientifique 87, 162 (1947).
(5) E. SEGRE, Phys. Rev. 71, 274 (1947)

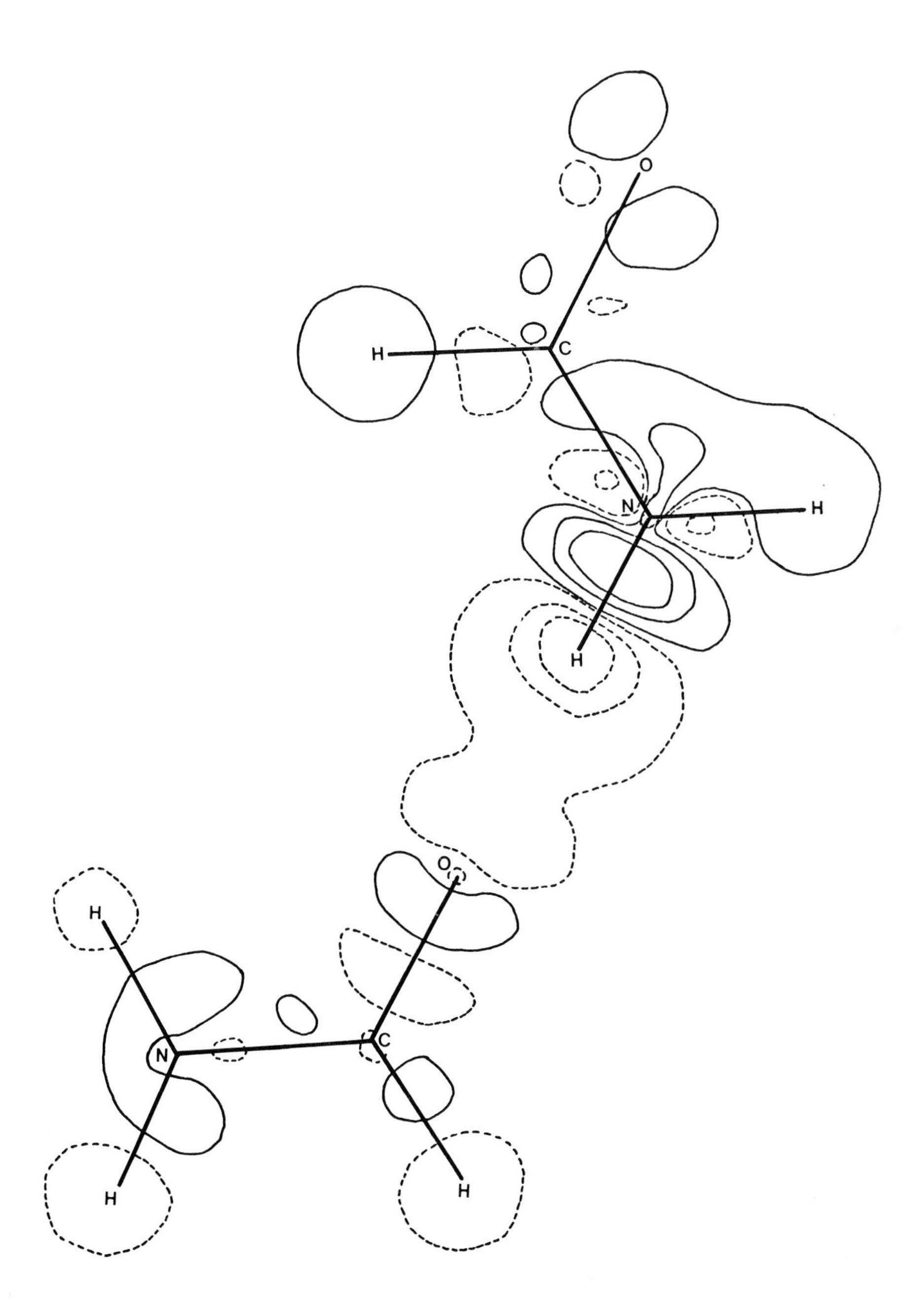

Figure 5 - Density difference function in the formamide dimer (after A. PULLMAN in R. DAUDEL et A. PULLMAN, Aspects de la Chimie Contemporaine, C.N.R.S, Paris, 1971) reproduced by permission.

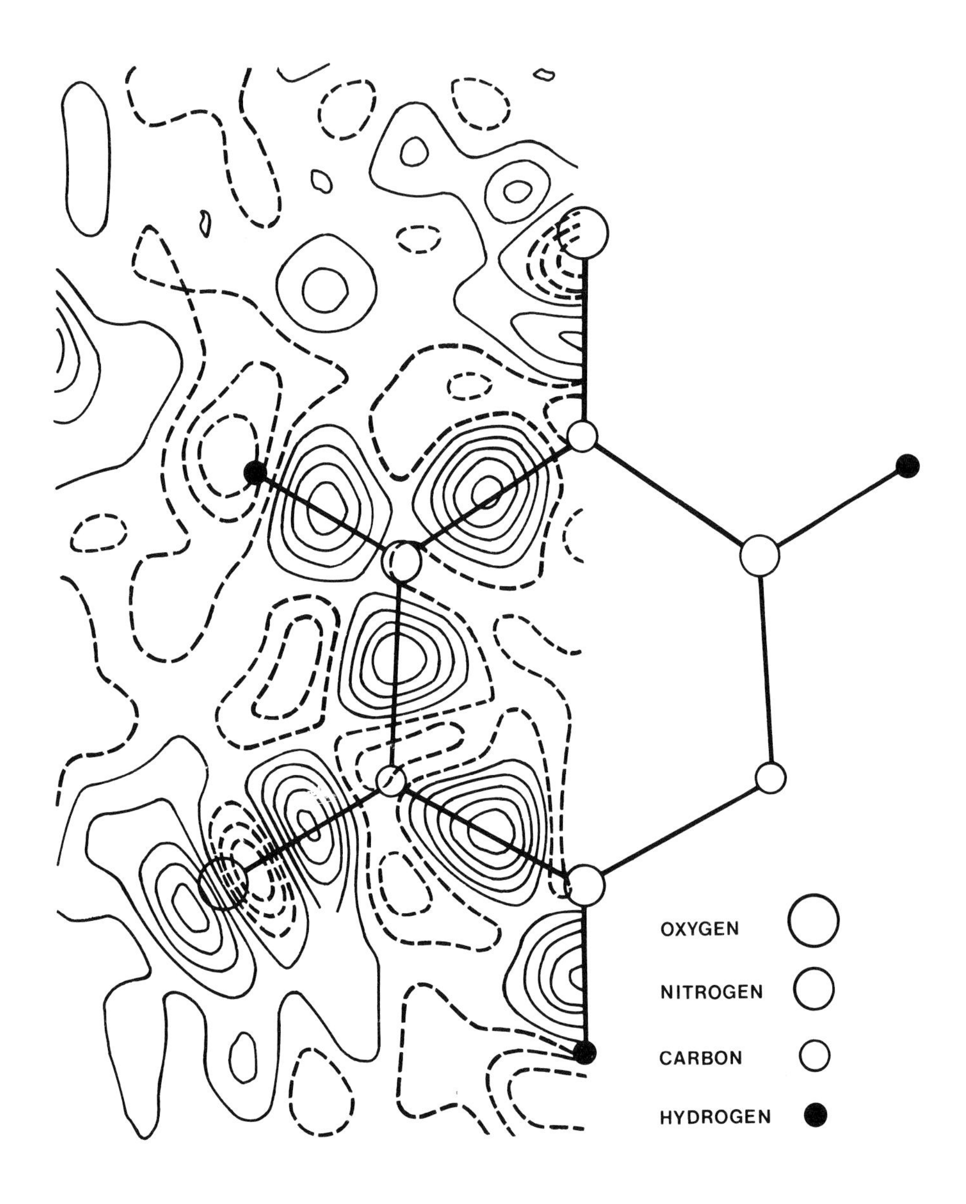

Figure 6 - Density difference in cyanuric acid : measurement (reproduced by permission from P. COPPENS M.T.P. International Review of Science Physical Chemistry Series II, BUTTERWORTHS, LONDON (1975))

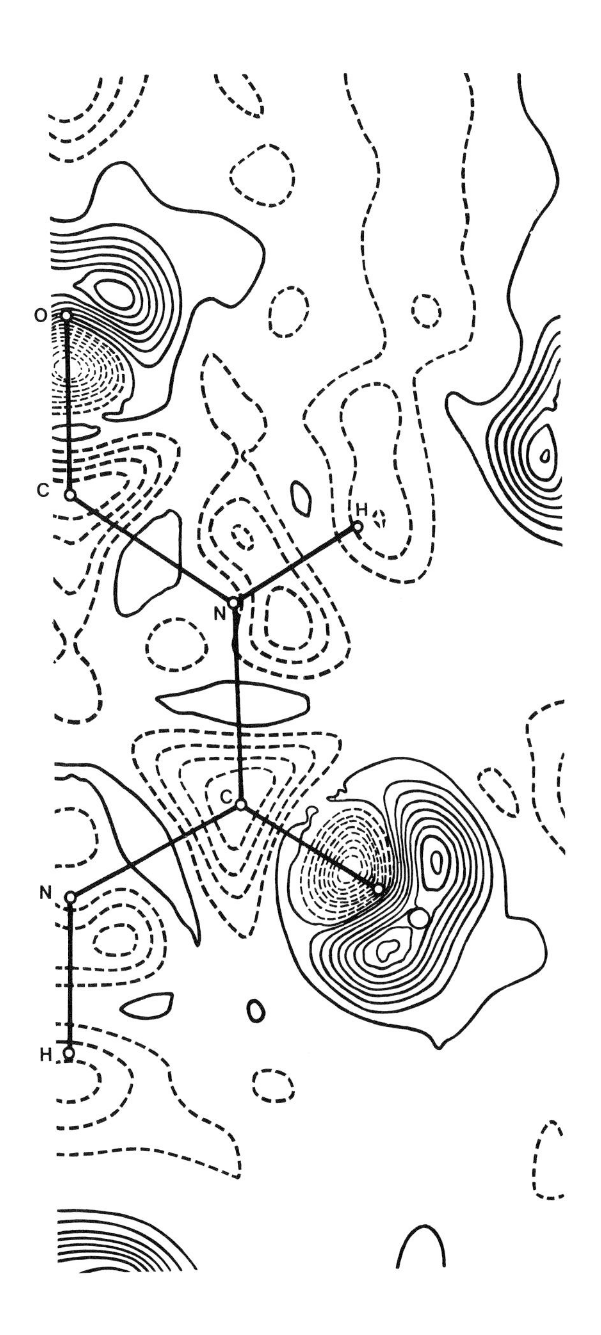

Figure 7 - Density difference in cyanuric acid : calculation (after COPPENS loc. cit.)

the measurement of decay is now a powerful tool for chemical analysis of very small amounts of radioactive compounds.[1]

THE PARTITIONING FACET

Partitioning the molecular space in fragments associated predominantly with bonds, cores, lone pairs and so on is another way to draw some light on the concept of bond. Various kinds of partitioning are possible : one is the loge theory, the other the virial partitioning method.

The loge theory is a procedure of analyzing the localizability of the electrons of an atom or a molecule. A good loge is a volume in which there is a high probability of finding a certain number $\underline{n}$ of electrons (and that number only).
For atom a simple relationship :

$$p^{3/2} v = \text{cste}$$

has been found[2] between the volume which an electron occupies when it visits a loge and the mean value p of the electric potential acting on that electron during the visit.
Let us consider the BH molecule. It is a 6 electron problem. Consider a sphere of arbitrary radius R centered at the boron nucleus. Finding 0, 1, 2, 3, 4, 5 or 6 electrons in that sphere are seven different possible electronic events. From a wave function associated with the BH molecule it is possible to calculate the probability P_i of the electronic event i. Figure 8 shows the variations of the various P_i's as a function of R.[3] It appears that the probability P_2 of finding 2 electrons (and 2 only) in the sphere reaches a high probability (0.85) for a certain value of R. The other probabilities remain above 0.5 for all values of R. It is said that finding two electrons in the sphere is the only leading event. It can be anticipated that there exists a good spherical loge in which there is a high probability of finding two electrons.

---

(1) J.I. VARGAS, M.T.P. Int. Rev. Sci. Serie I, 8, 75 (1971)
(2) S. ODIOT and R. DAUDEL, C.R. Acad. Sci. 238, 1384 (1954)
(3) R. DAUDEL, R.F.W. BADER, M.E. STEPHENS and D.S. BORRETT J. Can. Chem. 52, 1310 (1974).

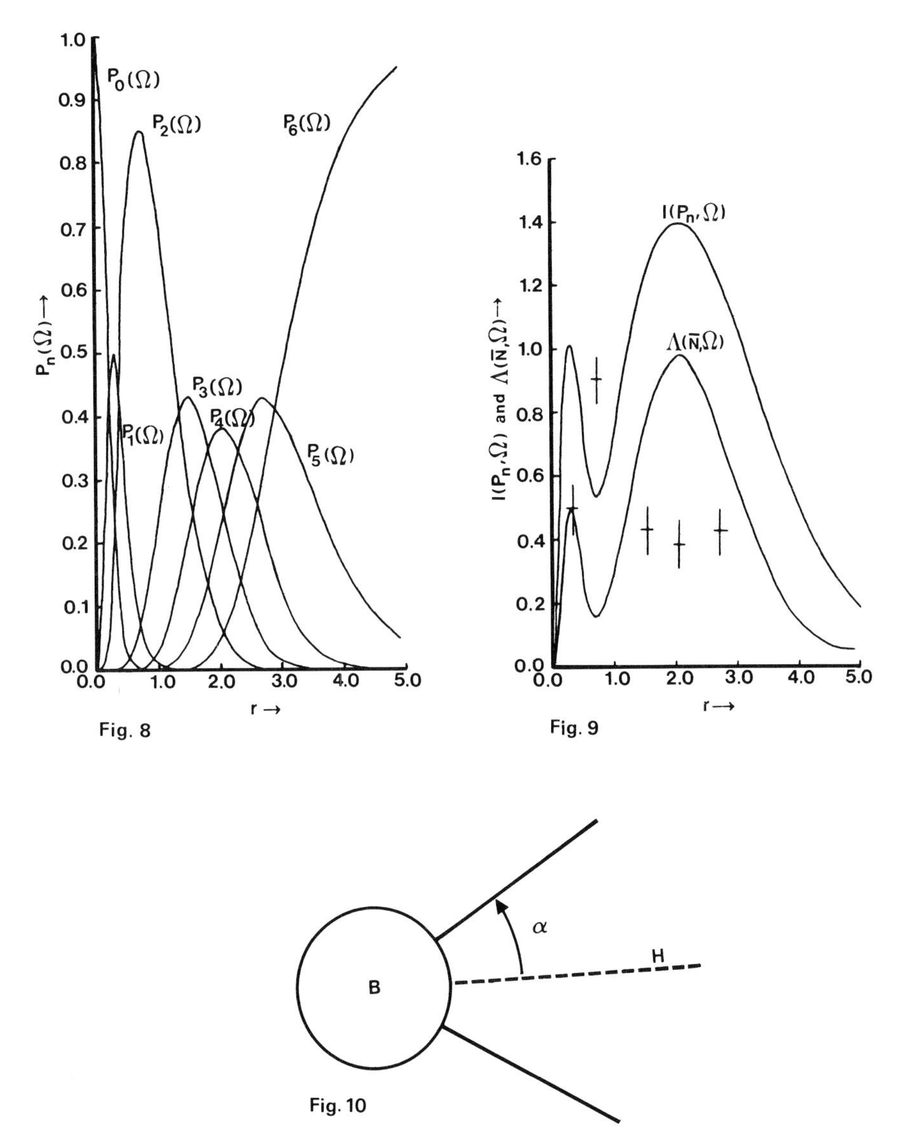

Figure 8 - Probability of occurence of the various electronic events.

Figure 9 - Missing information function and fluctuation of the number of electrons.

Figure 10 - Three loge partition (after R. DAUDEL, R.F.W. BADER, M.E. STEPHENS and D.S. BORRETT, J. Can. Chem. 52, 1310 (1974) reproduced by permission.

To show this fact by following a systematical procedure[1] we can compute the missing information function associated with the P_i's :

$$I = \sum_i P_i \log_2 P_i^{-1}$$

To obtain the maximum amount of information about the localizability of the electrons in the sphere we have to minimize the missing information function. Figure 9 shows that this minimum is reached approximately when P_2 is a maximum (R = 0.7 a.u.). This radius defines the best partition of the space into two loges : the sphere and the remaining part of the space. It can be said that the sphere is the boron core loge.

It is possible to go farther by considering a partition into three loges dividing the outer loge with a cone of angle α, the summit of the cone being the boron nucleus and it axis the BH line (Figure 10). The missing information function associated with that partition of the space reaches its minimum value for :

$$R = 0.7 \text{ a.u.} \qquad \alpha = 73°$$

Finding two electrons in each loge is the leading event. It can be said that the region inside the cone corresponds to the bond loge and that the outer part of the cone is the lone pair loge.

It is seen that the lone pair is more bulky than the bond loge. More precisely if an electronic density contour is taken as an arbitrary limit of a loge such that it contains 95% of its electronic charge the ratio of non bonded to bonded loge volumes equal to 2.11. This result is in accord with one of the basic postulates of the very powerful Gillespie theory of molecular geometry[2].

Instead of the missing information function it can be interesting and practical to study the fluctuation Λ of the number N of electrons which is found in a loge. Figure 9 shows the variation as a function of the radius R of that fluctuation :

$$\Lambda = \overline{N^2} - (\overline{N})^2$$

in the boron core loge.

---

(1) C. ASLANGUL, C.R. Acad. Sci. B272, 1 (1971)
(2) R.J. GILLESPIE, Molecular Geometry, Van Nostrand Reinhold Co., Ltd. London (1972).

It is seen that missing information function and fluctuation run parallel. Therefore a good loge is a region of the space in which the fluctuation of the number of electrons is small.

Recently Bader[(1)] has shown that there is a direct relationship between the frontiers of loges and electronic correlation. Let us recall the McWeeny definition of the correlation factor. Let $P_2(M_a, M_b)$ be the probability density of finding simultaneously one electron at point M_a and another one at point M_b. If $P_1(M_a)$ is the probability density of finding one electron at point M_a and $P_1(M_b)$ the probability density of finding one electron at point M_b the correlation factor $f(M_a, M_b)$ is defined by the equation :

$$P_2(M_a, M_b) = P_1(M_a)\, P_1(M_b) \left[1 + f(M_a, M_b) \right]$$

It vanishes when there is no correlation between the movement of the electrons. It has usually a negative value and measures "the Fermi hole" when a significant correlation occurs. Bader has shown that the fluctuation Λ of the number N of electrons in a loge V may be written as :

$$\Lambda = \bar{N} + \int_V P_1(M_a)\, P_1(M_b)\, f(M_a, M_b)\, dv_a\, dv_b$$

It follows that when the integral has the largest local negative value (that is when the correlation inside of the loge is maximized) the fluctuation is minimized. Therefore a good partition of the space into loges maximizes the correlation between the movement of the electrons in each loge and as a consequence minimizes that correlation between different loges.

An other important property of loges is their transferability from a molecule to another one.
Figure II permits the comparison of the BeH bond loge in BeH and the BeH bond loge in H BeH. It is seen that there are very similar. Their frontiers correspond approximatively to steepest descent lines on the electronic density surface.

It is possible to demonstrate rigorously[(2)] that any expectation value $\langle\psi|\Omega|\psi\rangle$ associated with a two-electron (or less) observable Ω can be expressed as a sum of loge contribution Ω_1 and of pair loge contribution $\Omega_{1,1'}$

(1) R.F.W. BADER in Localization and Delocalization in Quantum Chemistry, Chalvet, Daudel, Diner, Malrieu ed, Reidel pub. (1975) p. 15.
(2) C. ASLANGUL, R. CONSTANCIEL, R. DAUDEL and P. KOTTIS, Adv. Quantum Chemistry 6, 93 (1972).

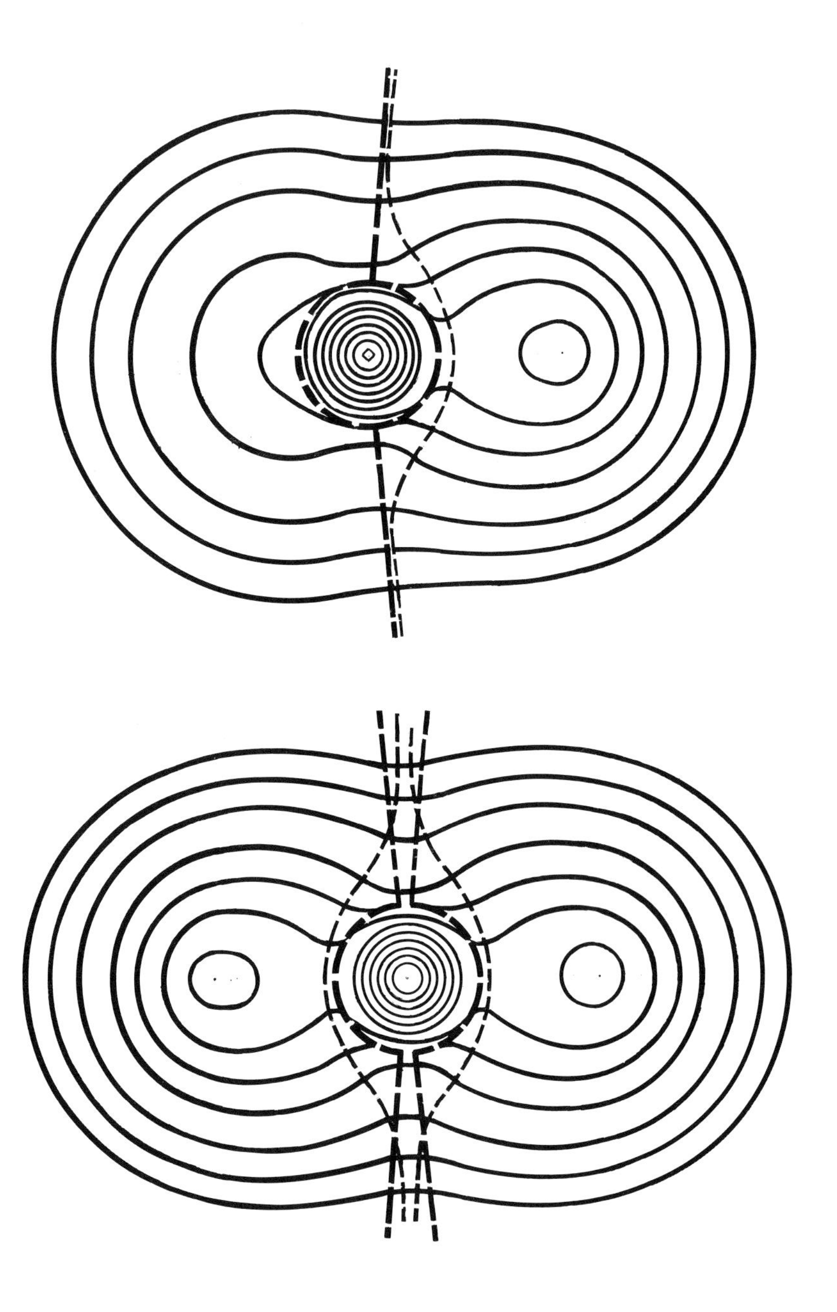

Figure 11 - Loge and virial partitioning in BeH and in HBeH.

$$< \psi \mid \Omega \mid \psi > \; = \sum_{1} \Omega_{1} + \sum_{1,1'} \Omega_{1,1'}$$

For some monoelectronic operators the $\Omega_{1,1'}$'s vanish and the formula reduces to

$$< \psi \mid \Omega \mid \psi > \; = \sum_{1} \Omega_{1}$$

If furthermore the loges are transferable the modulus Ω_1 associated with a given kind of loge 1 remains approximately the same in a certain family of molecules. Therefore the formula shows that the molecular property described by the operator Ω is simply the sum of moduli associated with each loge constituing the molecule. A kind of additivity law appears for that molecular property. Such is the case of the Faraday effect[1]. For a bielectronic operator as the energy a modulus like $\Omega_{1,1'}$ representing an interacting term between the loge 1 and the loge 1' does not vanish. The sum $\sum_{1,1'} \Omega_{1,1'}$ yields deviation to the additivity and is responsible for isomerization energies.[2]
Finally let us add that the loge theory has been used to suggest a criterion to distinguish between covalent and dative bonds.[3]
But one of the most interesting application of the loge theory is its use to compare the degree of localizability of the electrons in different molecules. Table 1 shows for example the percentage of localization of the electrons of bonded loges defined by BADER and STEPHENS[4] from the relative fluctuation of the number of electrons in the loges.

TABLE 1
Properties of bonded loges

Molecule	percentage localization
LiH	95,5
BeH_2	92,8
CH_4	68,9
OH_2	56,2
F_2	17,4

(1) R. DAUDEL, F. GALLAIS and F. SMET, Int. J. Quantum Chemistry 1, 873 (1967)
(2) R. DAUDEL, C.R. Acad. Sci. 270, 929 (1970)
(3) R. DAUDEL and A. VEILLARD in Nature et Propriétés des Liaisons de Coordination, CNRS pub. (1970)
(4) R.F.W. BADER and M.E. STEPHENS, J. Am. Chem. Soc. 97, 7391 (1975).

These results show that the two electron bond concept is very satisfactory for LiH, borderline for CH_4 and OH_2, vanishing for F_2, at least from the localizability viewpoint.

Bader has introduced a procedure of partitioning a molecule in fragments in which the virial theorem is locally satisfied. A partitioning surface is defined by that collection of all gradient paths, as traced out by the vectors $\nabla\rho$ (M), which originate and terminate at stationary points in the density distribution.[1] A stationary point being a point at which $\nabla\rho$ (M) = 0. Between each pair of bonded nuclei there is usually such a point : the saddle point. The partitioning surface through this point divide the system into two fragments. The mathematical requirement for a partitioning surface S is given by the zero flux requirement :

$$\nabla\rho\ (M)\ \vec{n} = 0 \qquad M \in S$$

where n is the vector normal to S.

It has been rigorously demonstrated[2] that a fragment bounded by a surface of zero-flux has a well-defined kinetic energy and that this kinetic energy satisfies the virial relationship.

The short dashed line on figure II gives the frontier of virial fragments in BeH. It is seen that BeH bond loge and BeH virial fragment are very similar. Therefore the virial theorem is approximately satisfied in the BeH bond loge. As the loges the virial fragments are transferable.

The similarity between loges and virial fragments is not general[x]

THE FUNCTIONAL FACET

The last section of this paper is devoted to the study of the functional aspect of chemical bonds. The loge theory can be used to establish a bridge between the partitioning facet and the functional facet of a bond. With each electronic event λ it is possible to associate an event function ψ_λ, identical to the

---

(1) R.F.W. BADER and P.M. BEDDALL, J. Chem. Phys. 56, 3320 (1972)
(2) S. SREBRENIK and R.F.W. BADER, J. Chem. Phys. 63, 2945 (1975)
(x) on this point see BADER, Localization and Delocalization in Quantum Chemistry (loc. cit.).

total wave function ψ when there is in each loge a number of points equal to the number of electrons which is found in the loge when this event occurs. The function ψ_λ vanishes for all other positions of the points.
It is readily seen that :[1]

$$\psi = \sum_\lambda \psi_\lambda \qquad (1)$$

Now each event function can be expressed in terms of completely localized loge functions $L_{i\lambda}$ and of a correlation function f_λ. A completely localized loge function for a given event λ is a function depending on a number of points identical to the number of electrons which are found in the loge for that event and which vanishes for other positions of the points.[2] The event function ψ_λ can be written as

$$\psi_\lambda = a \prod_i L_{i\lambda} f_\lambda \sigma \qquad (2)$$

where a is an antisymmetrizer and σ the spin function.

Finally any wave function can be rigorously written as :

$$\psi = \sum_\lambda a \prod_i L_{i\lambda} f_\lambda \sigma \qquad (3)$$

As it is assumed that a leading event exists an approximate wave function can be obtained by taking account of this event only. Furthermore as we said that the correlation between loges is minimized we can also neglect the correlation function. The approximate wave function becomes :

$$\psi_{ap} = a \prod_i L_{i\lambda_o} \sigma \qquad (4)$$

if λ_o denotes the leading event :

In the framework of that approximation a unique function is associated to each loge, and therefore to each bond, each lone pair ..... . This is the more natural functional aspect of chemical bonds.
A difficulty arises : the fact that completely localized loge functions are discontinuous. This is why in my first proposal[3]

---

(1) C. ASLANGUL, R. CONSTANCIEL, R. DAUDEL, L. ESNAULT, E. LUDENA, Int. J. Quantum Chemistry 8, 499 (1974)
(2) E.V. LUDENA and AMZEL, J. Chem. Phys. 52, 5923 (1970).
(3) R. DAUDEL. Les Fondements de la Chimie Théorique, Gauthier Villars (1956) p. 200 (english version : The Fundamentals of Theoretical Chemistry, Pergamon (1968)).

I suggested the use in equation (4) of loge functions not completely localized in the loges. Rather it was only orthogonalized and roughly localized functions. Klessinger and McWeeny[1] suggested the introduction of strongly orthogonalized loge functions (called group functions) and described a variational procedure to calculate them. Recently a program for IBM electronic computers has been written which permits the calculation of group or loge functions expanded on a basis set of gaussian functions. Figure 12 shows the density contour maps of the loge functions associated with some loges of isobutane. It has been observed[2] that the loge functions of a given type (CH functions, CC functions ...) are well transferable from a molecule to another one. Furthermore it is seen that in the region of a given loge the density contours of the corresponding loge function have higher values than the density contours associated with other loges. This fact suggests estimating the frontiers of the loges by the collection of points in which intersect the contour lines of same value.
Figure 13 gives the result of that simple procedure.

The powerful P.C.I.L.O method[3] of calculating wave functions amounts to taking account of the event functions not introduced in equation (4) but present in equation (3) by following a perturbation procedure : the leading term corresponding to the leading event.

Let us consider again equation (4). If the loge functions are approximated by simple products of orbitals, orbitals are associated with each bond. They are called bond orbitals. Such bond orbitals can be often generated from molecular orbital under the effect of a unitary transform. A "size" of such localized orbitals has been recently defined[4]. It is the expectation value of the spherical quadratic moment operator with origin at the orbital centroid of charge :

---

(1) M. KLESSINGER and R. McWEENY, J. Chem. Phys. 42, 3343 (1965)
(2) M. SANCHEZ, P.D. DACRE, R. DAUDEL, S. KWUN, C. VALDEMORO and R. McWEENY, Int. J. Quantum Chemistry (in the press)
(3) S. DINER, J.P. MALRIEU and P. CLAVERIE, Theoretica Chimica Acta 13, 1, 18 (1969) and 15, 100 (1969).
(4) M.A. ROBB, W.J. HAINES and I.G. CSIZMADIA, J. Am. Chem. Soc. 95, 42 (1973).

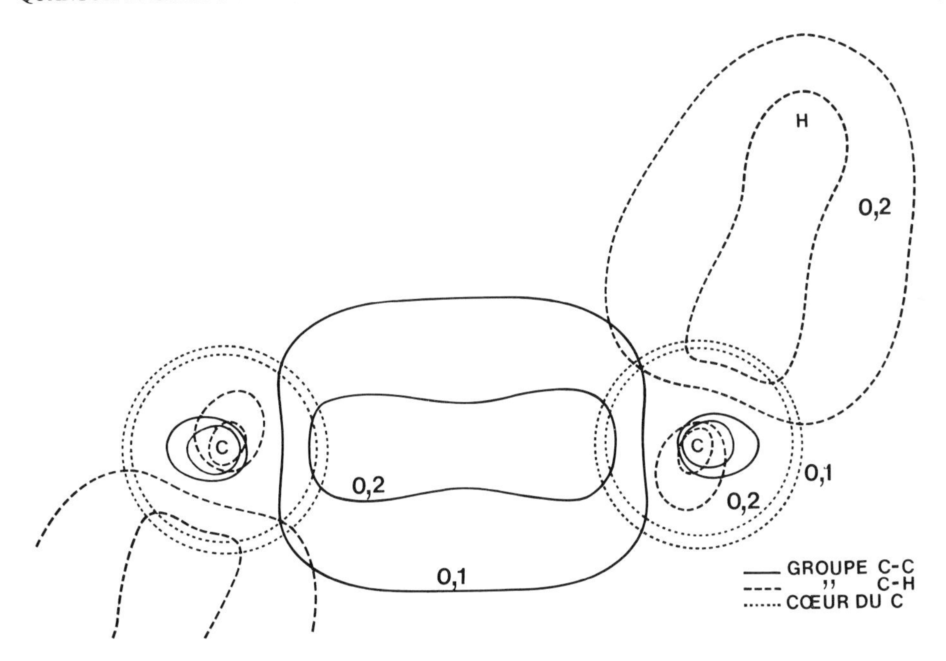

Figure 12 - Contour maps of group function in isobutane. (after M. SANCHEZ et al. Int. J. Quantum Chemistry) reproduced by permission.

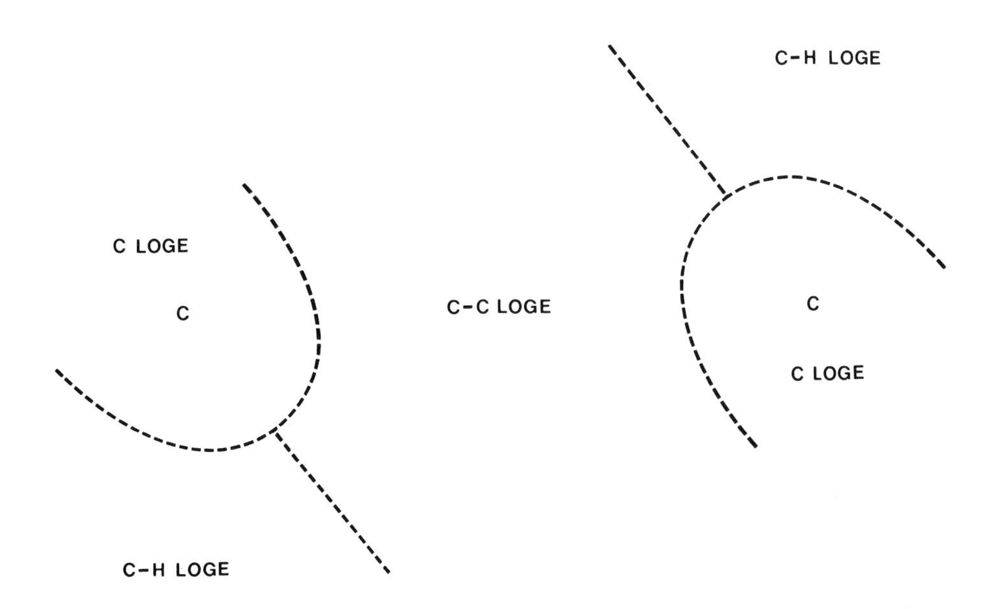

Figure 13 - Loges in isobutane (ibidem).

$$<r^2> = <x^2> + <y^2> + <z^2>$$

A more precise "shape is obtained if the three individual components $<x^2>$, $<y^2>$ and $<z^2>$ are considered separately.
Then, it is possible to represent the orbital by an ellipsoid with half-axes $<x^2>^{1/2}$, $<y^2>^{1/2}$, $<z^2>^{1/2}$. Figure 14 shows such ellipsoids for bond orbitals and lone pair orbitals in various molecules.[1] The angles contained by the rays between the end-points of the maximum width of the ellipsoid (at the centroid of charge) and the heavy nucleus provides a measure of the angular size of the various localized orbital.
It is seen that in some cases the size of a lone pair orbital can be smaller than the size $<r^2>$ of a bond orbital : this is for example the case of the molecule NH_3. This fact seems to be in contradiction with Gillespie postulate. But in fact the angular size of a lone pair is always larger than the angular size of a bond pair. The Gillespie bulk must be therefore interpreted as an angular size.

By using the size $<r^2>$ of an orbital it has been possible to extend to molecules the relation :

$$p^{3/2} v = cste$$

It takes a slightly different form[2]

Finally it must be emphasized that the consideration of the most localized orbitals can be confusing. Because in certain cases the most localized orbitals are not really localized.

To measure the degree of localizability of two orbitals f and g it is convenient to calculate the integral :[3]

$$S' = \int |f||g| \, dv$$

If the relative localization is complete the integral vanishes. If the orbitals completely overlap the integral is unity.

Between two localized bond CH orbital in methane :

$$S' = 0{,}404^{(4)}$$

(1) R. DAUDEL, M.E. STEPHENS, C. KOZMUTZA, E.KAPUY, J.D. GODDARD and I.G. CSIZMADIA, Int. J. Quantum Chem. (in press).
(2) R. DAUDEL, J.D. GODDARD and I.G. CSIZMADIA, Int. J. Quantum Chem. (in press).
(3) R. DAUDEL, Les Fondements de la Chimie Théorique, Gauthier Villars (1956).
(4) R. DAUDEL, M.E. STEPHENS, E. KAPUY and C. KOZMUTZA, Chem. Phys. Letters (in press).

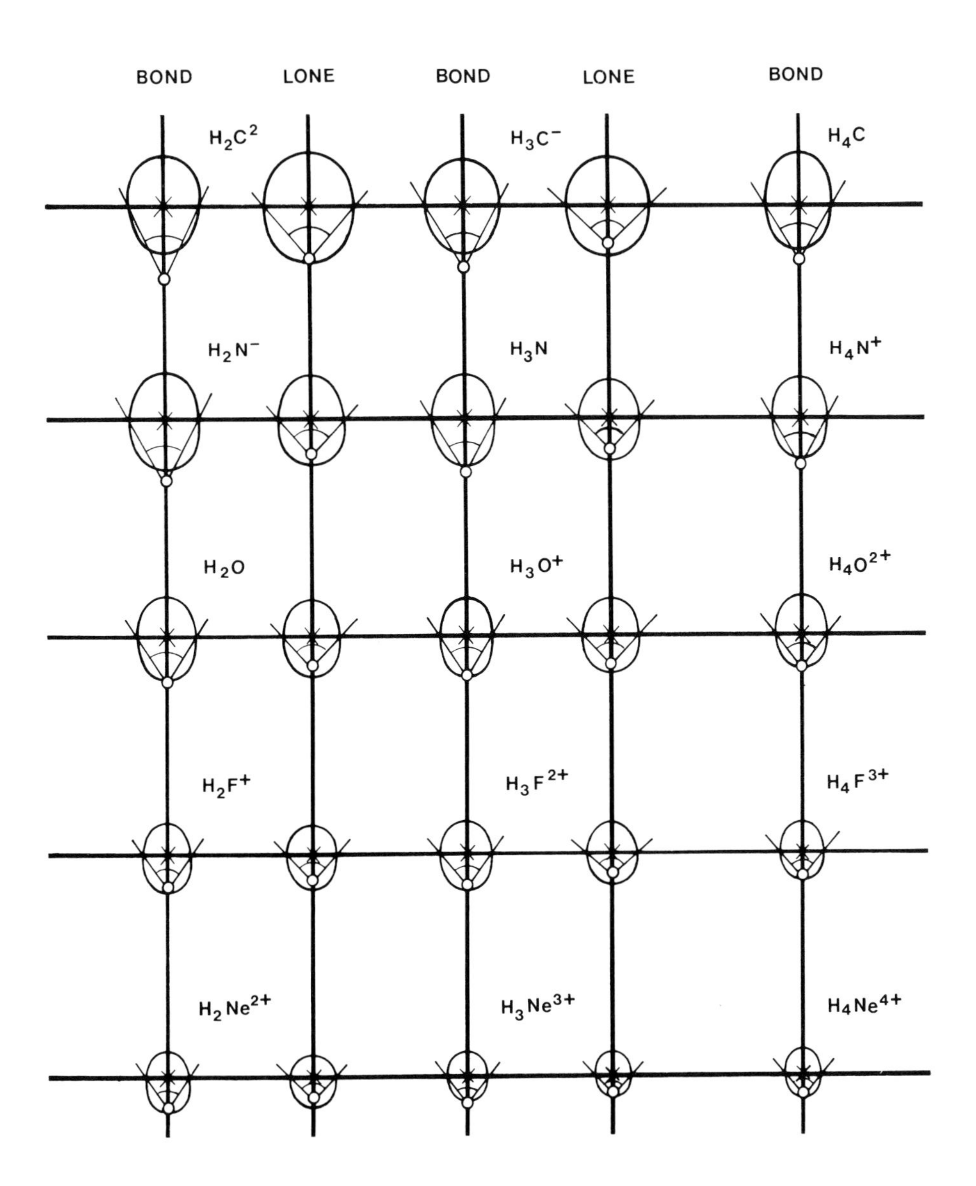

Figure 14 - Shapes of molecular orbital (after R. DAUDEL et al. Int. J. Quantum Chemistry) reproduced by permission.

It is seen that even in that classical example of two electron covalent bonds the localization of the most localized orbitals is far from being very significant.

There is an other procedure to analyze a wave-function in terms of group functions. It is based on functional separability (1). The group functions generated by following that method are not necessarily localized. The association of the obtained group functions with bond have to be made carefully.

(1) R. CONSTANCIEL, Phys. Rev. 11, 395 (1975).

PROPAGATOR THEORY OF ATOMIC AND MOLECULAR STRUCTURE

Yngve Öhrn

Quantum Theory Project
Department of Chemistry
University of Florida
Gainesville, Florida 32611

INTRODUCTION

When I accepted the invitation to speak at this conference, the organizers suggested that the title of my presentation should read "Methods Without Wavefunctions." Since I will present only some aspects of this subject employing the ideas of propagator theory, it might be appropriate to start this introduction with a few brief remarks on the general topic of methods without wavefunctions before concentrating on the central theme as given in the title of this paper.

It is not at all obvious which methods of use in quantum chemistry can be characterized as being "without wavefunctions." There are, however, a number of methods that purpose to calculate atomic and molecular electronic structure and properties without the explicit use of many-electron wavefunctions. Fully realizing the arbitrariness of any classification of such methods as well as the absence of sharp boundaries between any choice of groups, I find it useful to separate the three kinds of approaches: (i) Local Density Energy Functional Methods, (ii) Density Matrix Methods, (iii) Propagator or Green's Function Methods.

Local Density Energy Functional Methods

The first group of methods for the study of atomic and molecular electronic structure dates back to the Fermi-Thomas model[1], and to the Fermi-Thomas-Dirac model[2], which for atoms give a very simple representation of the potential in which an

B. Pullman and R. Parr (eds.), The New World of Quantum Chemistry, 57–78.

electron is moving. The electron charge density $n(\vec{r})$ plays a fundamental role in these methods and statistical treatments of the free electron gas yields energy expressions involving fractional powers of the electron charge density.[3] The Fermi-Thomas-Dirac total energy expression for a many-electron system thus in addition to the electron-nuclear attraction energy, linear in $n(\vec{r})$, and the electron-electron repulsion energy, quadratic in $n(\vec{r})$, consists of a total kinetic energy term, proportional to a volume integral over $n(\vec{r})^{5/3}$, and an exchange correction, proportional to a volume integral over $n(\vec{r})^{4/3}$. A proper quantum mechanical treatment of the kinetic energy uses the one-electron reduced density matrix[4] and the Laplacian rather than the density. Expressing the reduced one-matrix as a bilinear form of orbitals the total energy expression can be minimized by variation of these orbitals and a one-electron equation obtained for the determination of the orbitals. This kind of treatment was given by Gaspar[5], and by Kohn and Sham[6]. One-electron Schrödinger equations of this form were also studied by Slater[7].

The exchange potential in these one-electron equations is obtained to be proportional to the 1/3 power of the electron density, and is a local multiplicative potential rather than a non-local integral operator as in the Hartree-Fock treatment of a many-electron system. This simulation of the environment of an electron in a many-electron system by a single local potential is a simplifying feature of this approach, which has significantly contributed to its popularity. Early applications of this kind of models were limited to atoms[8], also with the thought of obtaining starting potentials for energy band calculations on perfect crystalline solids.

Recently applications of these kinds of methods to molecular systems have become more prominent. One of the more widely used methods is the MS-Xα method of Slater[9], and Johnson[10]. In this approach one has found that better agreement with experiment is obtained if the local exchange potential is multiplied by a semi-empirical scaling parameter, (α), the value of which in some applications has been chosen different for different elements and also for core and valence electrons of the same element. The multiple scattering formalism or the scattered wave model[11] is used for the self-consistent solution of these one-electron equations and the most common and simplest subdivision of space used in this connection has been the so called muffin-tin potential. The inadequacies of the muffin-tin assumptions have become obvious[12], and more recent attempts to solve the equations within a general cellular approach have been suggested[13], although many questions about the general feasability and efficiency are still unanswered. A very good review of the Xα energy functional method has recently been given by Connolly.[14]

These models also exist in spin unrestricted versions, where the charge density of the "spin up" electrons and the charge density of the "spin down" electrons are distinguished in the theory. The MS-Xα model has in this form been applied to a great variety of systems primarily with the idea of obtaining correlation between electron binding energies as obtained, say, by photoelectron spectra and the orbital energies calculated with the use of the transition state[15].

A related method to that of the Xα model for the calculation of electronic structure, is the spin density functional formalism of von Barth and Hedin[16]. This approach uses inherently systematic approximations of the formally exact expression of the exchange-correlation potential of a spin polarized electron gas to arrive at a set of coupled one-electron equations, which are solved self-consistently. The applications have mostly been to solid-state problems, particularly for transition metals. It is impressive that the total energy calculations within this model will predict correctly the relative stability of the para- and ferromagnetic phases of transition metals. The spin density functional formalism has also been applied[17] to the calculation of dissociation energies and bond distances for a number of small molecules and molecular ions involving H, and He atoms.

Another related model is the approach introduced by Gordon, and Kim[18] for the calculation of potential energy surfaces describing the interaction between closed shell atoms and molecules. In this approach the total energy functional is evaluated using a total electron charge density approximated as the sum of the charge densities of the closed shell constituents at interatomic distances greater than half that of the potential energy minimum. The nuclear-electron attraction energies and the electron-electron repulsion energies are evaluated by numerical quadrature using these additive charge densities obtained from Hartree-Fock atomic calculations. The kinetic energy, electron exchange, and electron correlation energy are evaluated using formulae from the electron gas theory. The correlation energy of an interacting electron gas is estimated using the well-known expansion formulae valid at high and low densities and employing an ad hoc interpolation formula at intermediate densities, which actually occur in these systems.

An interesting comparison could be made between the approach of Gordon and Kim and the method of Hedin and Lundquist[19], which is a precursor to the spin density functional method of von Barth and Hedin[16], and which uses more recent advances in the theory of the interacting electron gas to obtain electron correlation energies.

Almost all the more recent local density functional schemes

to calculate electronic structure refer to the work of Hohenberg and Kohn[20], Kohn and Sham[6], and Sham and Kohn[21]. The justification for studying total energy functionals is usually given in terms of a very general theorem[20] given by Hohenberg and Kohn stating that for a non-degenerate ground state and a given external potential, $v(\vec{r})$, the ground state energy can be expressed as

$$E[n] = \int v(\vec{r})n(\vec{r})d\vec{r} + F[n], \tag{1}$$

where F[n] is a universal functional of the electron charge density, $n(\vec{r})$, and E[n] is minimized by the correct charge density.

It has been mentioned many times that this theorem only states the existence of a universal energy functional of the density and does not provide any guide to the construction of such a functional. It is hence of limited value for actual calculations, and all the different schemes in use today rely heavily on physical arguments and ideas from the theory of the electron gas, for the construction of energy functionals of the electron charge density. Still the actual results of applications of these methods seem to indicate that they can be successfully used to describe electronic structure for a great variety of systems.

Formal objections to this kind of procedure can be raised. First of all, the constructed energy functionals are somewhat arbitrary and are most likely not universal. Secondly, the connection to established many-electron quantum mechanics is somewhat tenuous since one has given up the notion of a many-electron Hamiltonian and a well defined averaging procedure, which is present if one chooses to use many-electron wavefunctions or some explicit statistical density operator. The variational treatment of the energy functionals to obtain one-electron equations for the amplitudes which make up the density, has usually the only constraint that the density should be normalized to the correct number of electrons. The question of N-representability[22] is usually side stepped, something which is common for many of the methods which do not use many electron wavefunctions, or explicit ensemble averages[23]. The usefulness of these approximate total energy functionals is of course from a practical point of view dependent upon how efficiently they allow the calculation of ground state potential energy surfaces which closely simulate the exact ones. In this respect it is interesting to note the work of Trickey et al.[24] on rare gas crystals using the APW Xα model.

The usefulness of the one-article equations derived from these functionals can be manifold. The equations might be thought of as orbital generators which produce molecular orbitals well suited for many-electron wavefunction or propagator methods.

Whether this provides any advantages over standard Hartree-Fock methods is still an open question. The one-particle equations can also be used for the calculation of ionization and excitation energies, E_k. For applications of this type, the approach of Hedin and Lundquist has definite advantages since it attempts to find the exchange-correlation potentials from the formally exact equation for such quantities:

$$h(x)\Phi_k(x) + \int\Sigma(x,x';E)\Phi_k(x')dx' = E\Phi_k(x), \tag{2}$$

where x is a compound space spin variable, h(x) is the single-particle operator of kinetic energy and a common potential due to the presence of the nuclei, and Σ is a non-local, complex, and energy dependent potential describing exchange and correlation effects. The amplitudes, $\Phi_k(x)$ are used to calculate oscillator strengths. In order to obtain the single-particle reduced density matrix, and the density, one would calculate the complete Greens function or electron propagator, G(x,x';E) from:[25,26]

$$h(x)G(x,x';E) + \int\Sigma(x,y;E)G(y,x';E)dy = \delta(x-x'). \tag{3}$$

This is the topic of the main part of this paper. When G is obtained the total energy and other properties can also be readily calculated.

Density Matrix Methods

Husimi[4] seems to have been the first to consider the idea of a reduced density matrix. Löwdin[27], and McWeeny[28] have treated the many-electron problem with the use of density matrices. The question of the feasability of calculating the total energy by the direct variation of the second-order density matrix Γ was raised by Löwdin[27], who considered the following energy functional:

$$E[\Gamma] = \int[Nh(x) + \binom{N}{2}g(x,y)]\Gamma(xy|xy)dxdy/\int\Gamma dxdy, \tag{4}$$

where $g(x_1,x_2)$ is the potential energy of Coulomb repulsion between two electrons, say, $g(x_1,x_2) = |\vec{r}_1-\vec{r}_2|^{-1}$, and N is the number of electrons.

In order for this energy expression to correspond to a correct quantum mechanical expectation value, Γ must be derivable from a many-electron wavefunction. This means that direct variation of Γ over a class of functions must have a number of constraints, which would secure bounding properties of $E[\Gamma]$ and physically meaningful results. This problem has been referred to as the N-representability problem[22], and many years of research have produced a number of necessary and sufficient conditions[22,29] that density matrices, in particular the one-particle and the two-particle reduced density matrices, have to satisfy in order to be N-representable in the pure sence, i.e. correspond to a proper many-electron wavefunction, or to be N-representable in the statistical sense, i.e. to correspond to a statistical density operator.

At the 1972 Summer Research Conference at Boulder, Coleman[30] reported the formal solution of the N-representability problem, and Davidson[31] outlined a procedure for the direct variation of the second-order reduced matrix taking all the appropriate constraints into account. The complexity of such a procedure still seems too great to really make it a practical approach to the many-electron problem.

The idea that approximately N-representable density matrices might prove to be useful as well as the urge to understand in detail the implications of every condition has led many investigators to study special cases. Several significant papers in this field have been published in the last few years, and an excellent source of ideas and references is the proceedings from the Density Matrix Conference held at Queen's University, Kingston, Canada[32,33]. I find the work by Garrod, Mihailovic, Rosina[35] particularly interesting for applications to atoms and molecules. Some of the applications have been carried out for ground state energies of light nuclei[36], and the comparisons with the results from complete configuration interaction calculations have relevance also for applications to electron systems.

The use of density matrices[37] and transition matrices as useful quantities in wavefunction calculations is well established and of particular importance for comparative studies[38] of wavefunctions, which are obtained from widely different approximations and basis sets.

The direct calculations of the one- or two-particle density matrix is, however, still an extremely difficult problem for a general atomic or molecular system. Most progress has been made for the case of totally symmetric ground states, and the treatment of degenerate states and excited states in general is still very much in its infancy.

It is interesting to note that the one-electron density matrix $\gamma(x|x')$, is obtained as a contour integral of the electron propagator, $G(x,x';E)$:[26]

$$\gamma(x|x') = -(2\pi i)^{-1}\int_C G(x,x';E)dE \tag{5}$$

where the contour, C, in the complex E-plane consists of a vertical line at Re(E)=μ and an infinite semicircle to the left. The parameter μ plays the role of chemical potential and is defined such that the trace of γ equals the correct number of electrons, $\int\gamma(x|x)dx = N$. The electron propagator satisfies an equation of motion, which can be solved using well-defined approximations[39], and the density matrix obtained even from very simple, approximate propagators have proven to be quite adequate. For instance natural spin orbitals have been obtained[40] in good agreement with those obtained from full CI calculations for small atomic systems. Also charge and spin densities[41,42] of good quality can be obtained from approximate calculations of the electron propagator. Similarly, the two-electron reduced density matrix is obtained from the two-electron Green's function[43]. We will in particular consider the particle-hole Green's function or polarization propagator[26], from which the two-electron density matrix can be readily obtained for the many-electron ground state of quite complicated systems[44,45]. These methods for obtaining the reduced density matrices are usually not variational, but they can readily be systemetized by perturbation theory, and appropriate care in satisfying sum rules and other consistency criteria makes it possible at least for practical purposes to circumvent the N-representability problem.

Thus, there is ample evidence that propagator methods are important for the general approach to the many-electron problem which we can inadequately label "methods without wavefunctions", and which are inapired by the strive to "eliminate the redundancies" of the many particle coordinates and labels of the many-electron wavefunction. The propagators satisfy equations of motion, which can be solved for complicated many-electron systems, and in addition to ground-state properties they give information on transition energies and oscillator strengths.

In the next sections I attempt to present a unified approach to the calculation of double-time Green's functions or propagators[26], using the language of super-operators and perturbation theory. The literature on this topic is already quite extensive and I will try to compare some of the major approaches to calculate atomic and molecular structure from propagator theory.

PROPAGATORS

The concept of a double-time Green's function[46] addresses the very general problem of calculating the effect on an observation at time t represented by an operator, $A^\dagger(t)$, by another observation at time t' represented by the operator $B(t')$.[26] The double-time Green's function is defined as

$$\langle\langle A^{\dagger}(t);B(t')\rangle\rangle = -i\theta(t-t')\langle A^{\dagger}(t)B(t')\rangle \pm i\theta(t'-t)\langle B(t')A^{\dagger}(t)\rangle, \tag{6}$$

where the Heaviside step function, $\theta(t)$, gives an explicit time ordering, and the average,

$$\langle A^{\dagger}(t)B(t')\rangle = \mathrm{Tr}\{\rho A^{\dagger}(t)B(t')\} \tag{7}$$

is defined relative a density operator, ρ, which might be chosen as the density operator for a statistical ensemble or a pure state. The operators A, and B will in the general case be expressed as linear combinations of products of electron field operators, $\psi(x,t)$ and their adjoints, $\psi^{\dagger}(x,t)$. The plus sign in the second term of Eq. (6) applies only to the case of both A, and B containing products of odd numbers of electron field operators only, i.e. being fermion-like.

Introducing a set of orthonormal spin orbitals, $\{u_s(x)\}$ yields the electron field operators

$$a_s = \int u_s^*(x)\psi(x,t)dx$$

and

$$a_s^{\dagger} = \int u_s(x)\psi^{\dagger}(x,t)dx \tag{8}$$

associated with that basis of spin orbitals. The many-electron hamiltonian, H, can then be expressed as[26]

$$H = \sum_{uv}\sum h_{uv}a_u^{\dagger}a_v + \frac{1}{4}\sum_{uvxy}\sum\sum\sum(uv||xy)a_u^{\dagger}a_x^{\dagger}a_y a_v \tag{9}$$

where

$$h_{uv} = \int u_u^*(x)h(x)u_v(x)dx \tag{10}$$

and

$$(uv||km) = (uv|km) - (um|kv) \tag{11}$$

with

$$(uv|km) = \int u_u^*(x)u_v(x)g(x,y)u_k^*(y)u_m(y)dxdy \tag{12}$$

We are interested in the case when the density operator, ρ, and the many-electron hamiltonian, H, are constant in time. Then $\langle\langle A^{\dagger}(t);\ B(t')\rangle\rangle$ is a function of the time interval $(t-t')$, and we can readily study the Fourier transform ($\hbar=1$):

$$\langle\langle A^{\dagger};B\rangle\rangle_E = \int_{-\infty}^{+\infty} d(t-t')\langle\langle A^{\dagger}(t);B(t')\rangle\rangle e^{iE(t-t')} \quad (13)$$

which satisfies the algebraic equation:[26]

$$E\langle\langle A^{\dagger};B\rangle\rangle_E = \langle A^{\dagger}B \pm BA^{\dagger}\rangle + \langle\langle [A^{\dagger},H]_-;B\rangle\rangle_E \quad (14)$$

The commutator, $[A^{\dagger},H]_-$, comes from the fact that $A^{\dagger}$ (and B) is a Heisenberg operator.[26]

The spectral representation[47] of the Fourier transform of the double-time Green's function,[26]

$$\langle\langle A^{\dagger};B\rangle\rangle_E = \sum_{mn}\sum \langle n|A^{\dagger}|m\rangle\langle m|B|n\rangle\left[\frac{\rho_n}{E-E_m+E_n+i\eta} \pm \frac{\rho_m}{E-E_m+E_n-i\eta}\right] \quad (15)$$

exhibits the possible physical interpretation of such a propagator. Singularities in terms of simple poles (or cuts) occur at values of the energy parameter equal to differences of stationary state energies, E_m , corresponding to eigenstates, $|m\rangle$, of the many-electron hamiltonian, H, of Eq. (9). The corresponding residues would give transition probability amplitudes. Expression (15) is derived assuming that the density operator, ρ, is diagonal in the eigenstates of the hamiltonian.

Some approximate treatments of the equation of motion (14) of the propagators starts with the iterated form of Eq. (14):

$$\langle\langle A^{\dagger};B\rangle\rangle_E = E^{-1}\langle[A^{\dagger},B]_{\pm}]\rangle + E^{-2}\langle[[A^{\dagger},H]_-,B]_{\pm}\rangle + E^{-3}\langle[[[A^{\dagger},H]_-,H]_-,B]_{\pm}\rangle + \ldots \quad (16)$$

which has been discussed by Linderberg and Öhrn[26], and given the name of moment expansion for the corresponding spectral density function.[26]

A somewhat abstract but very convenient representation of propagators is achieved by introducing a linear space of field operators, spanned by independent products of simple electron field operators. If we label by X, and Y, two general elements of this linear space, we can define a scalar product,

$$(X|Y) = \langle[X^{\dagger},Y]_{\pm}\rangle = \mathrm{Tr}\{\rho[X^{\dagger},Y]_{\pm}\} \quad (17)$$

The superoperator hamiltonian, $\hat{H}$, and the superoperator identity, $\hat{I}$, are introduced and defined by the operations:

$$\hat{H}X = [X,H]_-, \quad \text{and} \quad \hat{I}X = X \tag{18}$$

where H is the many-electron hamiltonian from Eq. (9). The superoperator resolvent, $(E\hat{I} - \hat{H})^{-1}$, is then used to express the propagator.

The idea of superoperators has been used in statistical physics[48], and was later used by Goscinski, and Lukman[49] for propagators. Decoupling of the electron propagator equations has been extensively discussed by Pickup and Goscinski[50], by Purvis and Öhrn[51,52,53] in terms of this language, and by Jørgensen[54] for the polarization propagator.

One uses the identity,

$$(E\hat{I} - \hat{H})^{-1} = E^{-1}\hat{I} + E^{-1}\hat{H}(E\hat{I} - \hat{H})^{-1} \tag{19}$$

to obtain the following expression for the propagator:

$$\langle\langle A^\dagger;B\rangle\rangle_E = (A|(E\hat{I} - \hat{H})^{-1}B) \tag{20}$$

which is easily verified by referring to the moment expansion in Eq. (16).

Approximations to the propagator is then arrived at by using the idea of inner projection, emphasized by Löwdin[55,56], and by considering statistical or pure state density operators, ρ, which are given correct through some order of perturbation theory[57]. In the next section we will consider this problem for the electron propagator.

Applications of propagator techniques leads to large arrays of numbers, which originate from expressions $(X|\hat{H}Y)$, or $(\hat{H}X|Y)$, for a variety of operators X, and Y. The question arises if such arrays can be obtained from one another by transposition or by taking the adjoint, and thus if great savings in generation and storage of these arrays could be achieved. This question is directly related to whether

$$(X|\hat{H}Y) = (\hat{H}X|Y), \tag{21}$$

i.e. is the superoperator hamiltonian, $\hat{H}$, hermitian. Nehrkorn, Purvis, and Öhrn[57] have studied this problem in some detail, and Linderberg[58] has given a proof, which shows that when the density operator is correct through order n in perturbation theory, then $\hat{H}$ is hermitian to at least order n + 1. This can be seen from the identity

$$(X|\hat{H}Y) - (\hat{H}X|Y) = \langle[X^{\dagger},Y]_{\pm},H]_{-}\rangle = Tr\{[H,\rho]_{-}[X^{\dagger},Y]_{\pm}\}. \tag{22}$$

When the density operator is correct through order n, i.e.

$$\rho = \rho_o + \delta\rho + \delta^2\rho + \ldots + \delta^n\rho \tag{23}$$

and for $H = H_o + V$, we have

$$[\rho_o,H_o]_{-} = 0, \text{ and } [\delta^k\rho,H_o]_{-} + [\delta^{k-1}\rho,V]_{-} = 0,$$
$$\text{for } k = 1,2,3,\ldots,n,$$

then

$$(X|\hat{H}Y) - (\hat{H}X|Y) = Tr\{\delta^n\rho,V]_{-}[X^{\dagger},Y]_{\pm}\} \tag{24}$$

which completes the proof.

In the next section, the implications of these results for approximate treatments of the electron propagator are discussed.

The electron propagator

The particular choice of operators

$$A = a_r, \text{ and } B = a_s \tag{25}$$

results in the propagator, $\langle\langle a_r^{\dagger};a_s\rangle\rangle_E$, which is called the electron propagator in a discrete representation. Comparing with Eqs. (3) and (5), we would have the general connection

$$G(x,x';E) = \langle\langle\psi^{\dagger}(x);\psi(x')\rangle\rangle_E$$
$$= \sum_{rs}\sum u_r^*(x)\langle\langle a_r^{\dagger};a_s\rangle\rangle_E u_s(x') \tag{26}$$

with the configuration space.

The propagators,

$$\langle\langle a_r^{\dagger};a_s\rangle\rangle_E = (a_r|(E\hat{I} - \hat{H})^{-1}a_s) \tag{27}$$

can be arranged in a matrix:

$$\langle\langle \mathbf{a}^{\dagger};\mathbf{a}\rangle\rangle_E = (\mathbf{a}|(E\hat{I} - \hat{H})^{-1}\mathbf{a}), \tag{28}$$

and an inner projection manifold chosen from the field operator space. We then have

$$(\mathbf{a}|\mathbf{h})(\mathbf{h}|(E\hat{I} - \hat{H})\mathbf{h})^{-1}(\mathbf{h}|\mathbf{a}) = \mathbb{G}(E) \tag{29}$$

as an approximate expression for the electron propagator matrix. Any particular approximation is achieved by choosing the inner projection manifold, $\mathbf{h}$, and choosing the density operator, ρ, with respect to which the scalar product, in the field operator space, is defined. When the inner projection manifold, $\mathbf{h}$, is complete, and we have the exact density operator, then

$$\mathbb{G}(E) = \langle\langle \mathbf{a}^{\dagger};\mathbf{a}\rangle\rangle_E. \tag{30}$$

The particular choice, $\mathbf{h} = \mathbf{a}$, leads to

$$\mathbb{G}^{-1}(E) = (\mathbf{a}|(E\hat{I} - \hat{H})\mathbf{a}) \tag{31}$$

where

$$\begin{aligned}(a_r|(E\hat{I} - \hat{H})a_s) &= E\langle[a_r^{\dagger},a_s]_+\rangle - \langle[a_r^{\dagger},[a_s,H]_-]_+\rangle \\ &= E\delta_{rs} - F_{rs}\end{aligned} \tag{32}$$

with $\{F_{sr}\}$ being the Fock matrix[26] for the particular choice of density operator, as given in Eqa. (34, 35 or 36). If we simultaneously diagonalize this matrix and the first order reduced density matrix, with elements $\langle a_x^{\dagger}a_y\rangle$, we obtain

$$F_{sr} = h_{sr} + \sum_{xy}\sum (sr||xy)\langle a_x^{\dagger}a_y\rangle = \epsilon_r\delta_{sr} \tag{33}$$

where $\langle a_x^{\dagger}a_y\rangle = \delta_{xy}\langle n_x\rangle$. The spin orbitals, $u_r(x)$, corresponding to the spin orbital energies, ε_r , i.e. the functions that diagonalize $\{F_{sr}\}$ for a particular choice of occupation numbers n_x , can now be introduced as the spin orbital basis that defines the electron field operators in Eq. (8)[50]. In practice this is done by solving the atomic or molecular SCF problem, and finding the occupied and the virtual spin orbitals with their spin orbital energies. Depending on our choice of occupation numbers, we have a restricted Hartree-Fock problem or an unrestricted one. We can also have other special cases, even fractional occupation numbers as for the transition state,[59,60,61] or the grand canonical and canonical Hartree-Fock methods[23,23].

This is consistent with the average formation with respect to a pure state

$$|HF\rangle = a_N^\dagger a_{N-1}^\dagger \ \ldots \ a_1^\dagger|\text{vacuum}\rangle \tag{34}$$

or for the case of a transition state, the mixture[62]:

$$|N-\tfrac{1}{2}\rangle = 2^{-\frac{1}{2}}(a_N^\dagger+1)a_{N-1}^\dagger a_{N-2}^\dagger \ \ldots \ a_1^\dagger|\text{vacuum}\rangle \tag{35}$$

Both these situations are special cases of the density operator[26]

$$\rho = \prod_k[1 - \langle n_k\rangle + (2\langle n_k\rangle - 1)n_k] \tag{36}$$

with $n_k = a_k^\dagger a_k$, which applies to a variety of choices of fractional and integer fermion occupation numbers. The Hartree-Fock pure state of Eq. (34) occurs for the following choice of occupation numbers:

$$\begin{aligned} &\langle n_k\rangle = 1 \text{ for } k \le N,\\ \text{and } &\langle n_k\rangle = 0 \text{ for } k > N, \end{aligned} \tag{37}$$

while the transition state of Eq. (35) occurs when

$$\begin{aligned} &\langle n_k\rangle = 1 \text{ for } k < N,\\ &\langle n_k\rangle = \tfrac{1}{2} \text{ for } k = N,\\ \text{and } &\langle n_k\rangle = 0 \text{ for } k > N. \end{aligned} \tag{38}$$

The propagator can now be improved by enlarging the inner projection manifold and also by improving the description of the density operator. This can be done in a systematic manner by using Rayleigh-Schrödinger perturbation theory[50-53]. The inner projection manifold can be chosen as

$$\mathbb{h} = \{\mathbb{a},\mathbb{f}\} \tag{39}$$

which results in a partitioning of the propagator matrix in Eq. (29). When $(\mathbb{a}|\mathbb{f}) = \mathbb{0}$, in addition to $(\mathbb{a}|\mathbb{a}) = \mathbb{1}$, one finds that

$$\begin{aligned} \mathbb{G}^{-1}(E) = &(\mathbb{a}|(E\hat{I} - \hat{H})\mathbb{a})\\ &- (\mathbb{a}|\hat{H}\mathbb{f})(\mathbb{f}|(E\hat{I} - \hat{H})\mathbb{f})^{-1}(\mathbb{f}|\hat{H}\mathbb{a}) \end{aligned} \tag{40}$$

from the inverse of partitioned matrix.[63]

The electron propagator was first presented in this partitioned form by Linderberg and Öhrn[64] using an operator expansion technique, and a self-consistent solution was given[65] with application to the Pariser-Parr-Pople model.

The simplest choice of inner projection manifold would be

$$\mathbb{f} = \mathbb{f}_3 = \{f_{klm}\}, \tag{41}$$

with

$$f_{klm} = N_{klm}^{-\frac{1}{2}}[a_k^{+}a_l a_m + \delta_{km}\langle n_k\rangle a_l - \delta_{kl}\langle n_k\rangle a_m]$$

and the normalization constant

$$N_{klm} = [\langle n_k\rangle(1 - \langle n_l\rangle - \langle n_m\rangle) + \langle n_l\rangle\langle n_m\rangle]$$

when the average is obtained from the density operator in Eq. (36). In terms of this independent particle ensemble density operator the first term in Eq. (40) is of zeroth order in electron interaction. The quantities $(\mathbb{a}|\hat{H}\mathbb{f})$, and $(\mathbb{f}|\hat{H}\mathbb{a})$ are then of first order, and the second term of Eq. (40) would be correct through second order if we include only the zeroth order terms of $(\mathbb{f}|(EI-H)\mathbb{f})^{-1}$. We find the following expressions:

$$(a_i|(E\hat{I} - \hat{H})a_j) = (E - \epsilon_j)\delta_{ij} \tag{42}$$

$$(a_i|\hat{H}f_{klm}) = N_{klm}^{\frac{1}{2}}(mi||lk)$$
$$(f_{\varkappa\lambda\mu}|\hat{H}a_j) = N_{\varkappa\lambda\mu}^{\frac{1}{2}}(j\mu||\varkappa\lambda) \tag{43}$$

$$(f_{\varkappa\lambda\mu}|(E\hat{I} - \hat{H})f_{klm}) =$$
$$\delta_{k\varkappa}\delta_{l\lambda}\delta_{m\mu}(E - \epsilon_m - \epsilon_l + \epsilon_k). \tag{44}$$

These approximations give a particularly simple and effective form for the self-energy, which we identify as the second term in Eq. (40). This is a discrete representation of the (exchange) correlation potential $\Sigma(x,x';E)$ of Eq. (3), but with the Hartree-Fock part subtracted out and included in the first term of Eq. (40). The choice of the transition state occupation numbers of Eq. (38), with the occupation number, $\langle n_N\rangle = 1/2$, yields an electron propagator with one pole in very good agreement with a principal ionization energy.[62] In Table 1 the binding energies of 1s, 2s and 2p-electrons of the neon atom corrected for relativistic effects[66] are compared to the results from electron propagator calculations[62] using the approximations described here, and a basis of twenty-four Slater-type orbitals.[62]

Table 1

Ionization energies of the neon atom (eV)

Method	2p	2s	1s
HF, orbital en.	23.13	52.52	891.64
HF, Trans. state, orbital en.	19.65	49.10	868.07
Propagator, trans. state. Eq. (42)-(44)	21.02	48.19	869.28
EXP. Non-rel.	21.56	48.21	869.12

Many other approximate schemes have been used in calculating the electron propagator for atomic and molecular systems. The ingredients of the approximation method are, in this formalism, the choice of density operator defining the scalar product in the superoperator space, and the choice of inner projection manifold in Eq. (40). The pure state density operator obtained from Eq. (34), can be augmented by single, double and higher order excitations out of the Hartree-Fock state, and the relative weights of these higher order terms determined by perturbation theory. The inner projection manifold can be chosen as $\mathbb{h} = \{\mathbb{f}_1, \mathbb{f}_3, \mathbb{f}_5\}$ where

$$\mathbb{f}_1 = \{a_k\},\ \mathbb{f}_3 = \{a_k^\dagger a_l a_m\},\ \text{and}$$
$$\mathbb{f}_5 = \{a_j^\dagger a_k^\dagger a_l a_m a_n\},$$

and where all elements are orthonormalized[53,57]. A systematic treatment can be obtained by referring to perturbation theory[67], writing $H = H_o + V$, where

$$H_o = \sum_{kl}\sum F_{kl} a_k^\dagger a_l = \sum_k \epsilon_k n_k$$

and including terms through a certain order in the electron interaction. We can quite generally rewrite Eq. (40) to read:

$$\{\mathbb{G}^{-1}(E)\}_{ij} = (E - \epsilon_i)\delta_{ij} - \Sigma_{ij}(E) \tag{45}$$

where ϵ_i is a spin orbital energy, and $\Sigma_{ij}(E)$ is a matrix element of the self-energy (except for the Hartree-Fock potential, the effect of which is included in the spin orbital energies) the most important part of which has a pole structure due to the inverse matrix $(\mathbb{f}|(E\mathrm{I} - H)\mathbb{f})^{-1}$. The solution of the equation

$$[E\mathbb{1} - \mathbb{F} - \Sigma(E)]\mathbb{G}(E) = \mathbf{1}, \tag{47}$$

by direct pole residue search has been described elsewhere[51,53] together with results for photoelectron spectra of small atomic and molecular systems[52].

Most applications have been using a pure state density operator for the definition of the scalar product of Eq. (17). This means that there is a reference state, $|N\rangle$, such that

$$\mathrm{Tr}\{\rho \; ...\} = \langle N| ... |N\rangle. \tag{48}$$

Starting from the Hartree-Fock ground state for the N-electron systems as the zeroth approximation, this reference state can be systematically improved:

$$|N\rangle = [1 + \Sigma\, K^{qr}_{ab} a^{\dagger}_{q} a^{\dagger}_{r} a_{a} a_{b} + \Sigma\, K^{q}_{a} a^{\dagger}_{q} a_{a} + \; ... \;]|HF\rangle, \tag{49}$$

where the sums run over occupied spin orbitals a, and b, of $|HF\rangle$ in Eq. (34), and unoccupied ones q, and r. The double excitations with coefficient K^{qr}_{ab} contributes[57] already in first-order of Rayleigh-Schrödinger perturbation theory, while single, triple and quadruple excitations start contributing to $|N\rangle$ in second order. The approximations used by other workers in the field can now be described by the choice of reference state $|N\rangle$ which is employed, the inner projection manifold used, and what additional approximations are deviced to calculate the inverse of $(\mathbb{f}|(EI-H)\mathbb{f})$, which can be an extremely large matrix.

Cederbaum[68,69] at Munich has developed a powerful scheme for the calculation of the electron propagator. He has used diagrammatic perturbation theory to arrive at the working equations. This approach can be described in our language in the following manner:

1. $|N\rangle$ is chosen as $|HF\rangle$ plus single and double excitations.
2. $\mathbb{h} = \{\mathbb{f}_1, \mathbb{f}_3\}$
3. We write $(\mathbb{f}_3|(E\hat{I} - \hat{H})\mathbb{f}_3) = m(E) + M$, where only the diagonal part, $m(E)$, depends on the energy parameter, and then iterat the identity $(m + M)^{-1} = m^{-1} - m^{-1}M(m + M)^{-1}$ only keeping terms through third order in the electron interaction[67] for the self-energy.

This method and slight modifications of it have been applied to the calculation of ionization energies and one-electron properties of a variety of molecular systems[70]. The results are extremely impressive and the computational work of Cederbaum and his co-workers is now being applied to molecules of chemical interest also including vibrational structure.[73]

Simons[71] has developed a so-called third-order equation of motion (EOM) method and applied it to the calculation of electron binding energies in small molecules, particularly electron affinities. His approach can be described as follows:

1. $|N\rangle$ is chosen as $|HF\rangle$ plus double excitations.
2. $\mathbb{h} = \{\mathbb{f}_1, \mathbb{f}_3\}$.
3. We approximate $(\mathbb{f}_3|(E\hat{I} - \hat{H})\mathbb{f}_3)$ by its diagonal part m(E), and keep terms through third-order in electron interaction for the self energy. (This does not include all third-order terms).[67]

Quite reliable results seem to be obtainable with this method although some sensitivity to the choice of orbital basis might be found in some cases.

Öhrn and co-workers[51,52,53] have developed an approach to the electron propagator which can be described in the following terms:

1. $|N\rangle$ is chosen as $|HF\rangle$ or $|HF\rangle$ plus double excitations.
2. $\mathbb{h} = \{\mathbb{f}_1, \mathbb{f}_3\}$, or $\mathbb{h} = \{\mathbb{f}_1, \mathbb{f}_3, \mathbb{f}_5\}$.
3. $(\mathbb{f}_3|(EI-H)\mathbb{f}_3)$ or the more complicated expression coming from the inclusion of $\mathbb{f}_5$ is kept in closed form, symmetry blocked, and diagonalized before the pole and residue search in the solution of Eq. (47) is carried out. Approximate versions of this procedure are implemented.

There are certain computational advantages to this procedure in spite of the transformations which are necessary[67]. The particular approximations for the hole part of the electron propagator which have been employed in the calculation of the inverse, have been tested[72] within a very limited basis for the nitrogen molecule and compared with configuration interaction results for both poles and residues of the electron propagator. The results show that the approximate propagator calculations adequately simulate the exact result of the model, which is a full configuration interaction calculation of the neutral ground state and also for all the positive ion states in the energy range of interest.

In addition to the increasing activity in the application of the electron propagator calculations to molecular electronic spectra and properties, there is a growing interest in the application of polarization propagator techniques to molecular spectra and properties. In the next section I outline very briefly the most important development in this area.

The polarization propagator

When we consider a pure Hartree-Fock state we can identify occupied spin orbitals $u_a(x)$, or $u_b(x)$, and unoccupied spin orbitals $u_n(x)$ or $u_m(x)$ with their corresponding electron field operators. It is then possible to define particle-hole operators $q_k^\dagger = a_m^\dagger a_a$, and hole-particle operators $q_1 = a_b^\dagger a_n$ and to consider the polarization propagator or particle-hole propagator[26]:

$$\mathbb{P}(E) = \begin{bmatrix} \langle\langle \mathbb{q};\mathbb{q}^\dagger\rangle\rangle_E & \langle\langle \mathbb{q};\mathbb{q}\rangle\rangle_E \\ \langle\langle \mathbb{q}^\dagger;\mathbb{q}^\dagger\rangle\rangle_E & \langle\langle \mathbb{q}^\dagger;\mathbb{q}\rangle\rangle_E \end{bmatrix} \tag{50}$$

i.e. a blocked matrix of propagators with, say, the upper left block being the array

$$\langle\langle \mathbb{q};\mathbb{q}^\dagger\rangle\rangle_E = (\mathbb{q}^\dagger|(E\hat{I} - \hat{H})^{-1}\mathbb{q}^\dagger), \tag{51}$$

defined in terms of the superoperator resolvent of Eq. (19) and the particle-hole operators arranged in row and column matrices analogously to what was done to the electron field operators in Eq.(28).

The operator space on which the superoperators act will now consist only of boson-like operators and the commutator should be used in Eq. (17) defining the scalar product. Completely analogously to the case of the electron propagator, approximations can be found by invoking the idea of inner projection and density operators which are either pure state or referring to some suitable ensemble. We can choose the inner projection manifold to consist of some or all of the parts of $\mathbb{h}=\{\mathbb{f}_2,\mathbb{f}_4,\mathbb{f}_6\}$ where $\mathbb{f}_2=\{q_k,q_l^\dagger\}$, $\mathbb{f}_4=\{qq,q^\dagger q^\dagger\}$, and $\mathbb{f}_6=\{qqq,q^\dagger q^\dagger q^\dagger\}$, with all element properly orthonormalized.[53,57] The reference state, $|N\rangle$, is chosen as in Eq. (49), and the many-electron Hamiltonian expressed as $H = H_o + V$, with H_o being the Fock operator. This allows a systematic treatment of the polarization propagator through certain orders in perturbation theory.

Oddershede and Jørgensen[74] have shown that we need only consider $|N\rangle$ through single and double excitations and an inner projection manifold $\mathbb{h} = \{\mathbb{f}_2,\mathbb{f}_4\}$ provided we are interested in representing the propagator only through third order in electron interaction. The propagator, $\mathbb{P}(E)$ can then be written as

$$\mathbb{P}(E) = \begin{bmatrix} (\mathbb{q}^\dagger|\mathbb{f}_2) \\ (\mathbb{q}\ |\mathbb{f}_2) \end{bmatrix} (\mathbb{f}_2|P\mathbb{f}_2)^{-1}[(\mathbb{f}_2|\mathbb{q}^\dagger),\ (\mathbb{f}_2|\mathbb{q})] \tag{52}$$

following the inner projection idea of Eq. (29), where

$$\hat{P} = (E\hat{I} - \hat{H}) - \hat{H}\mathbb{f}_4(\mathbb{f}_4|(E\hat{I} - \hat{H})\mathbb{f}_4)^{-1}\mathbb{f}_4\hat{H}.$$

In terms of the explicit partitioning of $\mathbb{f}_2=\{\mathbb{q}^\dagger,\mathbb{q}\}$ this becomes:

$$\mathbb{P}(E) = \begin{bmatrix} 1 & 0 \\ 0 & -1 \end{bmatrix} \begin{bmatrix} (\mathbf{q}^{\dagger}|\hat{P}\mathbf{q}^{\dagger}) & (\mathbf{q}^{\dagger}|\hat{P}\mathbf{q}) \\ (\mathbf{q}|\hat{P}\mathbf{q}^{\dagger}) & (\mathbf{q}|\hat{P}\mathbf{q}) \end{bmatrix}^{-1} \begin{bmatrix} 1 & 0 \\ 0 & -1 \end{bmatrix} \tag{53}$$

The four different blocks of the polarization propagator in Eq. (50) all contain the same physical information, as can be seen from the spectral representation in Eq. (15). We thus concentrate on one block, say

$$\langle\langle \mathbf{q};\mathbf{q}^{\dagger}\rangle\rangle_E = [(\mathbf{q}^{\dagger}|\hat{P}\mathbf{q}^{\dagger}) - (\mathbf{q}^{\dagger}|\hat{P}\mathbf{q})(\mathbf{q}|\hat{P}\mathbf{q})^{-1}(\mathbf{q}|\hat{P}\mathbf{q}^{\dagger})]^{-1} \tag{54}$$

for the calculation of excitation energies, oscillator strengths and molecular electronic properties. We can write this even more in detail by following the analysis of Oddershede and Jørgensen[74] :

$$\langle\langle \mathbf{q};\mathbf{q}^{\dagger}\rangle\rangle_E^{-1} = \mathbb{A} - \mathbb{C}^{\dagger}\mathbb{D}^{-1}\mathbb{C} - \mathbb{B}^{*}(\mathbb{A}^{*})^{-1}\mathbb{B}, \tag{55}$$

where

$$\begin{aligned} &\mathbb{A} = (\mathbf{q}^{\dagger}|(E\hat{I} - \hat{H})\mathbf{q}^{\dagger}), \quad \mathbb{B} = (\mathbf{q}|\hat{H}\mathbf{q}^{\dagger}), \quad \mathbb{C} = (\mathbf{q}^{\dagger}\mathbf{q}^{\dagger}|\hat{H}\mathbf{q}), \\ &\mathbb{A}^{*} = (\mathbf{q}|(E\hat{I} - \hat{H})\mathbf{q}), \quad \mathbb{D} = (\mathbf{q}^{\dagger}\mathbf{q}^{\dagger}|(E\hat{I} - \hat{H})\mathbf{q}^{\dagger}\mathbf{q}^{\dagger}), \text{ and} \\ &\mathbb{B}^{*} = (\mathbf{q}^{\dagger}|\hat{H}\mathbf{q}). \end{aligned} \tag{56}$$

The time-dependent Hartree-Fock approximation[75] is obtained as

$$\langle\langle \mathbf{q};\mathbf{q}^{\dagger}\rangle\rangle_E^{-1} = \mathbb{A} - \mathbb{B}^{*}(\mathbb{A}^{*})^{-1}\mathbb{B} \tag{57}$$

i.e. the $\mathbf{f}_4$ projection manifold is neglected. In addition the reference state is $|N\rangle = |HF\rangle$, which means that $\mathbb{A}$ and $\mathbb{A}^{*}$ have terms of zeroth and first order in electron interaction, and $\mathbb{B}$ contains only first-order terms. Merely considering $\mathbb{A}$ in Eq (57) constitutes the Tamm-Dancoff approximation. The time-dependent Hartree-Fock approximation has also been applied by employing the statistical density operator of Eq. (36) with fractional occupation numbers.[76]

A number of second-order theories[77] can be recognized from studying Eq. (55) with $\mathbb{A}$ calculated through second order, which means that double excitations out of the Hartree-Fock ground state is included through first order in $|N\rangle$. $\mathbb{B}$ and $\mathbb{C}$ are of first order in electron interaction, $\mathbb{D}$ of zeroth order, and $\mathbb{A}^{*}$ is calculated through first order.

A comprehensive and very instructive discussion of the interconnection of a number of different second-order theories in terms of this superoperator formalism and also in terms of diagrammatic perturbation theory has been given by Oddershede and Jørgensen[74], with a quite complete set of references on this topic. It is rather evident from their study that there is a need to computationally explore the third-order polarization propagator.

REFERENCES

1. E. Fermi, Z. Physik 48, 73 (1928).
   L. H. Thomas, Proc. Camb. Phil. Soc. 23, 542 (1927).
2. P. A. M. Dirac, Proc. Camb. Phil. Soc. 26, 376 (1930).
3. See. e. g. J. C. Slater, Int. J. Quantum Chem. S9, 7 (1975).
4. K. Husimi, Proc. Phys. Math. Soc. Japan 22, 264 (1940).
5. R. Gaspar, Acta Physica Hungarica 3, 263 (1954).
6. W. Kohn, and L. J. Sham, Phys. Rev. 140, A1133 (1965).
7. J. C. Slater, Phys. Rev. 81, 385 (1951).
8. F. Herman and S. Skillman, Atomic Structure Calculations (Prentice Hall, 1963).
9. J. C. Slater, J. Chem. Phys. 43, S228 (1965).
10. K. H. Johnson, J. Chem. Phys. 45, 3085 (1966).
11. K. H. Johnson, Int. J. Quantum Chem. S1, 361 (1967).
12. See. e. g. J. B. Danese, J. Chem. Phys. 61, 3071 (1974).
13. A. R. Williams, and J. W. Morgan, J. Phys. C. 5, 1293 (1972); J. Phys. C. 7, 36 (1974).
14. J. W. D. Connolly, The X-α Model, Modern Theoretical Chemistry Vol. IV "Approximate Methods" (Edited by G. A. Segal) Plenum Press, 1976.
15. J. C. Slater, in Computational Methods in Band Theory, (P.M. Marcus, J. F. Janak, and A. R. Williams, editors) Plenum Press 1971, p. 447.
16. U. von Barth, and L. Hedin, J. Phys. C. 5, 1629 (1972).
17. O. Gunnarson, P. Johansson, S. Lundqvist, and B. I. Lundqvist, Int. J. Quantum Chem. S9, 83 (1975).
18. R. Gordon and Y. S. Kim, J. Chem. Phys. 56, 3122 (1972)
    Y. S. Kim and R. Gordon, ibid. 60, 1842 (1974).
19. L. Hedin, and B. I. Lundqvist, J. Phys. C. 4, 2064 (1971).
20. P. Hohenberg and W. Kohn, Phys. Rev. B. 136, 864 (1964).
21. L. J. Sham, and W. Kohn, Phys. Rev. 145, 561 (1966).
22. A. J. Coleman, Rev. Modern Phys. 35, 668 (1963).
23. S. F. Abdulnur, J. Linderberg, Y. Öhrn, and P. W. Thulstrup, Phys. Rev. 6, 889 (1972).
24. S. B. Trickey, F. R. Green Jr. and Frank W. Averill, Phys. Rev. B. 8, 4822 (1973).
25. L. Hedin and S. Lundqvist, in Solid State Physics Vol. 23, H. Eherenreich, F. Seitz, and D. Turnbull, editors) Academic Press, New York, (1969), p. 1.
26. J. Linderberg, and Y. Ohrn, "Propagators in Quantum Chemistry", Academic Press, (1973).
27. P.-O. Löwdin, Phys. Rev. 97, 1474 (1955).

28. R. McWeeny, Rev. Modern Phys. 32, 335 (1960) and reference therein.
29. B. Ruskai, Phys. Rev. 183, 23 (1969).
30. A. J. Coleman, in Energy, Structure and Reactivity, Proceedings of the 1972 Boulder Summer Research Conference on Theoretical Chemistry, (D. W. Smith, and W. B. McRae, editors), John Wiley and Sons, New York, p. 231.
31. E. R. Davidson, p. 234 of reference 30.
32. "Reduced Density Matrices with Applications to Physical and Chemical Systems," (A. J. Coleman and R. M. Erdahl, editors) (1968) Queen's Papers in Pure and Applied Mathematics, No. 11.
33. "Reduced Density Operators With Applications to Physical and Chemical Systems - 11," (R. M. Erdahl, editor) Queen's Papers in Pure and Applied Mathematics, No. 40 (1974).
34. E. R. Davidson, J. Math, Phys. 10, 725 (1969).
35. C. Garrod, M. V. Mihailovic, and M. Rosina, J. Math. Phys. 16, 868 (1975).
36. M. V. Mihailovic, and M. Rosina, Nucl. Phys. A237, 221 (1975).
37. E. R. Davidson, Adv. Quantum Chem. 6, 235 (editor P.-O.Löwdin) (1972).
38. See. e. g. D. W. Smith, and J. C. Ellenbogen, p. 67 of reference 33, and references therein.
39. J. Linderberg and Y. Öhrn, Chem. Phys. Lett. 1, 295 (1967)
L. S. Cederbaum, G. Holneicher, and S. Peyerimhoff, Chem. Phys. Lett. 11, 421 (1971)
L. S. Cederbaum, G. Holneicher, and W. von Niessen, Chem. Phys. Lett. 18, 503 (1973)
L. Tyner Redmon, G. D. Purvis and Y. Öhrn, J. Chem. Phys. 63, 5011 (1975).
40. J. D. Doll and W. P. Reinhardt, J. Chem. Phys. 57, 1169 (1972).
41. W. P. Reinhardt and J. B. Smith, J. Chem. Phys. 58, 2148 (1972).
42. L. S. Cederbaum, F. E. P. Matschke, and W. von Niessen, Phys. Rev. A 12, 6 (1975).
43. Gy. Csanak, H. S. Taylor, and R. Yaris, in Advances in Atomic and Molecular Physics (D.R. Bates, editor), Academic Press, New York (1971), p. 287.
44. J. Simons, J. Chem. Phys. 55, 1218 (1971).
45. P. Jørgensen, J. Oddershede, M. A. Ratner, J. Chem. Phys. 61, 710 (1974), and references therein.
46. D. N. Zubarev, Soviet Physics Uspekhi 3, 320 (1960).
47. H. Lehman, Nuovo Cimento 11, 342 (1954).
48. R. Zwanzig, "Lectures in Theoretical Physics," Vol. 111 (W. E. Brittain)
49. O. Goscinski, and B. Lukman, Chem. Phys. Lett. 7, 573 (1970).
50. B. T. Pickup, and O. Goscinski, Mol. Phys. 26, 1013 (1973).
51. G. D. Purvis, and Y. Öhrn, J. Chem. Phys. 60, 4063 (1974).
52. G. D. Purvis, and Y. Öhrn, J. Chem. Phys. 62, 2045 (1975).
53. L. Tyner Redmon, G. D. Purvis and Y. Öhrn, J. Chem. Phys. 63, 5011 (1975).
54. P. Jørgensen, Ann. Rev. Phys. Chem. 26, 359 (1975).

55. P.-O. Löwdin, Phys. Rev. 139, A357 (1965).
56. P.-O. Löwdin, Int. J. Quantum Chem. S4, 231 (1971).
57. C. Nehrkorn, G. Purvis, and Y. Öhrn, J. Chem. Phys. 64, 1752 (1976).
58. J. Linderberg, private Communication.
59. P.-O. Löwdin, Phys. Rev. 97, 1490 (1955).
60. O. Goscinski, B. T. Pickup and G. Purvis, Chem. Phys. Lett. 22, 167 (1973).
61. O. Goscinski, M. Hehenberger, B. Roos, and P. Siegbahn, Chem. Phys. Lett. 33, 427 (1975).
62. G. Purvis, and Y. Öhrn, J. Chem. Phys. in press (1976).
63. P.-O. Löwdin, J. Mol. Spectry. 10, 12 (1963).
64. J. Linderberg and Y. Öhrn, Chem. Phys. Lett. 1, 295 (1967).
65. J. Linderberg and Y. Öhrn, J. Chem. Phys. 49, 716 (1968).
66. The relativistic effects on the electron binding energies are estimated using the calculated Hartree-Fock-Dirac results of J. P. Declaux, Atomic Data and Nuclear Data Tables 12, 311 (1973).
67. G. D. Purvis and Y. Öhrn, Chem. Phys. Lett. 33, 396 (1975).
68. L. S. Cederbaum, Theoret. Chim. Acta, 31, 239 (1973).
69. L. S. Cederbaum, J. Phys. B8, 290 (1975).
70. W. von Niessen, L. S. Cederbaum, and W. P. Kraemer, J. Chem. Phys. in press (1976), and references therein.
71. J. Simons and W. D. Smith, J. Chem. Phys. 58, 4899 (1973), J. Simons, Chem. Phys. Lett. 25, 122 (1974), K. M. Griffing and J. Simons, J. Chem. Phys. 62, 535 (1975); J. Kenney and J. Simons, J. Chem. Phys. 62, 592 (1975), and references therein.
72. H. Kurtz, G. D. Purvis and Y. Öhrn, Int. J. Quantum Chem. S10, 000 (1976).
73. L. S. Cederbaum, and W. Domcke, J. Chem. Phys., 60, 2878 (1974), ibid, 64, 612 (1976).
74. J. Oddershede, and P. Jørgensen, Mol. Phys. (in press) 1976, plus references therein.
75. A. D. McLachlan and M. A. Ball, Rev. Modern Phys. 36, 844 (1964) P. Jørgensen and J. Linderberg, Int. J. Quantum Chem., 4, 587 (1970).
76. P. Jørgensen, and Y. Öhrn, Chem. Phys. Lett. 18, 261 (1972).
77. D. J. Rowe, Rev. Modern Phys. 40, 153 (1968)
T. Shibuya, and V. McKoy, Phys. Rev. A2, 2208 (1970)
E. A. Sanderson, Phys. Lett. 19, 141 (1965)
J. Linderberg, P. Jørgensen, J. Oddershede, and M. A. Ratner, J. Chem. Phys. 56, 6213 (1972).

SYMPOSIUM II. MOLECULAR SCATTERING

Chairman : J.O. Hirschfelder

Theoretical Chemistry Institute
University of Wisconsin, Madison,
Wisconsin, U.S.A.

QUANTIZED VORTICES IN MOLECULAR SCATTERING

Joseph O. Hirschfelder

Theoretical Chemistry Institute
University of Wisconsin
Madison, Wisconsin 53706

Since the period 1925-35 is known as the Golden Age of Quantum Chemistry, the 1970's should be called the Platinum Age. Thanks to the miracle of electronic technology, we can now perform all sorts of sophisticated experiments with precisions of one part in 10^7 being common; and, thanks to the same electronic technology, the theoretical chemists can solve all sorts of problems by making use of available subroutines and giant computers. The changes in chemical physics are occurring so rapidly it is hard to believe that only ten or twenty years ago, we believed that if only we could solve the Schrödinger equation with the electrostatic Hamiltonian, we would know the answers to most chemical problems. Most theoretical chemists regarded the relativistic corrections which lead to fine and hyperfine structure as being important in spectroscopy but not in chemistry. At the First Boulder Symposium, Coulson distinguished between Class A quantum chemistry which used a priori theory to treat small molecules, and Class B quantum chemistry which used semi-empiricism to understand the large molecules. Coulson speculated that it would be a very long time before a priori methods could be used to consider the complexities of large molecules. Thus, we have come a long way in a very short time. The present generation of quantum chemists must be topflite theoretical physicists well versed in all manner of sophisticated techniques.

There is probably no field which has undergone more change during the past few years than Reaction Kinetics. Now, instead of being limited to a consideration of questionable experimental data involving thermal averages, Dudley Herschbach and others are providing us with a myriad of detailed information regarding each of the quantum states involved in the reactive and nonreactive

B. Pullman and R. Parr (eds.), The New World of Quantum Chemistry, 81–86.

scattering processes. Bill Miller, Raphy Levine, and other theorists are busily engaged in, not only trying to explain the new data, but also trying to predict what will happen under widely different experimental conditions.

In 1935, Henry Eyring, Nathan Rosen, and I calculated the first a priori potential energy surface for the chemical reaction $H+H_2$ (with the three atoms in a line, using a minimum basis set and evaluating all of the integrals). There were four three-center integrals involved; each of which took me three months to evaluate on my Monroe calculating machine. The driving force was Wigner's remark that it could not be done. Then, Eyring suggested that it would be nice if I constructed a classical trajectory on the potential energy surface. Each day when I came to work, it would take me thirty minutes to calculate one new point; after six months and with the trajectory only half completed, I politely suggested to Eyring that it would be nice for him to complete the trajectory. In the winter of 1937-8, I used the Vannevar Bush mechanical Differential Analyzer at the University of Pennsylvania to calculate the potential energy for the three hydrogen atoms being nonlinear. The good old days were a lot of fun, but nowadays we get far more interesting results per gallon of sweat!

Like 90% of all academic scientists, I still work on the problems considered in my doctor's thesis. However, instead of calculating classical trajectories, I now determine quantum mechanical streamlines. If we think of a quantum mechanical system as a fluid, it has streamlines which follow the direction of the flux, $\underset{\sim}{J}$. Thus, a streamline is a particular solution to the set of coupled differential equations

$$dx/J_x = dy/J_y = \dots .$$

The particular feature of the quantum mechanical streamlines which has aroused my curiosity, are the quantized vortices which are due to wave interferences. These vortices may appear around (topologically suitable) nodal regions of the wavefunction and each vortex contains at least one quantum of angular momentum. I wish to discuss briefly two idealized examples which are related to molecular scattering:

I. SPHERE-SPHERE ELASTIC SCATTERING.[1]

One of the simplest examples corresponds to the elastic scattering of two particles which collide with a spherically symmetric potential, $V(r) = 0$ when $r > a$ and $V(r) = C$ when $r < a$. If the constant C is positive, the interaction potential corresponds to a barrier; whereas a negative value of C corresponds to a well. The greatest interest is associated with resonant collisions. Fig. 1 shows the streamlines and probability density

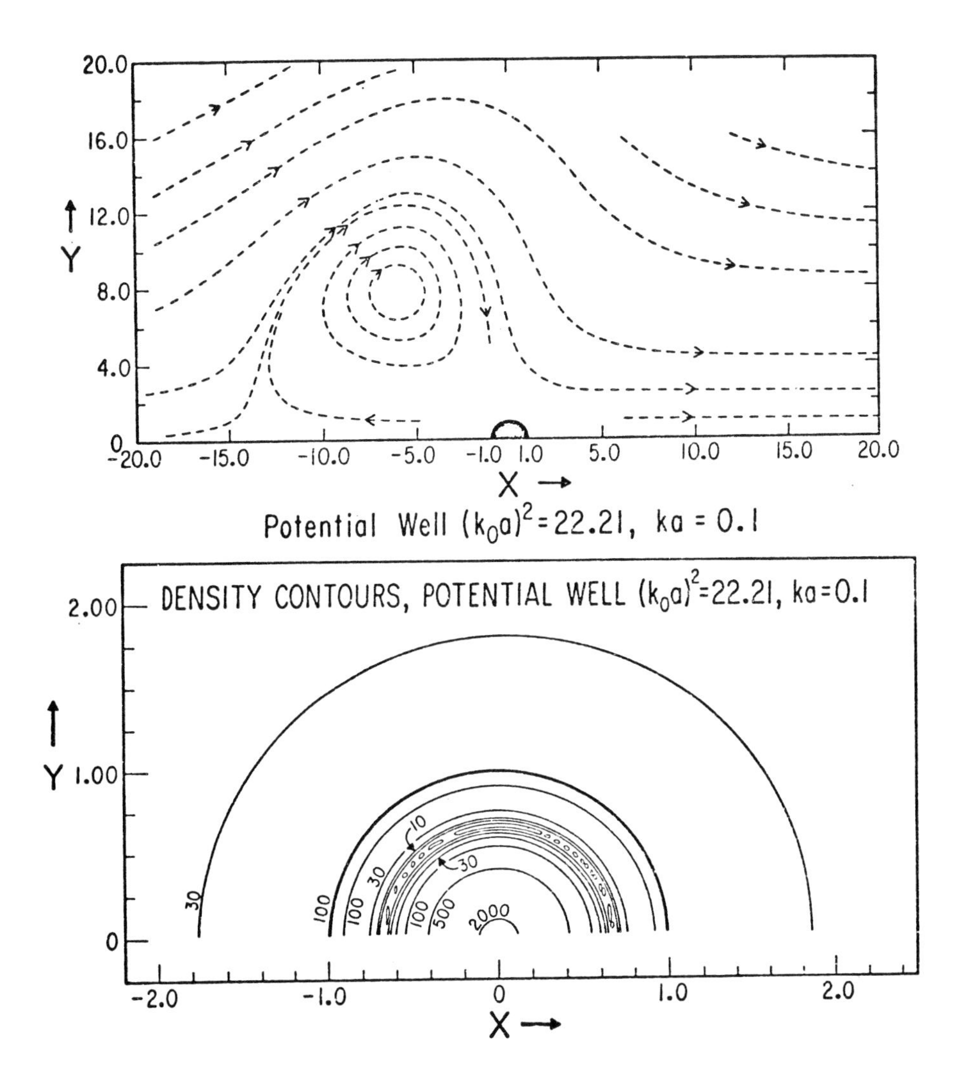

FIG. 1 Streamlines and Density Contours for an S-wave Resonance.

contours which occur near an S-wave resonant collision $[C = -22.21(\hbar^2/2\mu a^2),\ ka = 0.1]$. The incident wave corresponds to particles moving in the X direction; Y is the distance from the X axis; and the angle ϕ (not shown) measures the distance around the X axis. In this case, there is only one vortex outside of the interaction zone (r=a). Since the streamlines are independent of the angle ϕ, this vortex is really toroidal and shaped like a smoke ring. It is easy to prove that if r_0 is the distance from the center of the vortex to the origin, then $4\pi r_0{}^2 = \sigma_0$ where σ_0 is the S-wave partial cross section. Thus, r_0 is essentially the Fermi scattering length. For P-wave resonances, we find $2\pi r_0{}^2 = \sigma_1$ and for D-wave resonances, $1.5\pi r_0^2 = \sigma_2$ (where r_0 is the distance of the outermost vortex to the origin). Inside of the interaction zone, the density contours tend to correspond to the shape of the dominant partial wave, whereas the shape of the streamlines is largely influenced by the minor components of the wavefunction. Under resonant conditions, there is a strong focussing of the probability density into the interaction zone. Thus, for our S-wave example, the density at the center of the sphere is 2212 times the density of the incoming wave. Fig. 2 shows the streamlines corresponding to a Ramsauer-Townsend Effect $[C = -20.1906(\hbar^2/2\mu a^2),\ ka = 0.2]$. Outside the interaction zone the streamlines remain very nearly constant. However, inside the interaction zone, the streamlines are concentrated near the X axis. This results in the probability density at the origin being 21 times as large as the density of the incoming wave. Near the top of the interaction zone, the probability density is very small and three little vortex rings appear in the streamlines.

II. REACTIVE COLLINEAR ATOM-DIATOMIC MOLECULE COLLISION.[2]

The potential energy surface for a collinear atom-diatomic molecule collision can be idealized into a two-dimensional channel with a right angle bend. Inside the channel, the potential energy is zero and outside it is infinite. Fig. 3 shows the streamlines which we obtained as we gradually increased the energy. There is a threshold at $E^* = 2$ below which, no vortex can form. At $E^* = 2$, a small vortex appears in the upper right corner and as the energy is increased, moves towards the lower left corner and increases in size. Fig. 3d shows the streamlines when the energy is just a little smaller than the reflection energy $E^* = 3.535$. As the energy approaches 3.535 all of the streamlines disappear. At slightly higher energy, the streamlines reappear together with the same vortices as before, but the vortices are rotating in the opposite direction. We can prove rigorously that the streamlines are symmetric as long as there is only one open energy channel. However, for energies greater than $E^* = 4$ where more than one energy channel is open, the streamline pattern is very irregular. The vortex phenomena which we observe here is similar to the results of both Wyatt and Kuppermann (and their colleagues) who have

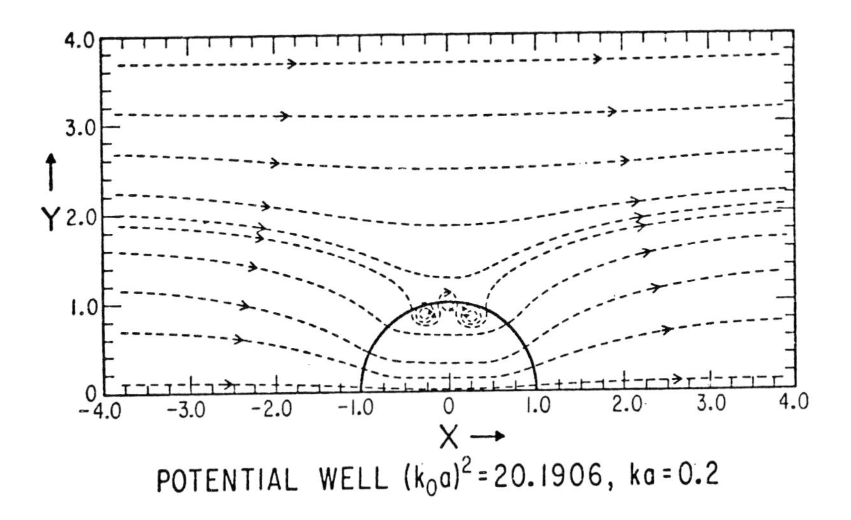

FIG. 2 Streamlines Corresponding to a Ramsauer-Townsend Effect.

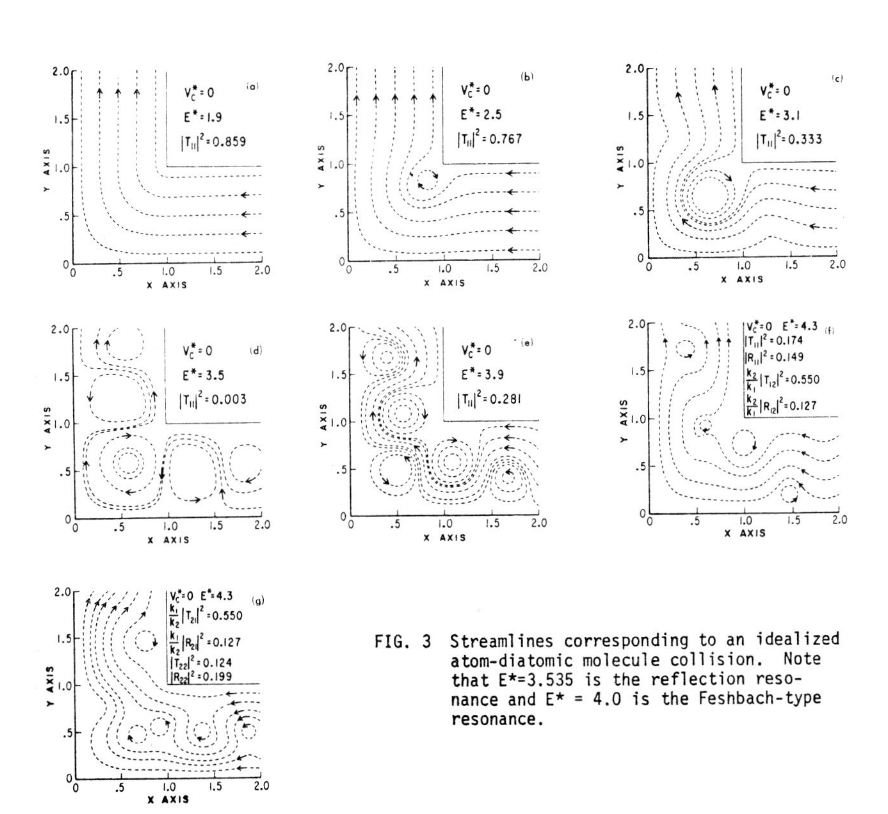

FIG. 3 Streamlines corresponding to an idealized atom-diatomic molecule collision. Note that E*=3.535 is the reflection resonance and E* = 4.0 is the Feshbach-type resonance.

studied atom-diatomic molecule collisions on realistic potential energy surfaces.

The vortex rings which occur in molecular collisions correspond to collisional magnetic moments. Jürg Riess[3] has shown that the induced paramagnetism of BH and AlH is due to the formation of vortices. The most important effects of the vortices probably occur in the presence of external magnetic fields. For example, ESR spectra and chemical shielding constants.

REFERENCES

1. J. O. Hirschfelder and K. T. Tang, J. Chem. Phys. 65 (July 1976).

2. J. O. Hirschfelder and K. T. Tang, J. Chem. Phys. 64, 760 (1976).

3. J. Riess, Annals of Phys. 57, 301 (1970) and 67, 347 (1971).

RECENT DEVELOPMENTS IN SEMICLASSICAL MECHANICS: EIGENVALUES AND REACTION RATE CONSTANTS*

William H. Miller†

Department of Chemistry and Materials and Molecular Research Division of the Lawrence Berkeley Laboratory, University of California, Berkeley CA 94720 and University Chemical Laboratory, Lensfield Road, Cambridge England

ABSTRACT

A semiclassical treatment of eigenvalues for a multidimensional non-separable potential function and of the rate constant for a chemical reaction with an activation barrier is presented. Both phenomena are seen to be described by essentially the same semiclassical formalism, which is based on a construction of the total Hamiltonian in terms of the complete set of "good" action variables (or adiabatic invariants) associated with the minimum in the potential energy surface for the eigenvalue case, or the saddle point in the potential energy surface for the case of chemical reaction.

I. INTRODUCTION

Semiclassical methods are the natural way of describing the effects of quantum mechanics in classical-like systems, and they have been an extremely powerful tool for describing quantum effects in atomic and molecular collision processes. Ford and Wheeler's papers[1] in 1959 showed how interference and tunneling effects in elastic scattering[2] could be described in a beautifully simple way: the collision dynamics are treated <u>via</u> classical mechanics but used to construct scattering <u>amplitudes</u> rather than the scattering probabilities (i.e., cross sections) themselves. This means that the quantum <u>principle of superposition</u> is included in the description, and this is the essential contribution of quantum mechanics. This approach provides a quantitative description of all quantum effects in elastic scattering; recently, for example, it has been shown[3,4] that a slight

B. Pullman and R. Parr (eds.), The New World of Quantum Chemistry, 87–102.

extension of the Ford and Wheeler analysis (to include "classically forbidden" contributions to the amplitude) even describes diffraction from purely repulsive potentials, a quantum effect heretofore thought to defy a simple semiclassical treatment.

Beginning in 1970 this semiclassical idea of "classical dynamics plus quantum superposition" was generalized to inelastic and reactive collision processes.[5-7] Although originally designed to deal with molecular collision phenomena, this "classical S-matrix" theory has also found application to electron scattering,[8] diffraction of atoms from crystal surfaces,[9,10] and coulomb excitation and fission dynamics in nuclear physics.[11-13] One of the predictions of this theory, for example, is that analogous to interference structure in the differential cross section for elastic scattering, there is an interference structure in the distribution of final states in inelastic-reactive scattering processes;[14] e.g., the cross section for the product molecule being in final vibrational state v_f, plotted as a function of v_f, in general has an interference structure. Although these interference effects in final state distributions have not yet been seen experimentally, it is certain that they exist and will presumably be seen when more refined techniques for state analysis becomes available.

This paper reviews some of the more recent developments in "semiclassical mechanics"--the semiclassical approximation to eigenvalues for multidimensional non-separable systems and the semiclassical approximation for the rate constants of chemical reactions which have simple activation barriers. It is an interesting feature of semiclassical theory that these two problems are described by essentially the same formalism.

Section II first discusses the semiclassical eigenvalue problem for multidimensional systems using both periodic orbit theory and the more rigorous Hamilton-Jacobi theory. In parallel fashion, Section III then discusses the semiclassical description of reaction rate constants--first the periodic orbit version of the theory and then the more rigorous Hamilton-Jacobi formulation.

II. SEMICLASSICAL EIGENVALUES

a. Periodic Orbit Theory

One of the truly new approaches to a multidimensional semiclassical quantum condition is the recent work of Gutzwiller[15] (with the modification of Miller[16]). The beginning point is to consider the trace of the resolvent operator, or green's function

$$\mathrm{tr}\, G(E) \equiv \mathrm{tr}\, (E-H)^{-1} \quad , \tag{2.1}$$

which is given quantum mechanically by

$$\mathrm{tr}\, G(E) = \sum_n (E-E_n)^{-1} \quad , \tag{2.2}$$

indicating that it has simple poles where E equals one of the quantum mechanical eigenvalues $\{E_n\}$. The green's function can be expressed in terms of the propagator,

$$G(E) = (i\hbar)^{-1} \int_0^\infty dt\, e^{iEt/\hbar}\, e^{-iHt/\hbar} \quad , \tag{2.3}$$

and the trace can be evaluated in a coordinate representation, so that

$$\mathrm{tr}\, G(E) = (i\hbar)^{-1} \int_0^\infty dt\, e^{iEt/\hbar} \int d\underset{\sim}{q}\, \langle \underset{\sim}{q}| e^{-iHt/\hbar} |\underset{\sim}{q}\rangle \quad . \tag{2.4}$$

Eq. (2.4) is still formally exact quantum mechanics, but one now injects the semiclassical approximation for matrix elements of the propagator,[17]

$$\langle \underset{\sim}{q}| e^{-iHt/\hbar} |\underset{\sim}{q}\rangle \sim e^{i\phi(\underset{\sim}{q},\underset{\sim}{q};t)/\hbar} \quad , \tag{2.5}$$

where ϕ is the classical action integral along the trajectory that goes from position $\underset{\sim}{q} \equiv \{q_i\}$, $i=1,\ldots, f$ at time $t = 0$ (f = number of degrees of freedom) back to $\underset{\sim}{q}$ at time t, and consistent with semiclassical mechanics, the integrals over $\underset{\sim}{q}$ and over t in Eq. (2.4) are all evaluated via the stationary phase approximation. The stationary phase condition for the $\underset{\sim}{q}$ integrals, for example, is

$$\begin{aligned} 0 &= \frac{\partial\phi(\underset{\sim}{q},\underset{\sim}{q};t)}{\partial \underset{\sim}{q}} \\ &= \left[\frac{\partial\phi(\underset{\sim}{q}_2,\underset{\sim}{q}_1;t)}{\partial \underset{\sim}{q}_2} + \frac{\partial\phi(\underset{\sim}{q}_2,\underset{\sim}{q}_1;t)}{\partial \underset{\sim}{q}_1} \right]_{\underset{\sim}{q}_1=\underset{\sim}{q}_2=\underset{\sim}{q}} \\ &= \underset{\sim}{p}_2 - \underset{\sim}{p}_1 \quad ; \end{aligned} \tag{2.6}$$

i.e., the values of $\underset{\sim}{q}$ which are points of stationary phase must be such that the initial and final momenta, as well as coordinates, are the same, or in other words, $\underset{\sim}{q}$ must lie on a periodic orbit of the system. In this very simple way, therefore, one sees how periodic orbits play a central role in this approach to the semiclassical eigenvalue problem.

The algebraic details involved in carrying out the stationary phase evaluation of the integrals in Eq. (2.4) are quite tedious but straight-forward, and the result of the calculation is[15]

$$\mathrm{tr}\, G(E) = -\frac{i\Phi'(E)}{\hbar} \sum_{n=1}^{\infty} \frac{e^{in[\Phi(E)/\hbar - \lambda\frac{\pi}{2}]}}{\prod_{k=1}^{f-1} 2i\,\sin(\frac{n}{2}\,\omega_k(E)\Phi'(E))} \quad . \tag{2.7}$$

In Eq. (2.7) $\Phi(E)$ is the action integral for one pass about the periodic orbit with energy E,

$$\Phi(E) = \int_0^{T(E)} dt\; \underset{\sim}{p}(t)\cdot\dot{\underset{\sim}{q}}(t) \quad , \tag{2.8}$$

where $T(E) \equiv \Phi'(E)$ is the period of the periodic orbit, $\{\omega_k(E)\}$, $k=1,\ldots, f-1$, are the stability frequencies of the (stable) periodic orbit, and λ is the number of "conjugate points" experienced in one pass about the periodic orbit. (In most cases λ is 0 or 2). The sum over n in Eq. (2.7) is a sum of amplitudes for all possible numbers of passes about the periodic orbit.

In order to identify the polar singularities in Eq. (2.7)--and thus the eigenvalues via Eq. (2.2)--it is useful to expand the sine functions in the denominator according to the identity

$$\frac{1}{2i\,\sin(\frac{x}{2})} = \frac{e^{-i\frac{x}{2}}}{1-e^{-ix}} = \sum_{n=0}^{\infty} e^{-i(n+\frac{1}{2})x} \quad . \tag{2.9}$$

Since there are f-1 sine factors in Eq. (2.7), this gives[16]

$$\mathrm{tr}\, G(E) = -\frac{i\Phi'(E)}{\hbar} \sum_{n_1=0}^{\infty}\sum_{n_2=0}^{\infty}\cdots\sum_{n_{f-1}=0}^{\infty}\sum_{n=1}^{\infty} e^{inA} \quad , \tag{2.10a}$$

where

$$A = \Phi(E)/\hbar - \lambda\frac{\pi}{2} - \Phi'(E)\sum_{k=1}^{f-1}\omega_k(E)(n_k+\tfrac{1}{2}) \quad . \tag{2.10b}$$

The sum over n (i.e., over multiple passes about the periodic orbit) in Eq. (2.10a) is easily accomplished,

$$\sum_{n=1}^{\infty} e^{inA} = e^{iA}(1-e^{iA})^{-1} \quad , \tag{2.11}$$

whereby the polar singularities in tr G(E) are seen to be those values of E for which $e^{iA} = 1$, or

$$A = 2\pi\, n_f \quad , \tag{2.12}$$

where n_f is an integer. Eqs. (2.10b) and (2.2) thus give the quantum condition of periodic orbit theory as[16]

$$\Phi(E) - \Phi'(E)\sum_{k=1}^{f-1}\hbar\,\omega_k(E)(n_k+\tfrac{1}{2}) = 2\pi\hbar(n_f+\tfrac{\lambda}{4}) \quad . \tag{2.13}$$

The energy levels are thus characterized by f quantum numbers, n_f being the number of quanta in motion along the periodic orbit, and n_k, k=1, ..., f-1, being the number of quanta in the kth harmonic mode of perturbation about the periodic orbit.

Although physically realistic and meaningful in many ways, Eq. (2.13) nevertheless has some fundamental shortcomings, primarily the implicit assumption that most of the energy is in motion along the periodic orbit, with deviations from it being treated only within a harmonic approximation. To see this, consider a completely separable system where the periodic orbit is chosen to be along the fth degree of freedom. The correct semiclassical eigenvalue condition in this case is

$$\Phi\left(E - \sum_{k=1}^{f-1}\varepsilon_k(n_k)\right) = 2\pi\hbar(n_f+\tfrac{1}{2}) \quad , \tag{2.14}$$

where $\varepsilon_k(n_k)$ is the one-dimensional eigenvalue for the kth degree of freedom, and in this case Φ is the one-dimensional action integral for the fth degree of freedom. To convert this into the expression given by periodic orbit theory, Eq. (2.13), two approximations are necessary: (1) All degrees of freedom except the fth one must be assumed to be harmonic, i.e.,

$$\varepsilon_k(n_k) \simeq \omega_k(n_k+\tfrac{1}{2}) \quad , \tag{2.15a}$$

and (2) Φ must be approximated by the first two terms of a Taylor's series expansion,

$$\Phi\left(E - \sum_{k=1}^{f-1} \hbar\omega_k(n_k+\tfrac{1}{2})\right) \simeq \Phi(E) - \Phi'(E) \sum_{k=1}^{f-1} \hbar\omega_k(n_k+\tfrac{1}{2}) \quad .(2.15b)$$

Assumption (1), Eq. (2.15a), is in general only valid for small n_k, and (2), Eq. (2.15b), is in general valid only if

$$E >> \sum_{k=1}^{f-1} \hbar\omega_k(n_k+\tfrac{1}{2}) \quad , \tag{2.16}$$

both of which imply that most of the total energy E is in one mode, the $f^{\underline{th}}$ one. If it should be that the first f-1 modes are harmonic, then Eq. (2.15a) is of course not an approximation; and further, if the $f^{\underline{th}}$ mode is also harmonic, then Eq. (2.15b) is also exact. Thus the one (trivial) case in which periodic orbit theory is actually exact is the completely separable, completely harmonic case.

Since these limitations implicit in periodic orbit theory stem directly from the stationary phase approximation used to evaluate the integrals in Eq. (2.4), it is difficult to see how to avoid them and stay within the present formalism. What would be required is some sort of uniform asymptotic evaluation of the integrals in Eq. (2.4), but it is not clear how to go about this. Fortunately, the approach described in the next section by-passes these difficulties by taking a different approach to the problem.

b. Hamilton-Jacobi Theory

The "Old Quantum Theory", as described by Born,[18] provides an alternate (much older) approach to the semiclassical eigenvalue problem[19] that has the major advantage of treating all degrees of freedom on an equal footing. Recently it has been shown[20] that the equations which result in this formulation can be cast in a form that permit efficient numerical solution, so that this approach is no longer limited to the realm of perturbation theory.[18]

In this approach the classical Hamiltonian for the system is divided into a separable reference Hamiltonian H_o and a residual non-separable interaction V,

$$H(\underset{\sim}{p},\underset{\sim}{x}) = H_o(\underset{\sim}{p},\underset{\sim}{x}) + V(\underset{\sim}{x}) \tag{2.17}$$

with

$$H_o(\underset{\sim}{p},\underset{\sim}{x}) = \frac{p^2}{2\mu} + V_o(\underset{\sim}{x}) \quad , \tag{2.18}$$

where $V_o(\underset{\sim}{x})$ is separable,

$$V_o(\underset{\sim}{x}) = \sum_{i=1}^{f} v_i(x_i) \quad . \tag{2.19}$$

It is convenient then to change canonical variables from the cartesian variables $(\underset{\sim}{p},\underset{\sim}{x})$ to the "zeroth order" action-angle variables $(\underset{\sim}{n},\underset{\sim}{q})$ which refer to H_o, in terms of which the Hamiltonian has the form

$$H(\underset{\sim}{n},\underset{\sim}{q}) = H_o(\underset{\sim}{n}) + V(\underset{\sim}{q},\underset{\sim}{n}) \quad . \tag{2.20}$$

Often the one-dimensional reference potentials $\{v_i(x_i)\}$ in Eq. (2.19) are taken to be harmonic, as we will now do, but this is not necessary. $H_o(\underset{\sim}{n})$ is then given explicitly by

$$H_o(\underset{\sim}{n}) = \sum_{i=1}^{f} \omega_i(n_i+\tfrac{1}{2}) \quad , \tag{2.21}$$

units being used so that $\hbar = 1$.

The presence of the non-separable interaction in Eq. (2.20) prevents the action variables $\underset{\sim}{n}$ from being constants of the motion and thus the "good" quantum numbers of the system. The next step, therefore, is to carry out a canonical transformation from these zeroth order action-angle variables $(\underset{\sim}{n},\underset{\sim}{q})$ to the "good" action-angle variables $(\underset{\sim}{N},\underset{\sim}{Q})$ such that in terms of these latter variables the total Hamiltonian is a function only of the action variables $\underset{\sim}{N}$:

$$H(\underset{\sim}{N},\underset{\sim}{Q}) = E(\underset{\sim}{N}) \quad . \tag{2.22}$$

The $\{N_i\}$ are thus constants of the motion, and the semiclassical eigenvalues are given by $E(\underset{\sim}{N})$ with all the $\{N_i\}$ required to be non-negative integers.

The $(\underset{\sim}{n},\underset{\sim}{q}) \rightarrow (\underset{\sim}{N},\underset{\sim}{Q})$ transformation is effected by a generating function $F(\underset{\sim}{q},\underset{\sim}{N})$ which is required to be a solution of the Hamilton-Jacobi equation,

$$\begin{aligned} E(\underset{\sim}{N}) &= H\left(\frac{\partial F(\underset{\sim}{q},\underset{\sim}{N})}{\partial \underset{\sim}{q}},\ \underset{\sim}{q}\right) \\ &= \underset{\sim}{\omega}\cdot\left(\frac{\partial F}{\partial \underset{\sim}{q}} + \frac{1}{2}\right) + V\left(\underset{\sim}{q},\ \frac{\partial F}{\partial \underset{\sim}{q}}\right) \quad . \end{aligned} \tag{2.23}$$

F is taken to be of the form

$$F(\underset{\sim}{q},\underset{\sim}{N}) = \underset{\sim}{N}\cdot\underset{\sim}{q} + i\sum_{\underset{\sim}{k}}{}' B_{\underset{\sim}{k}}(\underset{\sim}{N})\, e^{i\underset{\sim}{k}\cdot\underset{\sim}{q}} \quad , \tag{2.24}$$

the first term being that which generates the identity transformation, which is correct in the limit $V \to 0$, and where the prime on the summation indicates that the term $k = 0$ is omitted. Substituting this Fourier expansion into Eq. (2.23) and projecting out individual Fourier components leads to the following algebraic equations:

$$E\,\delta_{\underset{\sim}{k},\underset{\sim}{0}} = \underset{\sim}{\omega}\cdot(\underset{\sim}{N}+\tfrac{1}{2})\,\delta_{\underset{\sim}{k},\underset{\sim}{0}} - (\underset{\sim}{\omega}\cdot\underset{\sim}{k})B_{\underset{\sim}{k}}$$
$$+ (2\pi)^{-f}\int_{0}^{2\pi} d\underset{\sim}{q}\, e^{-i\underset{\sim}{k}\cdot\underset{\sim}{q}}\, V(\underset{\sim}{q},\underset{\sim}{N} - \sum_{\underset{\sim}{k}'}{}' \underset{\sim}{k}'\, B_{\underset{\sim}{k}'}\, e^{i\underset{\sim}{k}'\cdot\underset{\sim}{q}}) \ . \tag{2.25}$$

For $\underset{\sim}{k} \neq \underset{\sim}{0}$, Eq. (2.25) is a set of (non-linear) equations for the Fourier coefficients $B_{\underset{\sim}{k}}$,

$$(\underset{\sim}{\omega}\cdot\underset{\sim}{k})B_{\underset{\sim}{k}} = (2\pi)^{-f}\int_{0}^{2\pi} d\underset{\sim}{q}\, e^{-i\underset{\sim}{k}\cdot\underset{\sim}{q}}\, V(\underset{\sim}{q},\underset{\sim}{N} - \sum_{\underset{\sim}{k}'}{}' \underset{\sim}{k}'\, B_{\underset{\sim}{k}'}\, e^{i\underset{\sim}{k}'\cdot\underset{\sim}{q}}) . \tag{2.26a}$$

and for $k = 0$ Eq. (2.25) gives the eigenvalue $E(\underset{\sim}{N})$ in terms of the Fourier coefficients,

$$E(\underset{\sim}{N}) = \underset{\sim}{\omega}\cdot(\underset{\sim}{N}+\tfrac{1}{2}) + (2\pi)^{-f}\int_{0}^{2\pi} d\underset{\sim}{q}\, V(\underset{\sim}{q},\underset{\sim}{N} - \sum_{\underset{\sim}{k}'}{}' \underset{\sim}{k}'\, B_{\underset{\sim}{k}'}\, e^{i\underset{\sim}{k}'\cdot\underset{\sim}{q}}) . \tag{2.26b}$$

One thus first sets N equal to the desired quantum numbers and solves Eq. (2.26a) for the Fourier coefficients $B_{\underset{\sim}{k}}$; the eigenvalue corresponding to this set of quantum numbers is then given by Eq. (2.26b).

For the non-degenerate situation--i.e., the case that $\underset{\sim}{\omega}\cdot\underset{\sim}{k}$ is not too small for all $\underset{\sim}{k}$ which contribute significantly to the Fourier expansion--Eq. (2.26a) can often be solved by a simple successive substitution algorithm. Eq. (2.26a) is written as

$$B_{\underset{\sim}{k}} = \frac{1}{\underset{\sim}{\omega}\cdot\underset{\sim}{k}}\,(2\pi)^{-f}\int_{0}^{2\pi} d\underset{\sim}{q}\, e^{-i\underset{\sim}{k}\cdot\underset{\sim}{q}}\, V(\underset{\sim}{q},\underset{\sim}{N} - \sum_{\underset{\sim}{k}'}{}' \underset{\sim}{k}'\, B_{\underset{\sim}{k}'}\, e^{i\underset{\sim}{k}'\cdot\underset{\sim}{q}}) , \tag{2.27}$$

which is seen to be of the form

$$B_{\underset{\sim}{k}} = \text{function}\ [B_{\underset{\sim}{k}}] \quad .$$

The iterative procedure is then

$$B_{\underset{\sim}{k}}^{(\ell+1)} = \text{function}\ [B_{\underset{\sim}{k}}^{(\ell)}] \quad ,$$

$\ell = 0,1,2...$, with the initialization

$$B_{\underset{\sim}{k}}^{(0)} = 0 \quad .$$

Application[20] of this procedure for solving Eq. (2.26a) to a model two-dimensional potential well has shown that the semiclassical eigenvalues so obtained are in good agreement with accurate quantum mechanical values--comparable to the agreement for one-dimensional WKB eigenvalues--even for very strong nonseparable coupling. Essentially the same procedure has also been used[21] to determine a number of vibrational eigenvalues of H_2O and SO_2 with realistic potential functions. This is the first application of a multidimensional semiclassical quantum condition to real molecular systems.

If there are low order degeneracies in $H_0(\underset{\sim}{n})$--i.e., if $\omega \cdot k=0$ for some $\underset{\sim}{k}$ that contribute significantly to the Fourier series--then the above procedure for solving Eq. (2.26a) will clearly not be applicable. In such cases other algorithms--such as the multidimensional Newton iteration--must be used to solve Eq. (2.26a). For the two dimensional potential functions

$$V(x,y) = \frac{1}{2}\,\omega_1^{\,2}\,x^2 + \frac{1}{2}\,\omega_2^{\,2}\,y^2 + \lambda(xy^2 - \frac{1}{3}\,x^3) \quad ,$$

and

$$V(x,y) = \frac{1}{2}\,\omega_1^{\,2}\,x^2 + \frac{1}{2}\,\omega_2^{\,2}\,y^2 + \lambda\,x^2\,y^2 \quad ,$$

with $\omega_1 = \omega_2$, a multidimensional Newton procedure has been used[22] successfully to solve Eq. (2.26a) and thus obtain semiclassical eigenvalues.

III. RATE CONSTANTS FOR CHEMICAL REACTION

For a bimolecular exchange reaction of the general type

$$A + BC \rightarrow AB + C \quad ,$$

the thermal rate constant is given quantum mechanically by[23]

$$k(T) = Q_r^{-1}\ \mathrm{tr}(e^{-\beta H}\ F\ \mathcal{P}) \quad , \tag{3.1}$$

where Q_r is the partition function (per unit volume) for reactants, H is the total Hamiltonian, F is a flux operator, and $\mathcal{P}$ is the projection operator which projects onto states that have evolved in the infinite past from reactants.

Since the Boltzmann operator $e^{-\beta H}$ can be expressed in terms of the green's function by

$$e^{-\beta H} = \frac{-1}{2\pi i} \int^{\infty} dE\ e^{-\beta E}\ G(E) \quad , \tag{3.2}$$

the rate constant can be expressed equivalently as

$$k(T) = (2\pi\hbar\ Q_r)^{-1} \int^{\infty} dE\ e^{-\beta H}\ N(E) \quad , \tag{3.3}$$

where N(E) is[24]

$$N(E) = i\hbar\ \mathrm{tr}[G(E)\ F\ \mathcal{P}] \quad . \tag{3.4}$$

Semiclassical approximations for N(E), and thus for the rate constant via Eq. (3.3), are then obtained by introducing semiclassical approximations for G(E) and for the evaluation of the trace.

a. Periodic Orbit Theory

The periodic orbit analysis of Section IIa can also be used to evaluate the trace in Eq. (3.4). The factor $F\mathcal{P}$ in Eq. (3.4) makes only minor difference to the calculation, and if there is a simple activation barrier separating reactants and products (as, for example, with $H + H_2$) one obtains[25]

$$N(E) = \sum_{n=1}^{\infty} \frac{-e^{in[\Phi(E)/\hbar\ -\ \lambda \frac{\pi}{2}]}}{\prod_{k=1}^{f-1} 2i\ \sin[\frac{n}{2}\ \omega_k(E)\ \Phi'(E)]} \quad , \tag{3.5}$$

which is the analog of Eq. (2.7). In this case, though, the periodic orbit is for an <u>imaginary</u> time increment (because $e^{-\beta H} \equiv e^{-iHt/\hbar}$ for $t = -i\hbar\beta$), so that the action integral $\Phi(E)$ is pure imaginary. (The periodic motion is basically an oscillation back and forth through the potential barrier.[25]) The real quantity $\theta(E)$ is defined by

$$\Phi(E)/\hbar = i2\theta(E) \tag{3.6}$$

and since $\lambda = 2$ in this case, Eq. (3.6) becomes

$$N(E) = \sum_{n=1}^{\infty} \frac{(-1)^{n-1} e^{-n2\theta(E)}}{\prod_{k=1}^{f-1} \sinh [-n\hbar\omega_k(E)\theta'(E)]} \tag{3.7}$$

The sinh functions in the denominator of Eq. (3.7) are expanded in the analogous way as Eq. (2.9), so that Eq. (3.7) becomes

$$N(E) = \sum_{n_1=0}^{\infty} \sum_{n_2=0}^{\infty} \cdots \sum_{n_{f-1}=0}^{\infty} \sum_{n=1}^{\infty}$$

$$(-1)^{n-1} e^{-n[2\theta(E) - 2\theta'(E) \sum_{k=1}^{f-1} \hbar\omega_k(E)(n_k+\frac{1}{2})]} \quad , \tag{3.8}$$

and the sum over n--the number of "tunneling oscillations" inside the barrier--is easily carried out to give[25]

$$N(E) = \sum_{\underset{\sim}{n}=0}^{\infty} [1 + e^{2\theta(E) - 2\theta'(E)\hbar\underset{\sim}{\omega}(E)\cdot(\underset{\sim}{n}+\frac{1}{2})}]^{-1} \quad , \tag{3.9}$$

where $\underset{\sim}{\omega}(E) = \{\omega_k(E)\}$, k=1, ..., f=1, $\underset{\sim}{n} = \{n_i\}$, i=1, ..., f-1, and

$$\sum_{\underset{\sim}{n}=0}^{\infty} = \sum_{n_1=0}^{\infty} \sum_{n_2=0}^{\infty} \cdots \sum_{n_{f-1}=0}^{\infty} \quad .$$

Eq. (3.9) is the final expression for N(E) within periodic orbit theory, the rate constant then being given by Eq. (3.3).

Although Eq. (3.9) does provide a qualitatively correct physical picture of the reaction rate constant, this periodic orbit limit of the theory suffers from the same quantitative shortcomings as those in the eigenvalue problem discussed in Section IIa, namely that most of the energy is implicitly assumed to be in one mode (motion along the "reaction coordinate") with deviations from it being treated only within a harmonic approximation. To see this, consider the situation that motion along a reaction coordinate is separable from the remaining f-1 degrees of freedom; it is then easy to show that the exact quantum mechanical expression for N(E) in this case is

$$N(E) = \sum_{\underset{\sim}{n}=0} P_{tun}(E-\varepsilon_{\underset{\sim}{n}}) \quad , \tag{3.10}$$

where $P_{tun}(E_t)$ is the one dimensional tunneling probability with an energy E_t in translational motion along the reaction coordinate. The semiclassical approximation for the tunneling probability is

$$P_{tun}(E_t) = [1 + e^{2\theta(E_t)}]^{-1} \quad , \tag{3.11}$$

and if the f-1 degrees of freedom of the "activated complex" are approximated as harmonic,

$$\varepsilon_{\underset{\sim}{n}} \simeq \underset{\sim}{\omega} \cdot (\underset{\sim}{n} + \tfrac{1}{2}) \quad , \tag{3.12}$$

then the correct semiclassical limit of Eq. (3.10) is

$$N(E) = \sum_{\underset{\sim}{n}} \{1 + e^{2\theta[E - \hbar\underset{\sim}{\omega} \cdot (\underset{\sim}{n} + \frac{1}{2})]}\}^{-1} \quad . \tag{3.13}$$

To obtain the result given by periodic orbit theory, Eq. (3.9), it is also necessary to approximate the exponent in Eq. (3.13) by the first two terms of its Taylor's series expansion,

$$\theta[E - \hbar\underset{\sim}{\omega} \cdot (\underset{\sim}{n} + \tfrac{1}{2})] \simeq \theta(E) - \theta'(E)\hbar\underset{\sim}{\omega} \cdot (\underset{\sim}{n} + \tfrac{1}{2}) \quad . \tag{3.14}$$

As for the eigenvalue problem discussed in Section IIa, the approximations in Eqs.(3.12) and (3.14) are in general valid only if most of the total energy E is in motion along the reaction coordinate. As there, too, the periodic orbit expression happens to be _exact_ for the (trivial) case that all the f-1 modes are harmonic and the barrier for the fth mode is parabolic (i.e., an inverted harmonic potential).

Just as for the eigenvalue problem, therefore, the periodic orbit approach gives a result that is qualitatively correct but which can be inaccurate quantitatively because of dynamical assumptions implicit in it. Here too, though, there exists a Hamilton-Jacobi approach that treats all the degrees of freedom on an equal dynamical footing.

b. Hamilton-Jacobi Theory

A Hamilton-Jacobi analysis,[26] similar to that of the eigenvalue problem described in Section IIb, can be used to overcome the limitations in periodic orbit theory described in the previous section. This is possible by realizing that there are "good" action variables associated with the _saddle point_ of a potential surface just as there are those related to a minimum in a potential surface.

Consider first, for example, the case discussed in the previous section that motion along the reaction coordinate (the $f^{\underline{th}}$ degree of freedom) is separable from the other f-1 degrees of freedom. The correct semiclassical expression for N(E) is given in this case by Eqs. (3.10) and (3.11):

$$N(E) = \sum_{\underset{\sim}{n}=0} [1 + e^{2\theta(E-\varepsilon_{\underset{\sim}{n}})}]^{-1} \quad , \tag{3.15}$$

where $\underset{\sim}{n} \equiv \{n_k\}$, k=1, ..., f-1. An action variable for the reaction coordinate, n_f, can be defined in the usual one-dimensional fashion,

$$\begin{aligned}(n_f+\tfrac{1}{2})\pi &= \int dq_f \sqrt{2\mu[E_t - v_f]} \\ &= i\int dq_f \sqrt{2\mu[v_f - E_t]} \\ &\equiv i\theta(E_t) \quad ,\end{aligned} \tag{3.16}$$

with $E_t = E-\varepsilon_{\underset{\sim}{n}}$, where θ is the (real) barrier penetration integral. Eq. (3.15) can thus be written equivalently as

$$N(E) = \sum_{\underset{\sim}{n}=0} [1 + e^{2\pi \, \mathrm{Im} \, n_f(E,\underset{\sim}{n})}]^{-1} \quad , \tag{3.17}$$

where $n_f(E,\underset{\sim}{n})$ is given as a function of the total energy E and the f-1 action variables (or quantum numbers) by Eq. (3.16).

Although deduced for the separable case, Eq. (3.17) is also the result of the more rigorous Hamilton-Jacobi theory[26] for a general non-separable potential function. In the general case one must first construct the total Hamiltonian in terms of the complete set of "good" action variables $\{n_i\}$, i=1, ..., f using, for example, the methods of Section IIb. Then the equation

$$E = H(n_1, n_2, \ldots, n_{f-1}, n_f) \tag{3.18}$$

is solved for $n_f(E,n_1, \ldots, n_{f-1})$, and N(E) is given by Eq. (3.17). This result of the Hamilton-Jacobi analysis is seen to have the same qualitative structure as that of periodic orbit theory, Eq. (3.9), but it does not incorporate any approximation about the distribution of energy among the various modes nor approximate any of the modes as harmonic.

Eq. (3.17) can be derived for the general non-separable case by evaluating the trace in Eq. (3.4) in the representation of the "good" action variables $\{n_i\}$, i=1, ..., f:

$$N(E) = i\hbar\ \mathrm{tr}[G(E)F\mathcal{P}]$$

$$= i\hbar \sum_{\underset{\sim}{n}=0} \int dn_f \ \langle \underset{\sim}{n}, n_f | G(E)F\mathcal{P} | \underset{\sim}{n}, n_f \rangle$$

$$= i\hbar \sum_{\underset{\sim}{n}=0} \int dn_f \ \frac{\langle \underset{\sim}{n}, n_f | F\mathcal{P} | \underset{\sim}{n}, n_f \rangle}{E - E(\underset{\sim}{n}, n_f)} \quad , \tag{3.19}$$

where $\underset{\sim}{n} \equiv \{n_i\}$, i=1, ..., f-1, and $E(\underset{\sim}{n},n_f) \equiv H(n_1,n_2, \ldots, n_f)$. Because of the simple pole in the integrand, the integral over n_f can be evaluated by contour integration, giving

$$N(E) = 2\pi\hbar \sum_{\underset{\sim}{n}=0} \langle \underset{\sim}{n}, n_f | F\mathcal{P} | \underset{\sim}{n}, n_f \rangle \Big/ \frac{\partial E(\underset{\sim}{n},n_f)}{\partial n_f} \tag{3.20}$$

with $n_f = n_f(E,\underset{\sim}{n})$ determined by the pole condition, Eq. (3.18). It is shown elsewhere,[26] along with more details of the calculation, that the flux integral is given by

$$\langle \underset{\sim}{n}, n_f | F\mathcal{P} | \underset{\sim}{n}, n_f \rangle = \frac{\partial E(\underset{\sim}{n},n_f)}{\partial n_f} \left(1 + e^{2\pi \mathrm{Im}\, n_f}\right)^{-1} \Big/ 2\pi\hbar \quad , \tag{3.21}$$

which, inserted into Eq. (3.20), gives Eq. (3.17).

Application[27] of the periodic orbit version of this rate theory, Eq. (3.9), has been seen to give reasonably good results for the collinear $H + H_2$ exchange reaction <u>if</u> the exponent in Eq. (3.9) is modified in the following manner:

$$\theta(E) - \theta'(E)\ \hbar\underset{\sim}{\omega}(E)\cdot(\underset{\sim}{n}+\tfrac{1}{2}) \rightarrow \theta(E_{\underset{\sim}{n}}) \quad , \tag{3.22}$$

with $E_{\underset{\sim}{n}}$ determined by the equation

$$E_{\underset{\sim}{n}} = E - \hbar\underset{\sim}{\omega}(E_{\underset{\sim}{n}})\cdot(\underset{\sim}{n}+\tfrac{1}{2}) \quad . \tag{3.23}$$

This modification has the effect of correcting, in an approximate fashion, for the assumption in periodic orbit theory that all but infinitesimal of the total energy is in motion along the reaction coordinate. Use of Eq. (3.9) without this modification gives poor results because $\hbar\omega(E)(n+\frac{1}{2})$ is not at all small compared to E in the range of interest for this system. Since the Hamilton-Jacobi approach involves no such approximations, it is anticipated that it will provide a more satisfactory description of the rate constant for this (and similar) chemical reaction.

IV. CONCLUDING REMARKS

One thus sees that within the framework of semiclassical mechanics eigenvalues in a multidimensional potential well and chemical reaction through a saddle point region of a potential energy surface can be described by essentially the same formalism. In both cases the periodic orbit version of the theory provides a simple, physically correct picture of the phenomenon but can be inaccurate quantitatively because of certain assumptions implicit in it.

The Hamilton-Jacobi approach avoids the limitations of periodic orbit theory. It provides a way of constructing the total Hamiltonian for the non-separable system in terms of a complete set of "good" (i.e., conserved) action variables. In the eigenvalue case these action variables are required to be integers (i.e., quantum numbers), and this discretizes the energy. For chemical reaction through a saddle point region one of the action variables is identified as a "generalized barrier penetration integral", and its imaginary part (as a function of the other action variables and the total energy) gives the reaction probability. Applications of this approach to real molecular systems, either for eigenvalues or for rate constants of chemical reactions, have only begun but promise to be a useful way of describing these phenomena.

REFERENCES

* Supported in part by the U.S. Energy Research and Development Administration, and by the National Science Foundation under grant GP-41509X.

† Camille and Henry Dreyfus Teacher-Scholar; J. S. Guggenheim Memorial Fellow on sabbatical leave from the Department of Chemistry, University of California, Berkeley, CA 94720.

1. K. W. Ford and J. A. Wheeler, Ann. Phys. N.Y. 7, 259, 287 (1959).
2. For reviews of semiclassical theory of elastic scattering, see R. B. Bernstein, Adv. Chem. Phys. 10, 75 (1966), and M. V. Berry and K. E. Mount, Rept. Prog. Phys. 35, 315 (1972).
3. G. E. Zahr and W. H. Miller, Mol. Phys. 30, 951 (1975).
4. R. J. Gordon, J. Chem. Phys. 63, 3109 (1975).
5. W. H. Miller, J. Chem. Phys. 53, 1949 (1970).
6. R. A. Marcus, Chem. Phys. Lett. 7, 525 (1970).

7. For reviews, see W. H. Miller, Adv. Chem. Phys. 25, 69 (1974); 30, 77 (1975).
8. F. T. Smith, D. L. Huestis, D. Mukherjee, and W. H. Miller Phys. Rev. Lett. 35, 1073 (1975).
9. J. D. Doll, J. Chem. Phys. 61, 954 (1974).
10. R. I. Masel, R. P. Merrill, and W. H. Miller, Surf. Sci. 46, 681 (1974); J. Chem. Phys. 64, 45 (1976).
11. S. Levit, U. Smilansky, and D. Pelte, Phys. Lett. 53B, 39 (1974).
12. H. Massmann and J. O. Rasmussen, Nucl. Phys. A243, 155 (1975); H. Massmann, P. Ring, and J. O. Rasmussen, Phys. Lett. 57B, 417 (1975).
13. R. A. Malfliet, Lecture Notes in Physics (Springer Verlag) 33, 86 (1975); T. Koeling and R. A. Malfliet, Phys. Reports (Section C of Phys. Lett.) 22C, 182 (1975).
14. W. H. Miller, J. Chem. Phys. 53, 3578 (1970).
15. M. D. Gutzwiller, J. Math. Phys. 12, 343 (1971).
16. W. H. Miller, J. Chem. Phys. 63, 996 (1975).
17. R. P. Feynman and A. R. Hibbs, Quantum Mechanics and Path Integrals, McGraw-Hill, New York, 1965.
18. M. Born, The Mechanics of the Atom, Ungar, New York, 1960.
19. Other recent work on the semiclassical eigenvalue problem that is related to that discussed here is (a) I. C. Percival, J. Phys. B 6, L229 (1973); J. Phys. A 7, 794 (1974); (b) W. Eastes and R. A. Marcus, J. Chem. Phys. 61, 4301 (1974); D. W. Noid and R. A. Marcus, J. Chem. Phys. 62, 2119 (1975). These authors basically follow the formulation of Keller [J. B. Keller, Ann. Phys. N.Y. 4, 180 (1958); J. B. Keller and S. I. Rubinow, Ann. Phys. N.Y. 9, 24 (1960)] which is based on the Hamilton-Jacobi equation in cartesian coordinates, rather than that in action-angle variables used by Born.
20. S. Chapman, B. C. Garrett, and W. H. Miller, J. Chem. Phys. 64, 502 (1976).
21. N. C. Handy, S. M. Colwell, and W. H. Miller, Faraday Disc. Chem. Soc. 62, to be published.
22. N. C. Handy and W. H. Miller, unpublished results.
23. W. H. Miller, J. Chem. Phys. 61, 1823 (1974).
24. The more conventional expression for N(E) is in terms of the S-matrix elements of scattering theory,

$$N(E) = \sum_{\mathbf{n}_a, \mathbf{n}_b} |S_{\mathbf{n}_b, \mathbf{n}_a}(E)|^2$$

where $\mathbf{n}_a$ and $\mathbf{n}_b$ are the internal quantum numbers of the reactants and products, respectively.
25. W. H. Miller, J. Chem. Phys. 62, 1899 (1975).
26. W. H. Miller, Faraday Disc. Chem. Soc. 62, to be published.
27. S. Chapman, B. C. Garrett, and W. H. Miller, J. Chem. Phys. 63, 2710 (1975).

ENERGY CONSUMPTION AND ENERGY DISPOSAL IN ELEMENTARY CHEMICAL REACTIONS*

R.D. Levine

Department of Physical Chemistry
The Hebrew University, Jerusalem, Israel.

A. STATE-SELECTED CHEMICAL KINETICS

a. Specificity and Selectivity of Chemical Reactions

Traditionally, chemical reactions were studied under conditions which insured thermal equilibrium; the reaction was maintained at a rate slow compared to that of collisional energy transfer. Any momentary depletion of excited reagents due to their high reaction rate was immediately restored by collisional activation. The reactants could be characterized as being in an effective thermal equilibrium at all times. The only control over the energy of the reactants was via their temperature. To characterize the variation of the reaction rate it was sufficient to study the temperature dependence of the rate constant. Similarly, the energy released by the reaction could be measured by the exothermicity, the mean excess of energy of the (thermally equilibrated) reactants over that of the products.

One can however consider probing chemical reactions on a more 'molecular' level [1]. Here one overcomes the limitations imposed by thermal equilibrium and considers such questions as 'How would the reaction rate be affected if energy is specifically pumped into the reagent vibration?' or 'Immediately after the reaction, and before any collisional relaxation, how much of the available energy is in vibrational excitation of the newly formed products?' Ultimately one could even link the two questions: 'How does reagent selective vibrational excitation influence the vibrational energy disposal in the products' [2,3]. This considerable current interest in the selectivity of energy consumption

*Work supported by the Stiftung Volkswagenwerk

B. Pullman and R. Parr (eds.), The New World of Quantum Chemistry, 103–129.

and the specificity of energy disposable in chemical reactions is motivated both by potential practical applications [4-9] and by the drive to improve our understanding of the dynamics of molecular collisions. In order to achieve these goals there is a need to increase the arsenal of concepts and techniques which are employed to provide a theoretical discussion and interpretation of the experimental (or computational) results and to suggest new directions of research.

Consider, for example, the influence of reagent excitation on the rate of reactions [1-3,6,9,10-14]. Figure 1 shows the rate constant, $k(v\rightarrow;T)$ for reactions of vibrationally excited HCl molecules with the different halogen atoms

$$I + HCl(v) \rightarrow Cl + HI \quad (1)$$

$$Br + HCl(v) \rightarrow Cl + HBr \quad (2)$$

$$F + HCl(v) \rightarrow Cl + HF . \quad (3)$$

A far greater effect is evident for the endoergic I+HCl reaction than for the exoergic F+HCl reaction

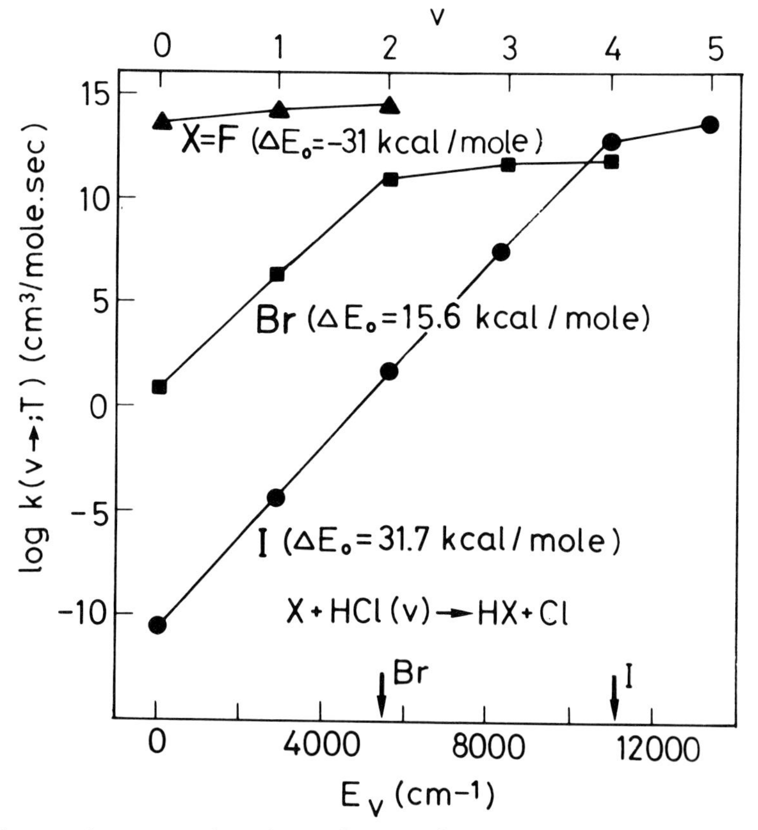

Figure 1. A compilation of experimental results for the effect of reagent vibrational excitation for endoergic (X = I, Br, ΔE_0 values

as indicated) and exoergic (X=F) reactions. (Adapted from Ref. 13. Experimental results from Refs. 14-18). The arrows indicate the threshold where reagents mean energy exceeds the endoergicity.

The activation energy for such reactions is often found to decrease linearly with increasing vibrational excitation [11,12]

$$E_a(v) = -k\,d\ln k(v\rightarrow;T)/d(1/T) \simeq E_o - (1-\lambda_v)E_v \, . \qquad (4)$$

Here E_v is the vibrational excitation of HCl and $E_o \equiv E_a(v=0)$ is the activation energy for vibrationally cold reagents. In the regime $E_v < E_o$, λ_v (read λ-vibration) is independent of v and is found to be negative and to have different values for different reactions. The results for the endoergic

$$H + HF(v) \rightarrow F + H_2 \qquad (5)$$

reaction [13] are shown in figure 2.

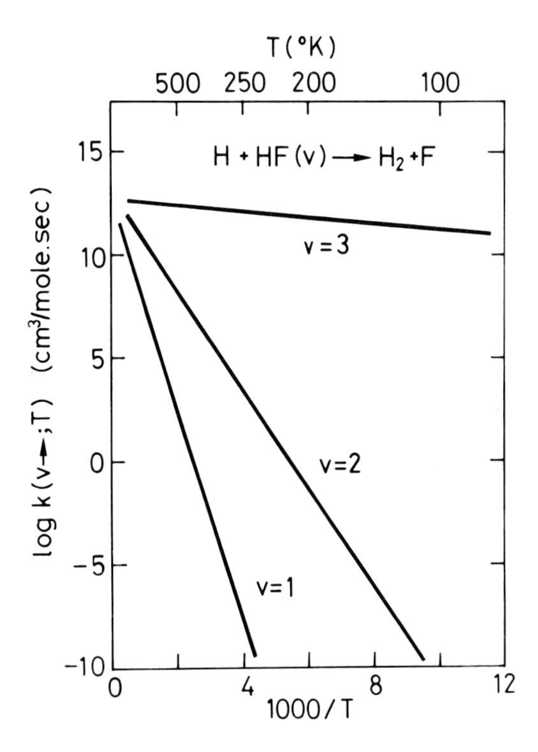

Figure 2. Temperature dependence of the reaction rate constant from specified reagent vibrational states for the endothermic H+HF(v) reaction. The 'Arrhenius' plot serves to show that the 'pre-exponential' factor, (the $T \rightarrow \infty$ limit), is nearly independent of v and the increase of the endothermic rate constant with reagent vibration is primarily due to the decrease in the activation energy.

The activation energy has a (rigorous) interpretation as the mean energy of those molecules that undergo reaction upon collision minus the mean energy of all molecules. The mean thermal energy of those reagents that react is thus $E_0 + \lambda_v E_V$. The mean thermal energy of the 'reactive' reactants is lower when they are vibrationally excited (recall that λ_v is negative). Reagent vibrational excitation is more effective than the thermal (translational and/or rotational) energy of the reactants in enhancing the reaction rate. Such selectivity in energy consumption [11,12] appears to be the rule rather than an exception for elementary chemical reactions.

For an endoergic reaction, the reagents need have a minimal excess energy (equal to the reaction endoergicity) before reaction can take place, on thermodynamic grounds. Thus, endoergic reactions will alwayshave an energy of activation and their reaction rates would be greatly enhanced by any excitation of the reagents. Even so, the wide ranges of behavior evident in figure 1 and in other reactions cannot be completely accounted for in this fashion.

Extreme variations are also evident in the energy disposal of chemical reactions. Figure 3 shows the populations of the vibrational states of HCl formed in a series of

$$Cl + HX \rightarrow X + HCl(v) \tag{6}$$

reactions. From extensive inversion in the highly exoergic Cl + HI reaction to a total lack of inversion in the endoergic Cl + HF reaction.

Comparison of figures 3 and 1 suggests that selectivity in energy consumption and specificity in energy disposal must be closely related. The $Cl + HI \rightarrow I + HCl(v)$ reaction preferentially populates the higher vibrational states of HCl and the rate of the $I + HCl(v) \rightarrow Cl + HI$ reaction is very strongly enhanced by HCl vibration excitation. For all reactions selectivity and specificity go hand in hand. Our task is to make this correlation explicit and then to devise a common measure to characterize it. While we shall not discuss this topic in detail here, the next step is obviously that of prediction [19-21].

The distributions P(v) shown in figure 3 (and in other figures, e.g. 5,7 and 8 below) can all be predicted, using information theoretic methods, given the exothermicity of the reaction and the spectroscopic constants of the reactants and products [29]. Of course, dynamical computations (e.g. classical trajectories, quantal close coupling) and models [1,2,13] have been quite successful in reproducing and predicting such distributions.

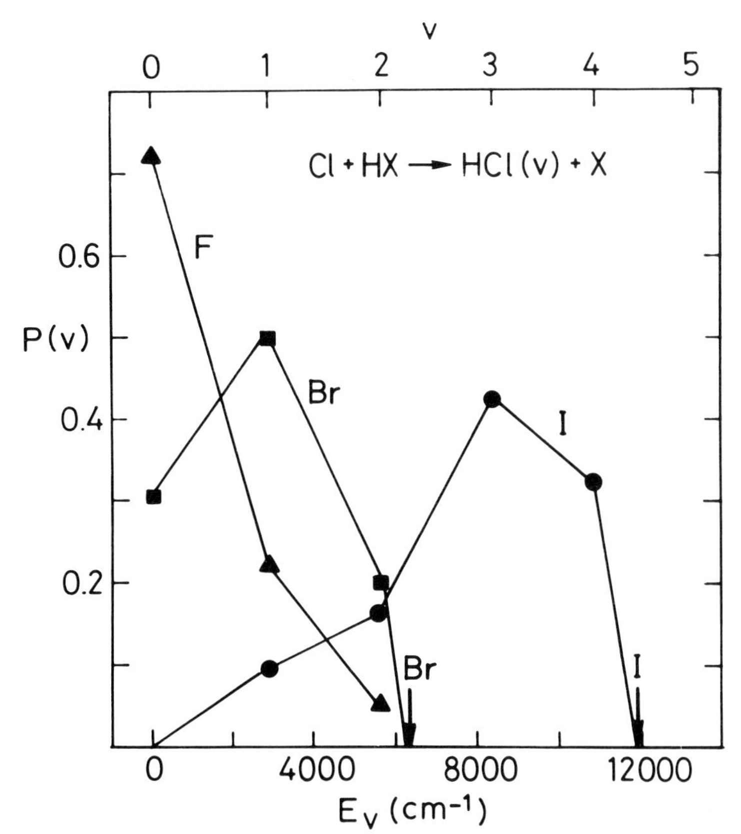

Figure 3. P(v), the fraction of HCl in the vibrational state v, produced in the Cl+ HX reaction vs. v. [Experimental results from Refs. 16 and 18; adapted from Ref. 13]. The arrows indicate the magnitude of the exothermicity, (the reaction is endothermic for X=F). The higher is the exothermicity the more extensive is the population inversion.

b. The Detailed Rate Constant.

For an elementary bimolecular reaction one can always introduce a reaction rate constant by writing down the reaction rate as a typical second order kinetics law. Thus for the reaction (1), $Cl + HCl(v) \rightarrow H + Cl_2$, the rate of disappearance of HCl molecules in the vibrational state v due to their reaction with Cl atoms is expressed as

$$-d[HCl(v)]/dt = k(v \rightarrow ;T)[Cl][HCl(v)] \quad . \tag{7}$$

Here the square brackets denote concentrations. We have assumed here, as before, that the rotational states of HCl and the kinetic energy have a thermal distribution at the temperature T. Hence the rate constant, $k(v \rightarrow ;T)$, can only depend on the vibrational state of HCl and on the temperature. As in ordinary chemical kinetics, the fact that the reaction is an elementary, bimolecular one insures that $k(v \to ;T)$ defined by (6) is independent of the concentration.

Chemical kinetics with state selected reactants is no different than ordinary chemical kinetics. We need simply regard HCl molecules in a specified vibrational state as a distinct chemical species. (Always provided that the other, non-selected degrees of freedom of the reagents remain in thermal equilibrium).

For completely thermal reactants, Cl atoms react with HCl molecules in different vibrational states. The total reaction rate is then just the sum of the rates of the separate processes that lead to disappearance of HCl molecules

$$\begin{aligned} -d[HCl]/dt &= \sum_v - d[HCl(v)]/dt \\ &= \sum_v k(v \to ;T)[Cl][HCl(v)] \\ &= \{\sum_v k(v \to ;T)[HCl(v)]/[HCl]\}[Cl][HCl] \ . \end{aligned} \tag{8}$$

Hence the thermal reaction rate $k(T)$ is related to $k(v \to ;T)$ by

$$k(T) = \sum_v p(v|T)k(v \to ;T) . \tag{9}$$

Here $p(v|T)$,

$$p(v|T) = [HCl(v)]/[HCl] \tag{10}$$

is the fraction of HCl molecules that are in the vibrational state v, at thermal equilibrium.

The thermal rate constant is obtained as an average of the detailed rate constants over the distribution of initial vibrational states. It is clear however that nothing in our discussion here limits it to vibrational states. v can equally well stand for any specification of the initial state. Hence the general rule is: when there is a distribution of the states of the reagents, the total rate constant is obtained as a weighted sum (i.e. an average) over the rate constants for the different initial states.

The asymmetric character of the definition of the rate constant is made explicit when we consider a resolution of the states of the products. Consider, say, the reaction

$$Cl + HCl(v) \to H + Cl_2(v') \ . \tag{11}$$

The rate of consumption of HCl molecules in the state v to form Cl_2 molecules in the state v' is expressed as $k(v \to v';T)[Cl][HCl(v)]$. But clearly the rate of consumption of HCl(v) is the sum of the rates to form $Cl_2(v')$ in the different vibrational states v' so that

$$k(v \to ;T) = \sum_{v'} k(v \to v';T) , \tag{12}$$

and, using (9)

$$k(T) = \sum_{v} p(v|T) \sum_{v'} k(v \to v';T) \quad (13)$$

when there is no resolution of the product states one need simply sum over the final states. The dichotomy between reagents and products states evident in (13) is summarized by the 'canon': sum over final states, average over initial states.

c. Detailed Balance

The reaction

$$H + Cl_2 \to Cl + HCl(v) \quad (14)$$

is simply the reaction (1) in the reversed direction. The rate constant $k(\to v;T)$ for formation of HCl in the vibrational state v,

$$d[HCl(v)]/dt = k(\to v;T)[H][Cl_2] , \quad (15)$$

may well therefore be related to the rate, $k(v \to ;T)$ for consumption of HCl(v) by the reversed, $Cl + HCl(v) \to H + Cl_2$ reaction, equation (1).

The principle of detailed balance is a statement about rates (not rate constants) of collisions which enables us to provide the connection. It states that at equilibrium the rate of a detailed process is equal to the rate of the reversed process. Hence, using equilibrium concentrations in (7) and (15)

$$k(v \to ;T)[Cl][HCl(v)] = k(\to v;T)[H][Cl_2] \quad (16)$$

or

$$k(\to v;T)/k(v \to ;T) = K(\to v;T). \quad (17)$$

Here $K(\to v;T)$,

$$K(\to v;T) = [Cl][HCl(v)/[H][Cl_2] \quad (18)$$

is the equilibrium constant for the $H + Cl_2 \rightleftarrows Cl + HCl(v)$ reaction, at thermal equilibrium. The concentrations in (18) as in (16) are those that obtain at equilibrium. Again we see that if we regard HCl(v) as a distinct chemical species then (17) is just the usual result that for an elementary process the equilibrium constant is the ratio of the forward and reversed rate constants.

The considerable utility of expressing the principle of detailed balance in the form (17) stems from our ability to compute the equilibrium constant without any kinetic input.

The simplest route to computing $K(\to v;T)$ is to relate it to K(T),

$$K(T) = \overrightarrow{k}(T)/\overleftarrow{k}(T) = [Cl][HCl]/[H][Cl_2] , \quad (19)$$

the equilibrium constant of the

$$H + Cl_2 \rightleftarrows Cl + HCl \quad (20)$$

reaction. Using the definitions (18) and (19)

$$K(\to v;T) = p(v|T)K(T) \tag{21}$$

Standard thermochemical tables enable one to compute $\Delta A^o(T)$, the standard free energy change of the reaction in terms of the standard free energies of the reactants and products

$$\Delta A^o(T) = \sum_{\text{products}} A_i^o(T) - \sum_{\text{reactants}} A_i^o(T) , \tag{22}$$

where

$$\Delta A^o(T) = -RT\ell nK(T) . \tag{23}$$

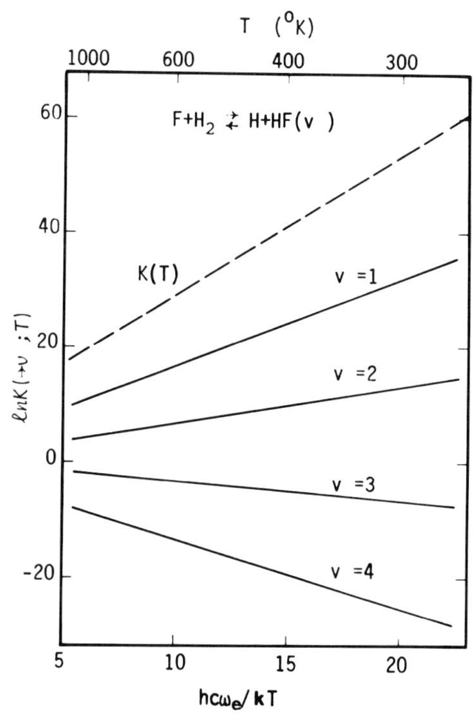

Figure 4. K(T) and K($\to$ v';T) vs. θ/T where $\theta = hc\omega_e/k$ is the characteristic vibrational temperature of HF, [22].

Alternatively, one can compute K(T) in terms of the partition functions of the reactants and products,

$$K(T) = (\mu'/\mu)^{3/2}(Q_{HCl}/Q_{Cl_2})\exp(-\Delta E_0/kT) . \tag{24}$$

Here μ' and μ are the reduced masses of the products, $\mu' = m_{Cl}m_{HCl}/(m_{Cl}+m_{HCl})$ and reactants respectively and the Q's are the partition functions for the <u>internal</u> degrees of freedom of the diatoms.

d. Energy Disposal and Energy Consumption

The principle of detailed balance cast in the form (16) clearly implies the essence. Those vibrational states that are preferentially populated in the forward reaction would react with a high rate in the reversed reaction. To compare the reactivity of two different vibrational states, say n and m we can write, using (17) and (21)

$$p(n|T)k(n\rightarrow;T)/p(m|T)k(m\rightarrow;T) = k(\rightarrow n;T)/k(\rightarrow m;T) \ . \qquad (25)$$

By definition, $p(n|T)$ is the fraction of HCl molecules in the vibrational state n at thermal equilibrium. Thus $p(n|T)k(n\rightarrow;T)$ is the actual rate of the reaction from the n'th vibrational state of HCl (at unit concentration) at thermal equilibrium. The relative rates of reaction of two different states equals the relative rates of populating the two states by the reversed reaction.

If we now allow a more general interpretation of the index n, as any state (or group of states) of the reactant molecule, then (25) remains valid.

Rather than considering the relative rates we can examine $P(v)$, the fraction of reactive collisions which populate the HCl vibrational state v in the reaction (14)

$$P(v) = k(\rightarrow v;T)/\overrightarrow{k}(T) \qquad (26)$$

$$\overrightarrow{k}(T) = \sum_v k(\rightarrow v;T) \qquad (27)$$

$$\sum_v P(v) = 1 \ . \qquad (28)$$

The principle of detailed balance can now be reexpressed by noting that $P(v)$ is also the fraction of reverse collisions which proceed from HCl in the vibrational state v,

$$P(v) = p(v|T)k(v\rightarrow ;T)/\overleftarrow{k}(T) \qquad (29)$$

$$\overleftarrow{k}(T) = \sum_v p(v|T)k(v\rightarrow ;T) \ . \qquad (30)$$

The two expressions, (26) and (29), do not appear to be equivalent. This is merely a reflection of the asymmetric canon: average over initial states sum over final states. By substituting (17) and (21) into (29) one readily verifies that (26) $\equiv$ (29).

The same probability distribution, $P(v)$, figure 4, characterizes the vibrational energy disposal in the forward reactic and the vibrational energy requirements in the reverse reaction. The results displayed in figures 1 and 3 are thus identical in their content. In fact, only one set of measurements was employed by us to draw both figures, using the equivalence of (26) and (29). Even so, one cannot conclude that the two routes to $P(v)$ are equivalent on practical grounds. Figure 5 [12], illustrates the difficulties. Consider the exothermic

$$O + CS \rightarrow S + CO(v) \qquad (31)$$

reaction. The measured [23] product vibrational state distribution has but a moderate v dependence. Hence, from (29), the exponential v dependence of $p(v|T)$ implies that $k(v \rightarrow ;T)$ for the endothermic

$$S + CO(v) \rightarrow O + CS \qquad (32)$$

reaction will increase essentially exponentially with v, figure 5. The exponential dependence is so dominant that it tends to wash-out the more moderate v-dependence of $P(v)$. Thus while both panels in figure 5 are mathematically equivalent, it is far easier to determine the v dependence of $P(v)$ via the energy disposal (i.e. using (26)) then through the energy consumption (i.e. using (29)). In general, one finds [12] that it is always easier to characterize $P(v)$ via measurements (or computations) in the exothermic direction. Similar conclusions apply to the role of reagent translational energy [23].

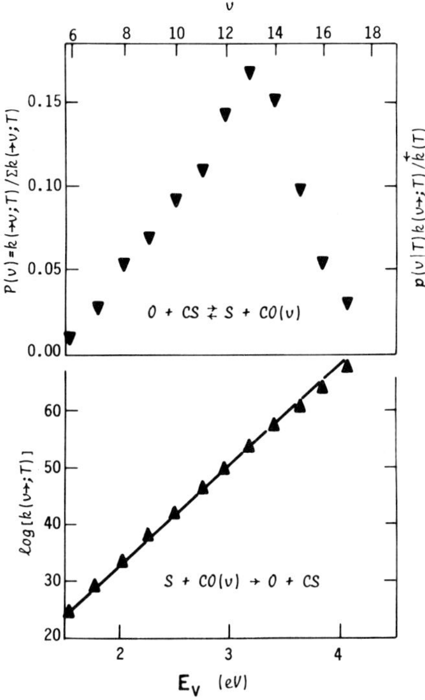

Figure 5. $P(v)$ as determined from an experimental [24] measurements of the relative rates of populating CO vibrational states in the O+CS reaction at 300°K. Using (29) the rates $k(v \rightarrow ;T)$ for the endothermic S+CO(v) were computed and are shown vs. the vibrational energy of CO. Since only relative rates were measured, $k(v \rightarrow ;T)$ can be determined only up to an arbitrary (but v independent) factor.

e. Thermodynamic Considerations

Using the principle of detailed balance we were able to introduce the equilibrium constant $K(\rightarrow v;T)$. Thereby the vibrational energy requirements of the reaction are determined if the vibrational energy disposal of the reversed process is known (or vice-versa). In this way one and the same distribution is sufficient to describe both the forward and reversed reactions. Can we do better, i.e. can we use equilibrium considerations to characterize the distribution $P(v)$ itself. The simple answer is no, we cannot. Any pair of forward and reversed reaction rate constants must be related as in (17). All that (17) implies is that if the forward reaction is highly specific in its vibrational energy disposal then the reversed reaction will be highly selective in its vibrational energy consumption. Alternatively, if the specificity is low, so is the selectivity. Equilibrium considerations in the form (17) provide no readily evident information on selectivity or specificity.

The equilibrium constant (and other equilibrium thermodynamic functions [22]) for state selected reactions does not therefore reflect the selectivity or the specificity of the reaction. Is there then any other thermodynamic property wherein a highly specific energy release is manifested? One suspects that there must be. Designers of chemical lasers are particularly interested in chemical reactions which release much of the energy as product vibration [13]. Why? Chemical lasers are devices which produce coherent light, i.e. work [25,26]. How is the specificity of energy disposal related to available work?

The change in free energy when thermal reactants form products in the vibrational state v is, from (21) and (23)

$$\Delta A^{o}(\rightarrow v;T) = -RT\ell nK(\rightarrow v;T) = \Delta A^{o}(T) - RT\ell np(v|T). \qquad (33)$$

Hence, if the fraction of product molecules in the vibrational state v is $P(v)$ the free energy available is

$$\begin{aligned} \Delta A(\rightarrow\{P(v)\};T) &= \Sigma P(v)\Delta A^{o}(\rightarrow v;T) + RT\Sigma P(v)\ell nP(v) \\ &= \Delta A^{o}(T) + RT\Sigma P(v)\ell n[P(v)/p(v|T)] . \end{aligned} \qquad (34)$$

Here $R\Sigma P(v)\ell nP(v)$ is the 'mixing entropy' of the distribution $P(v)$, [27], a term made familiar by the need to consider the free energy of separation of isotope mixtures.

As a check of the result (34) consider $\Delta A^{o}(\rightarrow\{p(v|T)\};T)$. This should come out to be just $\Delta A^{o}(T)$, the change in free energy from thermal reactants to thermal products. The substitution $P(v)=p(v|T)$ in (34) verifies that this is indeed the case.

Consider now a chemical reaction using thermal reagents where the vibrational relaxation of the products is arrested at some non-equilibrium distribution $P(v)$. If the vibrational relaxation were complete, the change in free energy would be $\Delta A^o(T)$. Now however it is $\Delta A^o(\to\{P(v)\};T)$. Hence, the work still available during a (reversible) isothermal relaxation of the distribution $P(v)$ to equilibrium (i.e. to $p(v|T)$) is

$$-\Delta A^o - [-\Delta A^o(\to\{P(v)\};T)] = RT\Sigma P(v)\ell n[P(v)/p(v|T)] \quad . \qquad (35)$$

In writing down (35) we noted that the work available (under reversible conditions) is minus the change in the free energy [27]. (Under realistic conditions a lesser amount of work is available).

On practical grounds what we have achieved is a way of computing the (maximal) work available when a non equilibrium distribution is notallowed to spontaneously relax to equilibrium but is made to do work instead. As in other spontaneous processes, here too one can harness the process and extract work. Concrete applications to the efficiencies of chemical lasers have been reported [28].

For our present purpose the theoretical implications of (35) are quite revealing. In particular, (35) shows that specificity of energy disposal is directly related to the (maximal) amount of work that can be extracted if the spontaneous relaxation of the nascent reaction products is arrested [29].

In the same vein selectivity in energy consumption is directly related to the (minimal) work required to drive the reaction (under isothermal conditions).

Equilibrium thermodynamics sufficed to establish the connection between energy requirements and energy disposal. We had to turn however to non equilibrium considerations [29,30] to find a quantitative measure for the specificity (or the selectivity). We proceed now to an actual analysis.

B. SURPRISAL ANALYSIS

a. The Prior Distribution

The distribution of vibrational states $p(v|T)$ of a diatomic molecule in a bulk, thermal equilibrium can be very accurately characterized by the canonical form $p(v|T) = \exp(-E_v/kT)/Q$. Given the temperature T the fraction of molecules in any state v can be readily computed. Equivalently, given the mean vibrational energy of the diatomic molecules one can find the corresponding temperature and hence determine $p(v|T)$. Can one now provide a similar procedure for the distribution of nascent products vibrational states?

Faced with the need to characterize the distribution $P(v)$ we must explicitly recognize that it reflects the distribution of outcomes in isolated binary reactive collisions. For any given collision the total available energy E is the sum of the products vibrational rotational and translational energies

$$E = E_V + E_R + E_T \quad . \tag{36}$$

The energy not deposited in the vibration, $E - E_V$ is necessarily either in the translation or the rotation. If E_V is high, less energy is available for the other degrees of freedom and vice versa. When we specify E and E_V there may be many collisions with the specified values but with different magnitudes of E_R and E_T (provided only that $E_R + E_T = E - E_V$). There are many possible outcomes of the collision all of which correspond to the same given vibrational state but which differ in the way the rest of the energy, $E - E_V$, is partitioned.

Next we need to recognize that energy is conserved in an isolated binary collision. Consider now an exoergic reaction where the relaxation is completely arrested. The point is that due to conservation of energy one can never expect to find the nascent products with an equilibrium, thermal-like, distribution. Never, no matter what. The reason why is obvious. Since no relaxation took place the distribution of total energy is precisely the same as the distribution of total energy in those collisions that [illegible] to reaction. Since, by assumption, the reaction is exoergic, the distribution of products total energy is bound to be zero for $E < -\Delta E_0$, where $-\Delta E_0$ is the reaction exoergicity. Such a distribution is necessarily non-thermal. It is only via subsequent collisions, either with the buffer gas or among molecules (or atoms) which are products of different collisions, that this strong correlation is dissipated. As long as we deal with nascent products we must learn to live with the implication of conservation of energy in isolated collisions. What we need in particular is a distribution, similar to $p(v|T)$ in that it can serve as a reference to compare $P(v)$ to, yet one which does take into account our specific concern with the results of isolated collisions. We shall refer to this distribution as the prior, $P^0(v)$, distribution. Like $p(v|T)$ it is to be the most random (i.e. most relaxed) distribution possible, under the appropriate conditions. Unlike $p(v|T)$, $P^0(v)$ will explicitly recognize the energy correlations that necessarily exist in the nascent products, i.e. that if E_V is low then considerable energy is present in translation and rotation and vice versa, figure 6.

To start off the discussion we consider first the simplest of all possible situations. A collision between an atom and a rigid diatomic molecule at a fixed total energy E, which can be partitioned between rotation and translation. The 'equilibrium' distribution at a given total energy is one where all quantum states are equally probable [35]. The number of quantum states (per unit volume) when the translational energy is in the range E_T to E_T+dE_T, $E_T = p^2/2\mu$ is $\rho_T(E_T)dE_T$,

$$\rho_T(E_T)\,dE = d\underset{\sim}{p}/h^3 = 4\pi p^2 dp/h^3 = 4\pi p\mu dE_T/h^3 \equiv A_T E_T^{\frac{1}{2}} dE_T \tag{37}$$

where $A_T = (2\mu)^{3/2}\pi/h^3$. The number of quantum states of the atom-diatom when out of the total energy E, the amount E_R present as rotation, $\rho(j;E)$ is

$$\rho(j;E) = (2j+1)\rho_T(E-E_R) = A_T(2j+1)(E-E_R)^{\frac{1}{2}} \tag{38}$$

and $E_R = B_e j(j+1)$ where B_e is the rotational constant.

The prior distribution of the final rotational states, assumed proportional to $\rho(j;E)$, is thus given by

$$P^o(j) = \rho(j;E)/\sum_j \rho(j;E) \quad , \tag{39}$$

and is independent of the initial rotational state in the collision. When we consider such collisions as

$$Ar + N_2(j) \rightarrow Ar + N_2(j') \tag{40}$$

then $P^o(j \rightarrow j') \equiv P^o(j')$.

Allowing for vibrational excitation requires only a modest extension of the formalism [31-33]. Now $E = E_T + E_V + E_R$. When the diatomic molecule is in a definite vibrotational state, the number of quantum states of an atom-diatom pair is

$$\rho(v,j;E) = (2j+1)\rho_T(E-E_v-E_R) = A_T(2j+1)(E-E_v-E_R)^{\frac{1}{2}} \tag{41}$$

and

$$P^o(v,j;E) = \rho(v,j;E)/\sum_{vj}\sum \rho(v,j;E) \; . \tag{42}$$

Clearly, the number of molecules in a given vibrational state is obtained by summing over all vibrotational states in a given vibrational manifold. Thus

$$P^o(v) = \sum_j P^o(v,j;E) = \rho(v;E)/\sum_v \rho(v;E) \, , \tag{43}$$

where (cf. (42))

$$\rho(v;E) = \sum_j \rho(v,j;E) \; . \tag{44}$$

The summation over j in (43) and (44) is over all rotational states allowed by conservation of energy, i.e. such that $E_R \leqslant E-E_v$.

Thus far we have considered collisions at a given total energy. To discuss reactions of thermal reagents it is necessary to average over the distribution of total energy in the reagents. The details [11,19,34] are somewhat tedious due to the fact that one needs to account for conservation of energy as discussed in figure 6. Explicit results are however available [11,19,34].

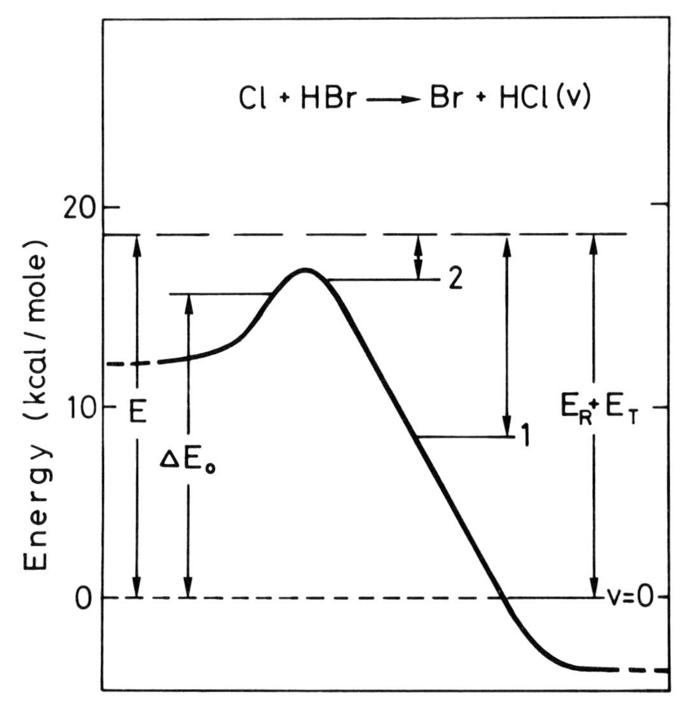

Figure 6. Energy balance in an exoergic reaction. Reactants with the total energy E are converted to products in different vibrational states with the energy balance (shown as arrows on the left hand side) present as rotation of HCl and relative translation of Br and HCl. The correlation between vibrational excitation and the amount of energy released as rotation or translation ($E_V = E - E_R - E_T$) is readily evident. Such correlation between E_V, E_R and E_T is absent at thermal equilibrium.

c. Surprisal Analysis

The considerations in section A.e suggest that we center attention on the mean deviation of the actual product state distribution from the prior distribution. The measure of the deviance of a particular final state is the 'surprisal'. Thus, for a vibrational state distribution

$$I(v) = - \ln[P(v)/P^{o}(v)] \; . \qquad (45)$$

Figure 7 shows P(v), $P^{o}(v)$ and the surprisal for the Cl+HI and Cl+DI reactions. Due to the high exoergicity of the reaction one can neglect the thermal spread in the energy of the reactants and use (43) to compute $P^{o}(v)$.

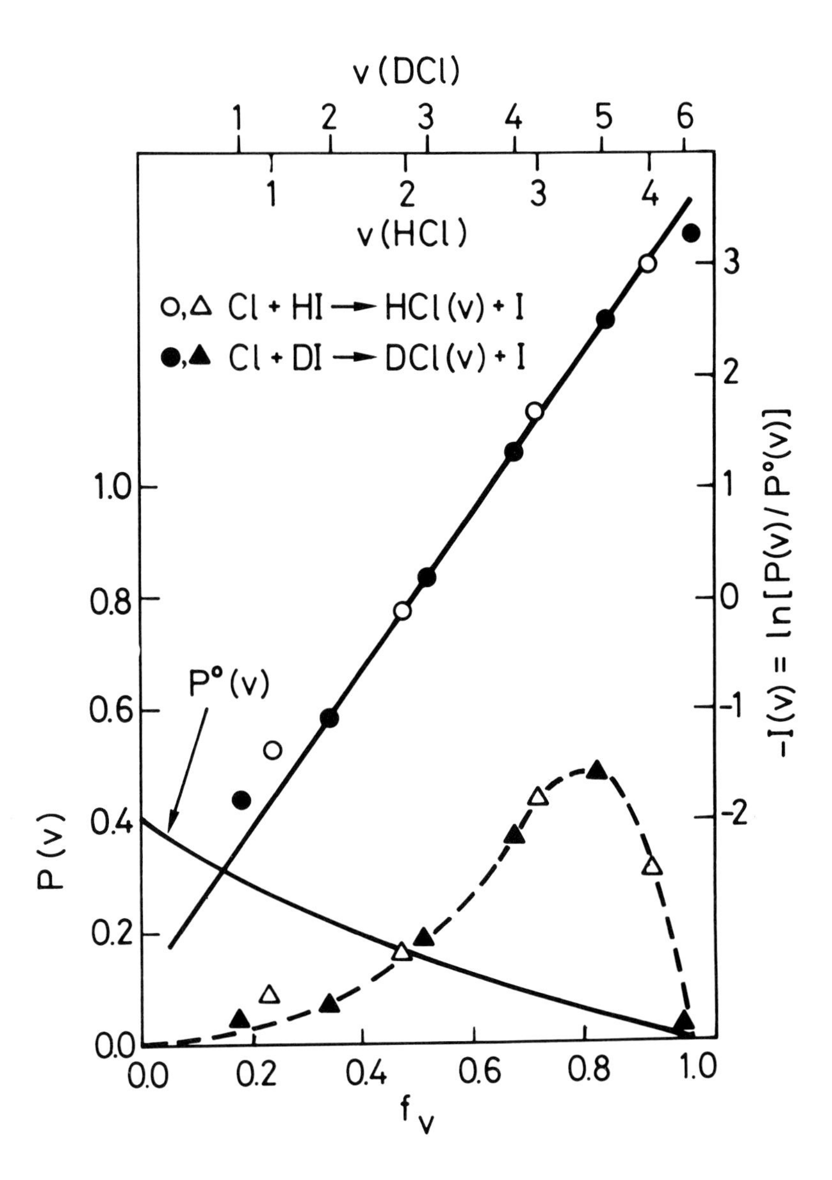

Figure 7. Observed [Ref. 16] and prior ($P^{o}(v)$) product vibrational state distribution in the Cl+HI (open triangles) and the Cl+DI (solid triangles) reactions vs. v. (Note that on prior grounds v=0 is the most probable final vibrational state). The abscissa shows $f_V = E_V/E$, the fraction of available energy present as product vibration. The circles show the surprisals of the different product vibrational states (adapted from Ref. 32).

While P(v) and $P^{o}(v)$ have a qualitatively different behavior, the surprisal, figure 7, is found to have a simple dependence on v. The same is found for other reactions, figure 8.

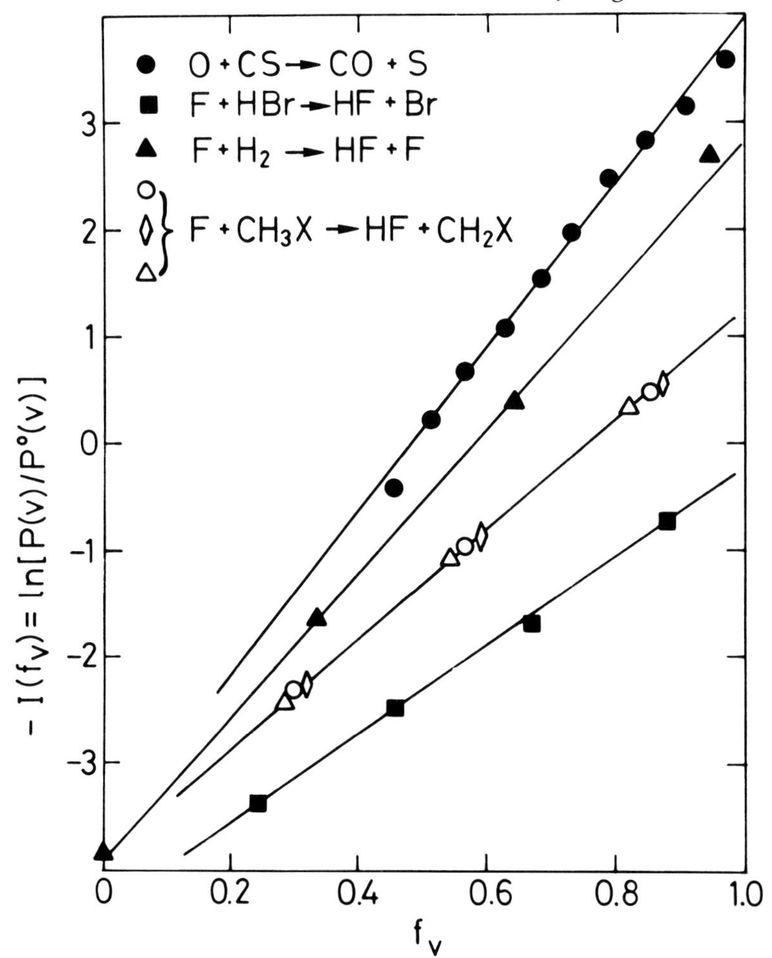

Figure 8. Surprisal analysis of the products vibrational state distribution for several exoergic reactions (adapted from Ref. 13. Experimental results: $F+CH_3X$ (X=Br, Cl, I), Ref. 36. $F+H_2$, Ref. 37. F+HBr, Ref. 38. O+CS, Ref. 24).

The method of surprisal analysis has been applied to other examples of vibrational energy disposal [36, 39-43] and vibrational energy consumption [10-13,19]. Translational energy disposal [32,44, 45] and consumption [21] and rotational energy disposal and consumption [11,12,20,32,34,46-50] were also considered. Figure 9 shows the analysis for the collision (40). The lines shown are not a fit but an independent prediction of the surprisal [19].

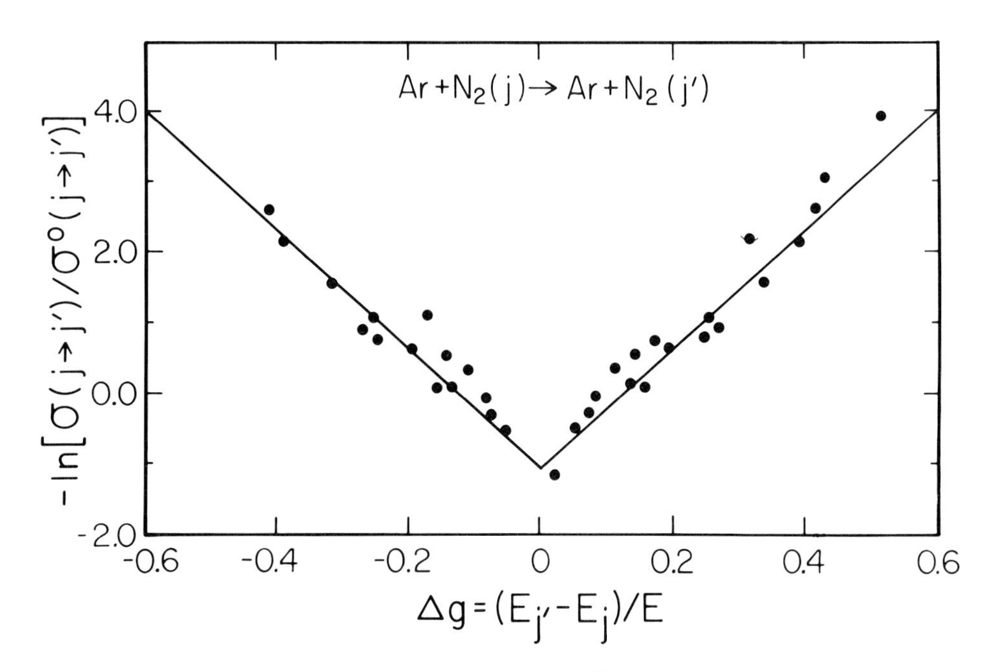

Figure 9. The surprisal, $-\ell n[P(j \to j')/P^{0}(j \to j')]$ for the collision (40), vs. the reduced variable $(E_j - E_{j'})/E$ (at a low energy (E=0.066 eV) where N_2 can be regarded as a rigid rotor). The dots are the results of quantum mechanical ('close coupling') computations [51] for the cross sections. All the different initial and final rotational states could be represented on the same surprisal plot. The solid line is the predicted [19] surprisal on the potential surface employed.

d. The Exponential Gap Representation

The efficiency of an energy transfer process typically shows an exponential dependence on the amount of energy transferred out of (or into) the translational motion [1]. The surprisal offers a quantitative expression of this rule of thumb. Consider, for example, the implications of figure 9. The probability of a $j \to j'$ transition can be well represented as

$$P(j \to j') = P^{0}(j \to j')\exp[-|\Delta E_R|(\theta_R/E) - \theta_0] \qquad (46)$$

where $\Delta E_R = E_R - E_R' = E_{T'} - E_T$ is the amount of translational energy released by (or required for) the transition, and θ_R (and θ_0) depend only on the total energy E. The probability of a final rotational energy is given by $P(E_R')d(E_R/B_e) = P(j')dj'$. Hence, using (38) and (39), $P^{0}(E_R') \propto \rho_T(E_T') = A_T(E-E_R')^{1/2}$. On prior grounds releasing energy into the products translation is the most favorable outcome (see also figure 7). The actual probability $P(E_R \to E_R')$ deviates exponentially from the prior

$$P(E_R \to E_R') = P^0(E_R \to E_R') \exp[-|\Delta E_R|(\theta_R/E) - \theta_0] \ . \qquad (47)$$

The more probable transitions are those where $|\Delta E_R|$ is small and the higher is θ_R the more confined is the distribution of E_R'. As $E \to \infty$, $(\theta_R/E) \to 0$, [19], so that the prior behavior obtains in the high energy ('sudden' [1]) limit.

The linear dependence of the surprisal on the 'energy gap' is not unique to rotational energy transfer. Figure 10 shows that the same applies to vibrational energy transfer.

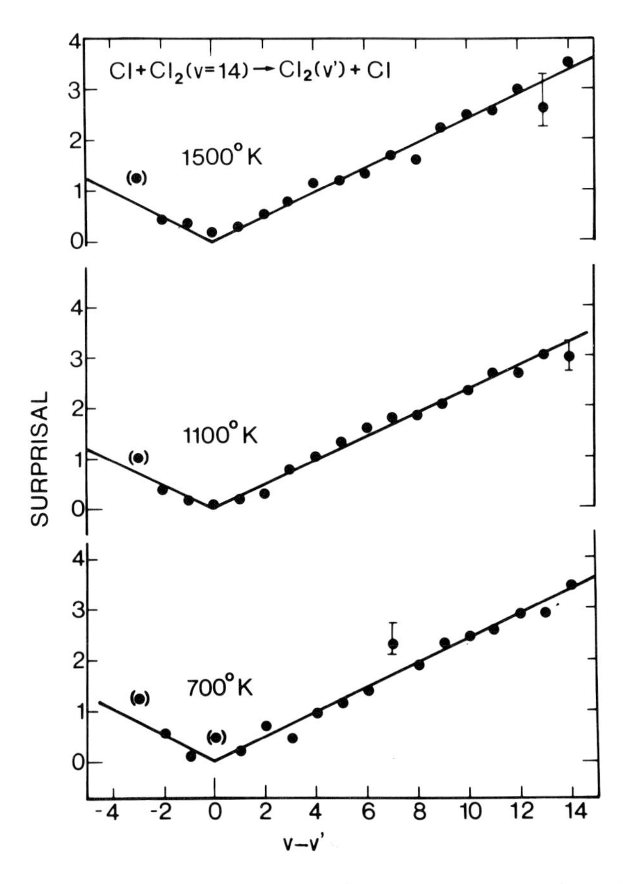

Figure 10. Surprisal analysis of the distribution of final vibrational states [adapted from Ref. 19, based on trajectory computations of Ref. 52]. The reagent Cl_2 molecule is in a definite vibrational state and the other degrees of freedom have a thermal distribution. v-v' is positive (negative) for deactivation (activation) collisions. The slopes of the surprisal plot can be predicted [19].

Even small energy gaps follow a similar rule. In the energy transfer process

$$DF(v) + F \rightarrow DF(v-1) + F \tag{48}$$

the energy gap depends on v only due to the anharmonicity of the DF molecule. The surprisal analysis is shown in figure 11.

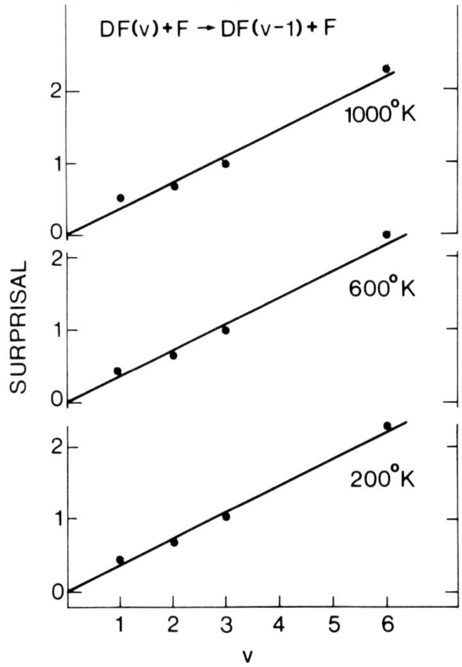

Figure 11. Surprisal analysis for the energy transfer process (48). [Adapted from Ref. 19, based on trajectory computations of Ref. 53].

The exponential gap representation is not limited to energy transfer collisions. Consider, for example, the reaction

$$Cl + HCl(v) \rightarrow H + Cl_2(v') \tag{49}$$

where the reagents are thermal except that the vibrational state of HCl is specified. If v and v' are zero, the reaction is endoergic. If now v is increased, the endothermicity is diminished. Less energy need be provided (by the translation or the rotation) to drive the reaction as v is increased. Finally, by taking v to be large enough (i.e. $E_V > \Delta E_0$) the reaction becomes exothermic, (i.e. the mean products E_R and E_T exceed those of the reagents). Figure 12 shows the surprisal plot where the arrows indicate the point where the energy balance changes. To the left of the arrow energy is consumed by the reaction; to the right, energy is released.

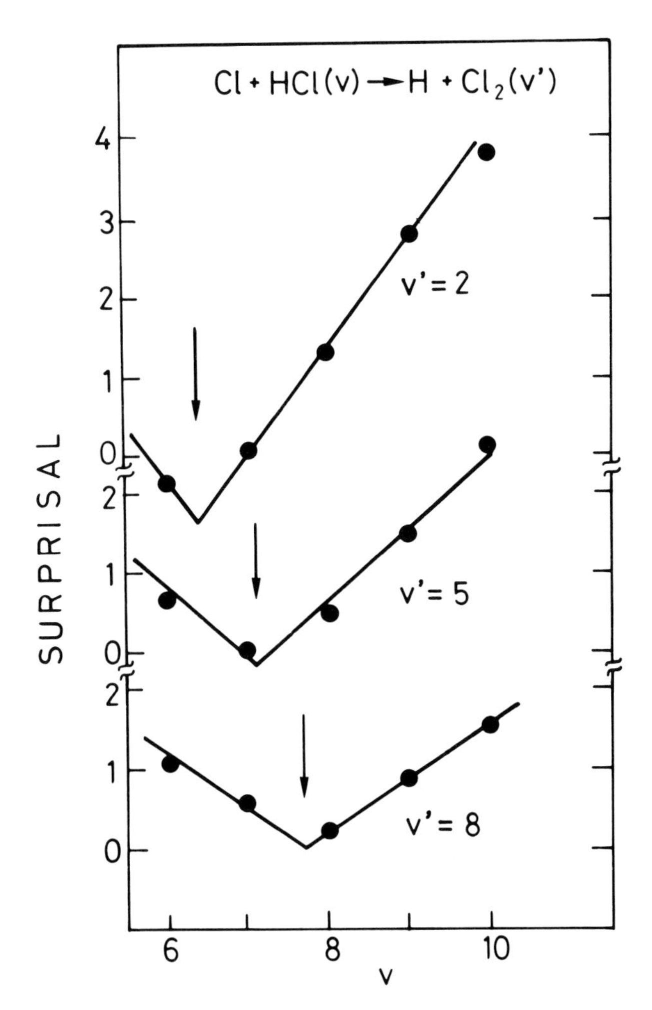

Figure 12. Surprisal analysis of vibrational energy consumption in the reaction (49). (For different final vibrational states of Cl_2. Adapted from Ref. 10). The transition from an endothermic to an exothermic behavior is quite evident.

We can summarize the results in figures 10-12 by

$$P(v \rightarrow v') = P^{O}(v \rightarrow v')\exp[-\lambda_{\mathbf{v}}|\Delta E_{V}|/kT - \lambda_{O}] \quad (50)$$

where $\Delta E_V = E_V - E_{V'}$ for a thermoneutral process (energy transfer or a symmetric atom exchange) and $\Delta E_V = E_V - E_{V'} - \Delta E_O$ for reactive collisions.

The representation (50) enables us to interpret the different selectivity and specificity of vibrational energy in exothermic and endothermic collisions which we noted already in section A.a and which are summarized in the table:

	Consumption	Disposal
Endothermic process	Exponential increase of the reaction rate constant upon vibrational excitation	P(v) declines with increasing product vibrational state
Exothermic process	Rate constant increases only very slowly with v. P(v) declines with increasing v.	Population inversion

As expected from the discussion of detailed balance, there are only two independent entries in the table (if the forward process is endothermic the reversed process is exothermic). Consider now say an exoergic process, i.e. $-\Delta E_0 > 0$. It then follows from (50) that as long as $E_{v'} \leqslant E_v - \Delta E_0$, the surprisal will increase linearly with v, as seen also in figures 7 and 8. The same argument shows that in the reversed endoergic process, the surprisal of $P(v' \rightarrow v)$ will increase linearly with v' as long as $E_{v'} \leqslant E_v - \Delta E_0$. Similarly, for an endothermic process, the surprisal of $P(v \rightarrow v')$ will decrease linearly with v', etc. The v and v' dependence of the probability itself is determined (as seen, say, in figure 7) both by the surprisal and by the prior.

A given reaction (or energy transfer) process can have an exothermic or an endothermic disposal and consumption depending on the sign of ΔE_v. This change of character is shown in figure 10 for an energy transfer collision and in figure 12 for a chemical reaction.

Even when we consider summing over all final states, the exponential gap rule continues to provide a useful guidance. Thus, for an endoergic process at low reagent excitation, some translational energy must be consumed if we are to reach even the ground internal state of the products. The reaction rate will thus increase exponentially with increasing reagent excitation (which implies diminishing the amount of energy provided by the translation). Once the reagent excitation equals the endoergicity the situation begins to change. Any further increase in reagent excitation requires energy disposal. The reaction rate will then increase by less than expected on prior grounds. Experimental and computational results are in accord with this interpretation, figure 13.

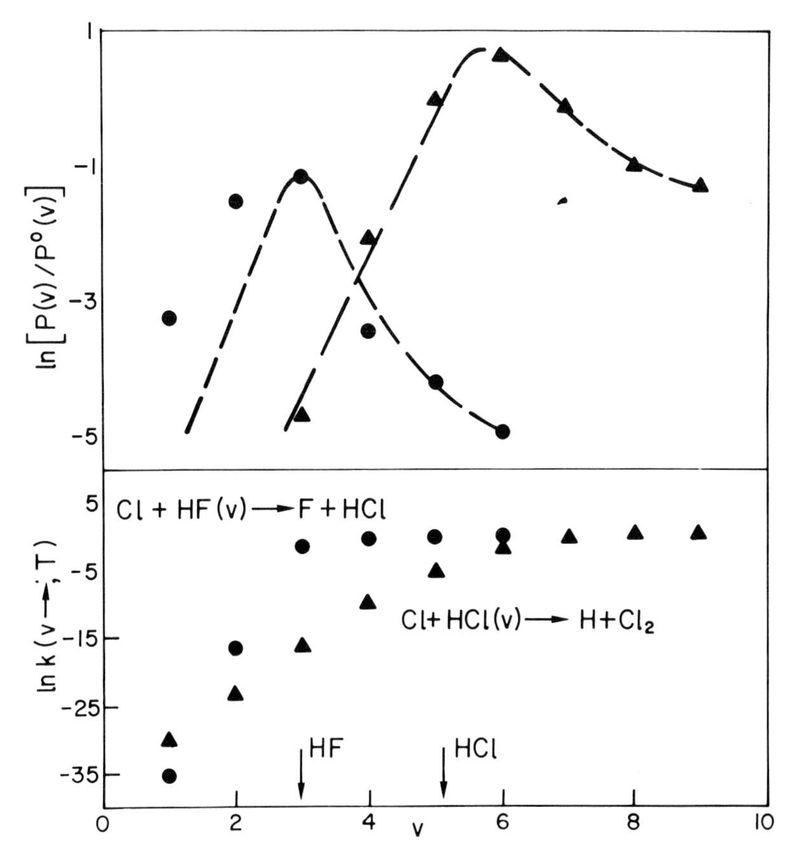

Figure 13. Surprisal analysis of the reaction rate $k(v \rightarrow ;T)$, [10]. Experimental results for Cl+HF(v) (dots) for $v \gtrsim 3$ from [18], for $v \lesssim 3$ from the measured [2] vibrational energy disposal in F+HCl and the use of detailed balance. Trajectory computations [10] for Cl+HCl(v) (triangles). The arrows indicate the threshold for exothermic behavior ($E_V + \Delta E_0 \gtrsim 0$). The broken lines in the surprisal plot (upper panel) are the results based on an exponential gap representation for $P(v \rightarrow v')$ where $P(v) = \Sigma P(v \rightarrow v')$. Note the linear behavior of the surprisal in the endothermic regime for consumption (which is equivalent to an exothermic regime for disposal)

C. CONCLUDING REMARKS

We have centered attention on the analysis of the specificity of internal energy disposal and the selectivity of internal energy consumption in elementary chemical reactions. Not considered was the important ability to predict such behavior, figures 14 and 15.

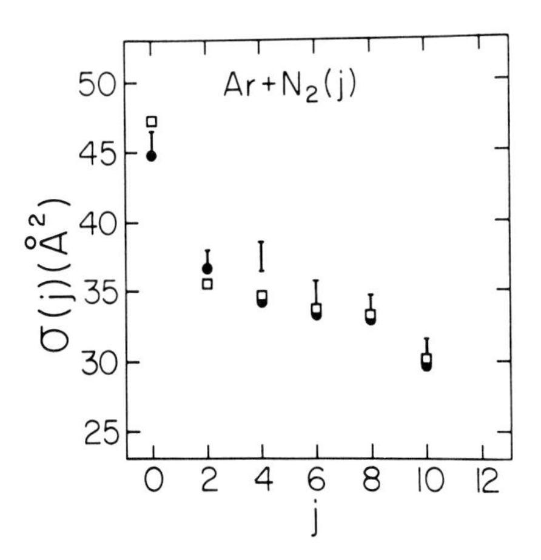

Figure 14. Surprisal synthesis (squares) cross section in $Ar+N_2(j)$ collision vs. j [20]. Shown for comparison are quantal scattering (dots, [51]) and classical trajectory (bars [54]) computations.

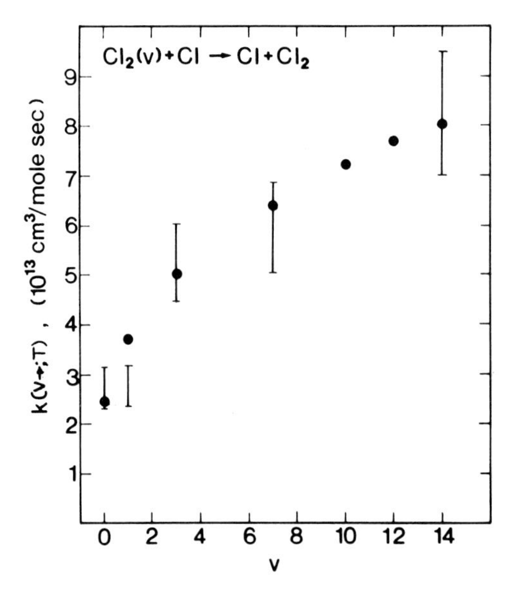

Figure 15. Surprisal synthesis (points) of the rate constant $k(v \rightarrow;T)$ in $Cl+Cl_2(v) \rightarrow Cl_2+Cl$ collisions, vs. v, [19]. Shown for comparison as classical trajectory (bars, [52]) computations.

Also not considered was the analysis of the role of translational energy, [10,21,44,45]. Here as well, surprisal synthesis can be employed to predict novel effects. A recent illustration, figure 16, is the prediction that (for some mass combinations) reagent translational energy will become more effective than internal excitation in promoting reaction at higher collision energies.

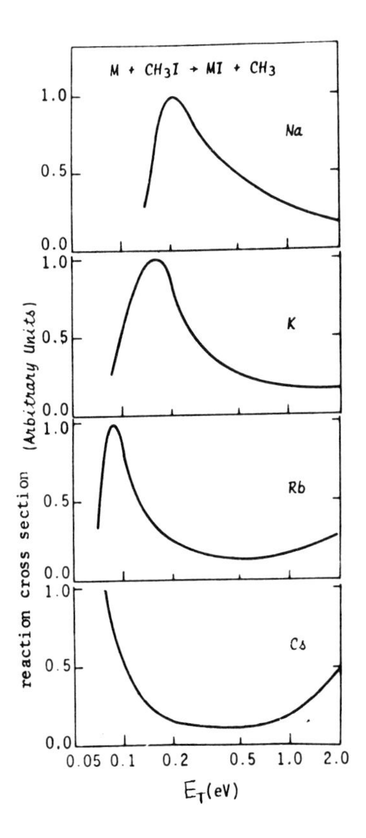

Figure 16. Surprisal synthesis [21,55] of the translational energy dependence of the reaction cross section in $M+CH_3I \rightarrow MI+CH_3$ reactions. Note the increase at higher E_T's for Rb and Cs.

Much of the current work is aimed at providing a 'mechanical' approach to the surprisal. How can one proceed directly from a knowledge of the potential (and kinetic) energy during the collision to the surprisal parameters? Many elements of this link are already available but considerable work remains to be done.

References

1. R.D. Levine and R.B. Bernstein, Molecular Reaction Dynamics (Clarendon Press, Oxford, 1974).
2. A.M. Ding, L.J. Kirsch, D.S. Perry, J.C. Polanyi and J.L. Schreiber, Far.Disc.Chem.Soc. 55, 252 (1973).
3. J.G. Pruett and R.N. Zare, J. Chem. Phys. 64, 1774 (1976).
4. R.B. Bernstein, Isr.J.Chem. 9, 615 (1971).
5. J.C. Polanyi, Appl. Optics 10, 1717 (1971).
6. C.B. Moore and P.F. Zittel, Science 182, 541 (1973).
7. R.N. Zare and P.J. Dagdigian, Science 185, 739 (1974).
8. B.F. Gordietz, A.I. Osipov, E.V. Stupechenko and L.A. Shelepin, Sov. Phys. Usp. 15, 759 (1973). V.S. Letokhov and A.A. Makarov, J. Photochem. 3, 249 (1974).
9. M.J. Berry, Ann. Rev. Phys. Chem. 26, (1975).
10. E. Pollak and R.D. Levine, Chem. Phys. Letters. (1976).
11. R.D. Levine and J. Manz, J. Chem. Phys. 63, 4280 (1975).
12. H. Kaplan, R.D. Levine and J. Manz, Chem. Phys. 12, 447 (1976).
13. A. Ben-Shaul, M.J. Berry, K.L. Kompa and R.D. Levine, Chemical Lasers (1976).
14. L.J. Kirsch and J.C. Polanyi, J. Chem. Phys. 57, 4498 (1972).
15. K.L. Kompa and J. Wanner, Chem. Phys. Letters 12, 560 (1972).
16. D.H. Maylotte, J.C. Polanyi and K.B. Woodall, J. Chem. Phys. 57, 1547 (1972).
17. K. Bergmann and C.B. Moore, J. Chem. Phys. 63, 643 (1975).
18. D.J. Douglas, J.C. Polanyi and J.J. Sloan, Chem. Phys. 13, 15 (1976).
19. I. Procaccia and R.D. Levine, J. Chem. Phys. 63, 4261 (1975).
20. I. Procaccia and R.D. Levine, J. Chem. Phys. 64, 808 (1976).
21. H. Kaplan and R.D. Levine, Chem. Phys. (1976).
22. R.D. Levine, Chem. Phys. Letters (1976).
23. R.D. Levine and R.B. Bernstein, J. Chem. Phys. 56, 2281 (1972).
24. G. Hancock, C. Morley and I.W.M. Smith, Chem. Phys. Letters 12, 193 (1973).
25. R.D. Levine and O. Kafri, Chem. Phys. Letters 27, 175 (1974).
26. R.D. Levine and O. Kafri, Chem. Phys. 8, 426 (1975).
27. W.J. Moore, Physical Chemistry (Longmans, London, 1962).
28. A. Ben-Shaul, O. Kafri and R.D. Levine, Chem. Phys. 10, 367 (1975).
29. I. Procaccia and R.D. Levine, J. Chem. Phys. (1976).
30. R.D. Levine, J. Chem. Phys. (1976).
31. J.L. Kinsey, J. Chem. Phys. 54, 1206 (1971).
32. A. Ben-Shaul, R.D. Levine and R.B. Bernstein, J. Chem. Phys. 57, 5427 (1972).
33. R.B. Bernstein and R.D. Levine, Adv. Atom. Mol. Phys. 11, 215 (1975).
34. M. Rubinson and J.I. Steinfeld, Chem. Phys. 4, 467 (1974).

35. R.C. Tolman, Statistical Mechanics (Clarendon Press, Oxford, 1938).
36. D.J. Bogan and D.W. Setser, J. Chem. Phys. 64, 586 (1976).
37. M.J. Berry, J. Chem. Phys. 59, 6229 (1973).
38. N. Jonathan, C.M. Melliar-Smith, S. Okuda, D.H. Slater and D. Timlin, Mol. Phys. 22, 561 (1971).
39. A. Ben-Shaul, Chem. Phys. 1, 244 (1973).
40. M.J. Berry, Chem. Phys. Letters, 27, 73 (1974) ibid 29, 323 (1974).
41. U. Dinur, R. Kosloff, R.D. Levine and M.J. Berry, Chem. Phys. Letters 34, 199 (1975).
42. D.M. Manos and J.M. Parson, J. Chem. Phys. 63, 3575 (1975); R.G. Shortridge and M.C. Lin, J. Chem. Phys. (1976).
43. K.J. Schmatjko and J. Wolfrum, Ber. Bunsenges. physik. Chem. 79, 696 (1975).
44. A. Kafri, E. Pollak, R. Kosloff and R.D. Levine, Chem. Phys. Letters 33, 201 (1975).
45. E. Pollak and R.D. Levine, J. Chem. Phys. (1976).
46. R.D. Levine, B.R. Johnson and R.B. Bernstein, Chem. Phys. Letters 19, 1 (1973).
47. R.E. Wyatt, Chem. Phys. Letters 34, 167 (1975).
48. R.D. Levine, R.B. Bernstein, P. Kahana, I. Procaccia and E.T. Upchurch, J. Chem. Phys. 64, 796 (1976).
49. R.B. Bernstein, Int'l J. Quant. Chem. (1976).
50. H. Kaplan and R.D. Levine, Chem. Phys. 13, 161 (1976).
51. R.T. Pack, J. Chem. Phys. 62, 3143 (1975).
52. D.L. Thompson, J. Chem. Phys. 60, 455 (1974).
53. R.L. Wilkins, J. Chem. Phys. 59, 698 (1973).
54. M.D. Pattengill, Chem. Phys. Letters 36, 25 (1975).
55. H. Kaplan and R.D. Levine, J. Chem. Phys. 63, 5064 (1975).

SYMPOSIUM III. QUANTUM ORGANIC CHEMISTRY AND BEYOND

Chairman : B. Pullman

Institut de Biologie Physico-Chimique,
Fondation Edmond de Rothschild,
Paris, France

INTRODUCTION

Bernard PULLMAN

During a long period organic chemistry was one of the preferred field of exploration for many quantum chemists. It still remains of major interest for a number of them, but altogether it seems as if the center of gravity of the quantum molecular theories became displaced towards different horizons. The displacement seems in fact to occur in two directions. On the one hand, we see a prominent development of very refined computations for relatively small and frequently inorganic molecular systems with the view of a better and better reproduction of their observable properties and thus a better understanding of the fundamental principles governing the electronic structure of molecules. On the other hand, there is a no less prominent development of frequently also no less refined computations towards the study of molecular systems which because of their dimensions or the complexity of the questions which they raise go beyond the usual treatment of organic molecules. These later studies involve in particular the penetration of quantum-mechanical concepts and methods into the realm of biochemistry, biophysics, and pharmacology. It so happens that because of the nature of the problems involved this penetration has taken up a double aspect. On the one hand it implies, of course, the exploration of the electronic and conformational properties of the fundamental molecular species involved in these branches of science and this in itself involves frequently the study of the specific and arduous problem of the relation between the structure of biological polymers and that of their constituents. On the other hand, it raises constantly, with more strenghts than in chemistry, the problem of the effect of environment on molecular structure and properties and in particular the problems of the solvent effect, cation binding, presence of counter ions and, from a broader viewpoint, that of intermole-

B. Pullman and R. Parr (eds.), The New World of Quantum Chemistry, 133–135.

cular interactions between the numerous partners which generally make up a biological unit. The extension of the quantum-mechanical studies to these large structures and these new problems represents an essential enrichment of theoretical science of the last ten to fifteen years.

Today's session will be devoted to a presentation of some of the essential aspects of this recent development.

Prof. George Hall will give us a general presentation and, I hope, an appraisal of the different computational methods used for the study of large molecular systems. This is a crucial problem in this field. The methods may be both semi-empirical and non-empirical. The first ones are less precise but also less expensive, the second more precise but also more expensive. In as much as one cannot for practical reasons use constantly for such systems the most refined procedures available it is essential that the procedure utilized be well chosen and its validity ascertained for the type of problem or structure investigated. Professor Hall who belongs to the prestigious school of English quantum-chemistry has always combined an interest for fundamental theory with an equal interest for the elucidation of properties of large molecular systems. He seems thus particularly suited for discussing with discrimination this rich but complex problem.

Our next speaker, Madame Pullman, will carry us to one of the foremost fronts of nowadays quantum-chemistry by presenting and discussing her and others efforts to include environmental effects into computations of molecular structures and properties. In fact, the elements of complexity which she has to face are particularly great because she applies this treatment to biostructures and bioproblems. Important in all quantum chemistry, the environmental effects are truly particularly significant in relation to biochemistry, biophysics and pharmacology, by the very nature of the phenomena and problems considered. If I would not be her husband, I would say that Madame Pullman is one of the leaders of this line of investigation, which by distinguishing clearly between intrinsic and induced properties, enrich our knowledge of the electronic structure and mechanisms in biological phenomena and brings us thus a step forward towards a better understanding of experimental reality. Her studies englobe solvent effects and cation or counterion binding.

Our last speaker today, Professor Christoffersen, will describe to us another new field of recent developments, namely the penetration of quantum-mechanical theories and computations into pharmacology. It is a fascinating field which to some extent is more puzzling even then quantum biology because it adds to the intrinsic difficulties of the latter the complementary effects of the interactions of biological substrates with exogenous substances. It is thus again a field in which intra- and intermolecular interactions play a predominant role. But it is also a field whose development, if successful, could be particularly beneficial to humanity. Experimental pharmacology is a very ad hoc science in which much efforts are frequently waisted because of the lack of an adequate rationalization. It is permitted to hope that the application of quantum-mechanical methods to the problem of structure-activity relationships of compounds of medicinal interest will sooner or later help to simplify and accelerate the production of new and powerful drugs. Anyway the goal to achieve is worthy of the biggest efforts.

Ladies and gentlemen, it is my pleasure to leave the podium to the speakers.

COMPUTATIONAL METHODS FOR LARGE MOLECULES

George G. Hall

University of Nottingham, Department of Mathematics,
Nottingham, England NG7 2RD

INTRODUCTION

The distinction between large molecules and small molecules is not one made by nature nor is it one that arises out of our present computational methods. We can recognize that most of the molecules of interest to biochemists and organic chemists are large and that most of the molecules of interest to most quantum chemists are small. This suggests a division in terms of the number of electrons somewhere in the region between 50 and 100 electrons. Nevertheless the distinction is not entirely a matter of size. Large molecules have properties not possessed by small molecules and are of interest in new ways. Small molecules may even become large molecules in both senses if they are surrounded by solvent molecules to form supermolecules.

Although the number of electrons is a convenient index for some of the computational problems it is not in itself the source of the most serious problems. Our colleagues in solid state physics are always treating systems with an infinite number of electrons and we can learn much from their experience. Our particular problem arises from the variety and complexity of molecules. An atom shows small differences both in bonding geometry and electronic structure when it is in different molecules and these small differences may be of the greatest interest to us.

Recently, survey articles dealing with large molecules have been published by Christoffersen (1972), Hinze (1974), and Schaefer (1976). There is a Conference report edited by Herman, McLean, and Nesbet (1973) and another by Saunders and Brown (1975). The books by Sinanoğlu and Wiberg (1970), Pople and Beveridge

B. Pullman and R. Parr (eds.), The New World of Quantum Chemistry, 137–148.

(1970), Murrell and Harget (1972), and Segal (1976) are also directly relevant. Since these references give detailed discussions of the theory, I shall use this lecture to focus attention on a number of issues which seem to me to require further thought.

DIFFERENT APPROACHES TO LARGE MOLECULES

In the contemporary literature three, or possibly four different approaches to large molecules can be distinguished. These differ not only in methodology but also in the philosophy of what they try to do. Because of this they are generally adopted exclusively by people who then find it difficult to appreciate the validity of other approaches. It is relatively easy, nowadays, to adopt a computer program that incorporates some other person's techniques but it is not so easy to adopt his assumptions and standards. We still seem to be divided between those who wish to provide more accurate and more detailed data on molecular wavefunctions and those who wish to explain why molecules behave as they do.

The first approach simply extends to large molecules the *ab initio* techniques that have been developed for small molecules. The most obvious problem is that of calculating, storing, and manipulating the large number of integrals. For a typical molecule with about 250 electrons this number reaches 10^9. To solve this problem Clementi (1972) has proposed two techniques. The first is to set a threshold, such as 10^{-7} Hartree, and omit entirely all smaller integrals. The second is to replace, in smaller integrals, the contracted set of functions representing an atomic orbital by a single Gaussian (the adjoined function) so that the evaluation is greatly speeded without essential loss of accuracy. By these techniques and by skilled computing Clementi has successfully demonstrated that *ab initio* calculations can be performed on systems with more than 200 electrons in a reasonable computing time. Table 1 gives some details.

Table 1. Calculations on Large Molecules.

Molecule	No. of Electrons	No. of Gaussians	No. of SCF Coefficients	No. of Integrals	Time*
†Carbazole-$(NO_2)_3$ fluorene	232	618	194	9×10^9	6.5
‡Ethyl-Chlorophyllide	340	277	241	0.4×10^9	3

*Computer time on IBM 360/195, in hours
†Clementi, *et al.* (1972)
‡Christoffersen, *et al.* (1975)

The second approach is the semi-empirical method in one of its modern forms. With the invention of CNDO by Pople and his colleagues (1965) this approach has acquired a new authority and a new flexibility. Its strength lies partly in its increased theoretical clarity and partly in its more reliable methods of parameterization. The resulting theory, despite its limitations, has proved its worth in many discussions of molecular geometry, orbital structure, and even spin properties. In a reparameterized form due to Del Bene and Jaffé (1968) it includes excited states. Another variant, MINDO, (Bingham, Dewar, and Lo, 1975) relaxes some of the integral approximations and makes greater use of experimental quantities in an attempt to improve the agreement with experiment. I would like to mention, in particular, PCILO (Diner, et al., 1969) partly because its formalism deliberately goes beyond MO theory and partly because it has been extensively applied to study the conformations of large molecules. Figure 1 is a phospholipid with 206 electrons and 23 torsion angles recently studied by Pullman and Saran (1975) using this technique.

The third approach is the FSGO method, due originally to Frost (1967). This is an ab initio approach and so has no parameterization problems. It becomes applicable to large systems,

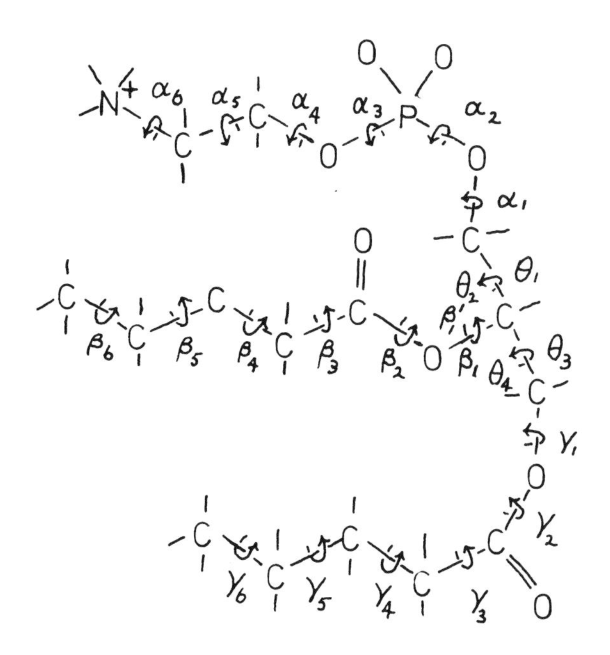

Figure 1. Phosphatidylethanolamine with 23 torsion angles.

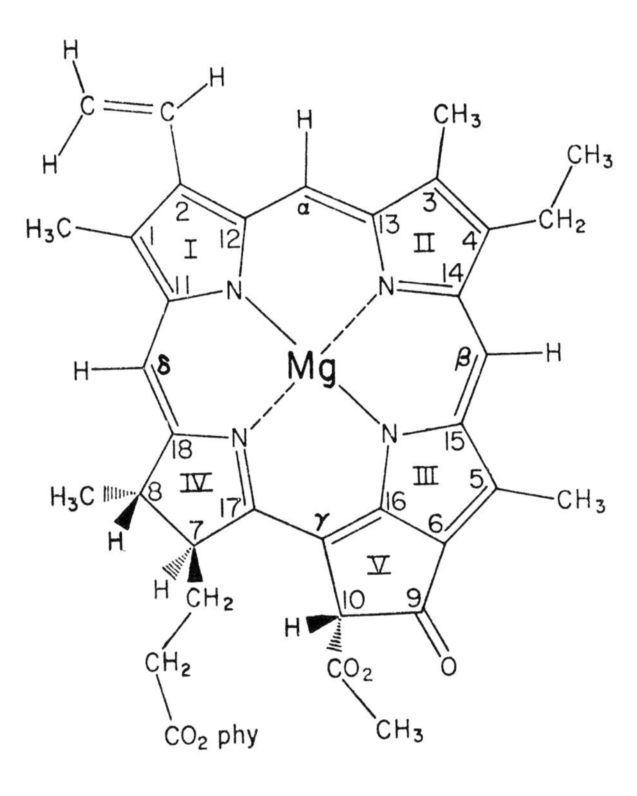

Figure 2. Ethyl-Chlorophyllide.

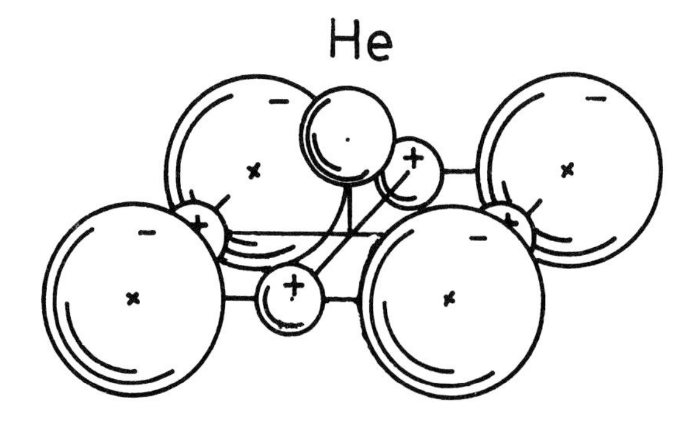

Figure 3. Conformation of He on [100] surface of LiH.

Table 2. Orbital Energies.

$$\epsilon_{FSGO} = A\,\epsilon_{DZ} + B.$$

	A	B	CORRELATION COEFF.
H_2O	1.01395	0.27853	0.99984
H_2O_2	1.00311	0.28355	0.99981
CH_3OH	0.99707	0.15583	0.99979
HF	1.01235	0.36947	0.99976
F_2	1.03123	0.29860	0.99954
CH_2CCH_2	0.99009	0.10729	0.99992
CHCH	0.98669	0.12825	0.99993

with computer demands comparable to those of a semi-empirical calculation, because the integral formulae remain simple and the number of integrals is kept small. The disadvantage is that, since no attempt is made to reproduce the correct behaviour of the wavefunction close to the nucleus, the total energies are poor. Much of this can be corrected by an adjustment after the calculation. Table 1 illustrates a recent calculation on ethyl chlorophyllide (Figure 2) by Christoffersen and his colleagues (Spangler, et al., 1975). Because this method is based on localized orbitals instead of MO it extends immediately to solids. Linnett and his students (Sutters, et al., 1974), have used it to study surface properties of LiH crystals. Figure 3 shows the conformation which they calculated to be the most stable for an He atom adsorbed on the [100] face of such a crystal. I find this approach attractive but tantalizing. Table 2, for example, shows some typical linear correlations (Brailsford, et al., 1975) with more accurate calculations. These correlations are often better than can at present be explained. Table 3 shows a recent calculation of static polar-

Table 3. Average Polarizabilities.

	Calc $(1/2 \sum_i 1/\alpha_i^2)$	Expt.
CH_4	24.4	25.6
C_2H_4	36.9	42.2
C_3H_6	57.5	56.4
H_2O	9.0	14.4
NH_3	18.2	22.2

izabilities (Amos and Yoffe, 1975a) which are also remarkably good and similar calculations of intermolecular forces are equally encouraging.

Finally, I must mention approaches in which the Hamiltonian itself is modified. The X_α method is the best known of these and I cannot do better than to refer you to the lecture on this topic by Slater in the proceedings of the Menton Congress (1973). There is also the pseudo-potential method which will be increasingly important as the heavier atoms become more studied. I am saying little about these methods solely because there are, as yet, few applications of them to large molecules.

THE NEED FOR BALANCE

The first general comment that I would like to make on these alternative computational methods concerns the need for a balanced treatment from one molecule to another and from one atom to another (Mulliken, 1962). There are some large molecules so important that an accurate calculation on each one is justified. More frequently the quantities of interest are the small differences between related molecules and to calculate these differences a balance of accuracy is more important than absolute accuracy. Table 4, whose first few columns are taken from a recent Pople paper (1975), illustrates this. He has proposed a series of tests for calculated molecular energies and shows that his balanced Gaussian bases satisfy these tests better than the MINDO/3 treatment. The double-zeta calculations of Snyder and Basch (1972) show reasonable balance and emerge well from the tests, but if we select the best available Hartree-Fock energies (Richards, et al., 1970) the results are very poor. It is much easier to calculate a good energy for a small molecule than for a large one but the energy differences then become quite poor.

Table 4. Energy Comparisons for Hydrocarbons.

Reaction	Energy Change (kcal/mole)				
	MINDO/3	6-31G/4-31G	DZ	"Best" HF	OBS (298K)
$CH_3CH_3+H_2 \rightarrow 2CH_4$	+7.1	-22.0	-25.0	-23.7	-15.6
$H_2C{=}CH_2+2CH_4 \rightarrow 2CH_3CH_3$	-46.2	-22.3	-16.4	-32.9	-17.0
$HC{\equiv}CH+4CH_4 \rightarrow 3CH_3CH_3$	-92.0	-55.4	-45.9	-31.4	-43.3
$CH_3CH_2CH_3+CH_6 \rightarrow 2CH_3CH_3$	-6.8	+1.1	---	-129.2	+2.3
$H_2C{=}C{=}CH_2+CH_4 \rightarrow 2H_2C{=}CH_2$	+2.7	-4.4	-5.1	-321.7	-3.2
$C_6H_6+6CH_4 \rightarrow 3CH_3CH_3+3H_2C{=}CH_2$	+7.2	+63.6	---	-40.9	+64.2

Table 5. Point Charge Models.

Distance(B)	Potential	Hall	Shipman	Amos and Yoffe
1.2	0.4901	0.482	0.48	0.47
3.6	0.0491	0.0490	0.0488	0.047
6.0	0.0124	0.0124	0.0123	0.0120

Hall $\rho^{\bullet}_{H} = 2 \sum_{st} S^{-1}_{st} S_{st} \; \delta(\underline{r} - \underline{R}_{st})$

Shipman $\rho^{\bullet}_{S} = 2 \sum_{st} S^{-1}_{st} S_{st} \frac{a_s}{a_s + a_t} \; \delta(\underline{r} - \underline{R}_s)$

Amos and Yoffe $\rho^{\bullet}_{AY} = 2 \sum_{s} \delta(\underline{r} - \underline{R}_s)$

THE NEED FOR ELECTROSTATICS

Electrostatics is the key both to successful computational methods for molecules and to objective interpretations of the results. This follows naturally from the Schrödinger equation in which the potential is wholly electrostatic. From this it can be deduced that, in a composite system, at least the charges and dipoles in each part of the system must be correctly accounted for to prevent the errors in the potentials from increasing with the size of the system. Any computational method involving the neglect of integrals has to take this requirement seriously. Wilhite and Euwema (1974) have given approximations which achieve this. We have been investigating an extreme form of this in which integrals are reduced to potentials between point charges. Table 5 shows some results for one-electron potentials due to Hall (1973), Shipman (1975), and Amos and Yoffe (1975b). Two-electron integrals behave in the same way. In particular, point charge models appear to give rapid and reliable estimates of the electrostatic forces between molecules (Christoffersen, Hall, McCreery, 1976).

Experience with the use of electrostatics in solids teaches one further lesson. To obtain correct bulk effects, dielectric polarizations must be included. Such effects are normally excluded in MO theory though they may be automatically included when parameters are obtained from experiment. The PCILO theory has a perturbation formalism which already includes these intramolecular dispersion effects explicitly.

Electrostatics is also important in the interpretation of the results of computation. The potential field around a molecule is objectively defined and is of considerable significance for a number of chemical purposes. Scrocco and Tomasi (1973) in a recent review have argued strongly for its greater use. I am sure that electrostatic forces play an even more important role in biology than in chemistry because they are the only forces which can be dramatically changed from attraction to repulsion, *e.g.*, by transfer of a charge. They may even be changed by rotation of one molecule through an angle. In a biological process it often happens that molecules are attracted to one another in one phase of the process and repelled in a later phase.

THE NEED FOR NUMERICAL ANALYSIS

In any large computation it is of vital importance that the design of the program and the choice of algorithm should take account of the realities of the numerical round-off and truncation errors. The literature contains many examples of naive optimism on this point for which there is no justification. I would like to mention just two simple examples of these numerical problems. In a recent paper (Ford and Hall, 1974) some more complicated examples have been discussed.

The first example concerns the cancellations between energy contributions. Everyone knows that some contributions to the total energy cancel but this example shows that, for a typical large molecule, the intermediate contributions may be ten times the total energy. If the contributions are summed carelessly one significant place can easily be lost from the result. The numbers in Table 6 illustrate the order of magnitude of the various contributions and the extent of the subtractions.

Table 6. Subtractions.

Energy Contributions		Hartrees
Electron-nuclear attractions		-16551.8325
Electron-electron repulsions		6782.1430
Nuclear-nuclear repulsions		6095.9439
Kinetic Energy		1782.0532
	TOTAL	-1891.6924

My second example concerns the variation principle. It is not sufficiently appreciated that once numbers are rounded off or neglected the variation principle is lost and the computed energy can fall below its true value. The results are most dramatic when

Table 7. H Atom Calculation.

TRIAL FUNCTION	$e^{-2r} + k\ re^{-1.5r}$		
ENERGY	-0.494888	RESCALED ENERGY	-0.496873
ARITHMETIC RESTRICTED TO 5 DECIMALS			
ENERGY	-0.49664	RESCALED ENERGY	-0.56126
ERROR RATIO	350	ERROR RATIO	12,800

the calculation involves some optimization since this tends to home in on the numerical error and uses any numerical instability to lower the energy. Table 7 shows a simple calculation on the hydrogen atom in which the arithmetic was rounded off to five decimals. The magnification of the error is here large but not untypical of the situation in many calculations we have performed where the energy diverged to $-\infty$.

THE NEED FOR FORMAL THEORY

No theory can be considered satisfactory until it can be defined unambiguously and its procedures justified rigorously. For semi-empirical theories this creates problems since the basis functions used are not defined in a way that justifies their assumed properties.

The justification for the transfer of an orbital from one molecule to another depends on its localization around one nucleus so that it remains characteristic of that atom in its bound state. This degree of localization cannot be achieved by a transformation of the occupied MO alone. I would like to reiterate a suggestion, which I made many years ago (1952), for one solution to this problem using the standard excited state. In this state the valence electrons are put into singly-occupied orbitals so that atomic localization can be achieved for the α spin orbitals. These functions are then used to set up the Fock matrix for the ground state. This SES also has a uniform distribution of charge which is desirable in separating-off the short range bonding from the two-electron electrostatic effects. One conclusion from this model is that atomic dipolar potentials are required to describe atoms having lone pairs (Gregson and Hall, 1969) and will contribute to the diagonal elements f_{rr} of the Fock matrix in a semi-empirical

Table 8. Standard Excited State.

		inner shells	lone pairs	valence electrons
Ground State	α	. . .	. . .	. . .
	β	. . .	. . .	. . .
S. E. S.	α	. . .	. . .	. . . . . .
	β	. . .	. . .	

Open shell calculations on SES defines enlarged basis set which can be localized to atoms. Atoms are in valence states with directed lone pairs.

$$f_{rs}\ (\text{gd. state}) = f_{rs}(\text{SES}) + 2\text{-electron electrostatic terms}$$

theory.

In some recent work (1976) I have developed a formalism in which the wavefunction is factorized into a correlation part and a determinant of orbitals. This is specially significant in a semi-empirical context since it may provide a much better form of theory. When simplified by invoking zero-differential overlap the additional terms in the Hamiltonian have the form:

$$\sum_{i<j<k} F_{ijk}$$

where F_{123} is a function of the coordinates of three electrons. This leads to an extra term in the Fock matrix. In the ZDO approximation this extra term has the form:

$$f_{uv} = 6 \sum_{wy} (2\, P^{ww}P^{yy} - P^{wy}P^{yw}) < uwy\ F_{123}\ uwy > \delta_{uv}$$

$$+ 6 \sum_{y} (P^{vy}P^{yu} - 2\, P^{yy}P^{vu}) < uvy\ F_{123}\ uvy >$$

CONCLUSION

The calculations already published prove that significant progress has been made in computing properties of large molecules so that there are genuine grounds for optimism. By using balanced basis sets in an MO calculation and comparing only closed shell molecules much better results have been obtained than was previously thought possible. Problems of integral estimation and numerical control remain.

The distinction between *ab initio* and semi-empirical approaches is now less clear than before with each adopting features of the other. For very large systems, some very simplified treatment will be needed and it is more likely to be of semi-empirical type since this obviates the most time-consuming steps in the computation.

The major problem before us remains that of generating really useful chemical concepts. The more accurate our computational methods become the less intelligible are the results. The task of the theoretician is to understand nature and not just to generate more unorganized facts. The concepts of molecular orbitals, density matrices, etc., have contributed much to the description and understanding of matter at the chemical level. One of our major future tasks is to develop concepts for describing matter at the biological level.

REFERENCES

Amos, A. T. and Yoffe, J. A.: 1975a, Chem. Phys. Letters 31, 57.

Amos, A. T. and Yoffe, J. A.: 1975b, Theor. Chim. Acta 40, 221.

Bingham, R. C., Dewar, M. J. S., and Lo, D. H.: 1975, J. Amer. Chem. Soc. 97, 1285.

Brailsford, D. F., Hall, G. G., Hemming, N., and Martin, D.: 1975, Chem. Phys. Letters 35, 437.

Christoffersen, R. E.: 1972, Adv. Quantum Chem. 6, 333.

Christoffersen, R. E., Hall. G. G. and McCreery, J. H.: 1976 in course of publication.

Clementi, E.: 1972, Proc. Nat. Acad. Sci. (US) 69, 2942.

Del Bene, J. and Jaffé, H. H.: 1968, J. Chem. Phys. 48, 1807, 4050.

Diner, S., Malrieu, J. P., and Claverie, P.: 1969, Theor. Chim. Acta 13, 1, 18.

Ford, B. and Hall, G. G.: 1974, Computer Phys. Comm. 8, 337.

Frost, A. A.: 1967, J. Chem. Phys. 47, 3707, 3714.

Gregson, K. and Hall, G. G.: 1969, Mol. Phys. 17, 49.

Hall, G. G.: 1952, Proc. Roy. Soc. A213, 102, 113.

Hall, G. G.: 1973, Chem. Phys. Letters 6, 501.

Hall, G. G.: 1976 in Quantum Science, Methods and Structure. A Festschrift in Honour of P. O. Löwdin (editor: J. Linderberg) Plenum.

Herman, F., McLean, A. D., and Nesbet, R. K.: 1973, Computational Methods for Large Molecules and Localized States in Solids Plenum.

Hinze, J.: 1974, Adv. in Chem. Phys. 26, 213.

Mulliken, R. S.: 1962, J. Chem. Phys. 36, 3428.

Murrell, J. N. and Harget, A. J.: 1972, Semi-empirical SCF MO Theory of Molecules Wiley.

Pople, J. A.: 1975, J. Am. Chem. Soc. 97, 5306.

Pople, J. A. and Beveridge, D. L.: 1970, Approximate Molecular Orbital Theory McGraw-Hill.

Pople, J. A., Santry, D. P., and Segal, G. A.: 1965, J. Chem. Phys. 43, S129.

Pullman, B. and Saran, A.: 1975, Int. J. Quantum Chem. QBS2, 71.

Richards, W. G., Walker, T. E. H., and Hinkley, R. K.: 1970, Bibliography of Ab Initio Molecular Wavefunctions Oxford.

Saunders, V. R. and Brown, J.: 1975, Quantum Chemistry: The State of the Art Science Research Council.

Schaefer, H. F.: 1976, Ann. Rev. of Phys. Chem. 27.

Scrocco, E. and Tomasi, J.: 1973, Topics in Current Chem. 42, 95.

Segal, G. A.: 1976, Approximate Methods for Molecular Structure Calculations Plenum.

Shipman, L. L.: 1975, Chem. Phys. Letters 31, 361.

Sinanoğlu, O. and Wiberg, K. B.: 1970, Sigma Molecular Orbital Theory Yale.

Slater, J. C.: 1973, in The World of Quantum Chemistry (editors: R. Daudel and B. Pullman) Reidel.

Snyder, L. C. and Basch, H.: 1972, Molecular Wave Functions and Properties Wiley.

Spangler, D., McKinney, R., Christoffersen, R. E., Maggiora, G. M., and Shipman, L. L.: 1975, Chem. Phys. Letters 36, 427.

Sutters, R. A., Linnett, J. W., and Erickson, W. D.: 1972, Surface and Defect Properties of Solids 3, 132.

Wilhite, D. L. and Euwema, R. N.: 1974, J. Chem. Phys. 61, 375.

THE SOLVENT EFFECT : RECENT DEVELOPMENTS

Alberte Pullman

Institut de Biologie Physico-Chimique
Laboratoire de Biochimie Théorique associé au C.N.R.S.
13, rue P. et M. Curie, 75005 Paris.

I. INTRODUCTION

The study of the solvent effect by actual quantum-mechanical computations was practically impossible for many years : neither the methods of quantum chemistry, nor the computational possibilities were apt to deal with the problems involved, if only because of their dimensions. The situation has entirely changed very recently due to the considerable developments of the methodology stimulated by the new computational facilities : it is now possible to treat non-empirically, with a reasonable accuracy, relatively large complex systems. The explicit incorporation of the solvent into quantum chemical computations was a logical development in this evolution and major steps in this direction have indeed been taken in the last few years. From the outset, these advances have followed two different strategies resulting in two different methodologies : one of them proceeds in the "traditional" way of dealing with the solvation problem, trying to account for the bulk effect of the medium surrounding a solute molecule by the use of a "continuum" model essentially based on the ideas of Born, Kirkwood, and Onsager, inserted into a quantum mechanical framework.

Early attempts to insert the potential of a dielectric medium into a quantum mechanical hamiltonian for the computation of solvation effects may be found in Brown and Coulson (1958), Coulson (1960), Jortner and Coulson (1961), Jortner (1962). After a reversion to the use of classical potentials to compute the various solvation terms to be added to the SCF energy of an isolated solute (e.g. Beveridge and Radna, 1974; Hopfinger, 1973) a return to the insertion of the continuum effect into proper SCF

B. Pullman and R. Parr (eds.), The New World of Quantum Chemistry, 149–187.

computations at the polyelectron level is observed. The essential lines followed in these most recent efforts are summarized in section II.

The second strategy recently developed for studying the solvent effect is based on a "discrete" treatment in which one tries first to establish, at the microscopic level, the individual effect of the solvent molecules upon the solute studied, through the utilization of the "supermolecule" model which treats as a single polyelectron system the solute and its neighbor solvent molecule(s), proceeding then by successive stepwise addition of individual solvent molecules up to the complete constitution of the solvation layer(s) and ultimately to the computation of the properties of the solute in the presence of this surrounding, bound solvent, still in a supersystem treatment.

In our laboratory we have essentially developed and used the supermolecule approach in its various steps and I shall illustrate it in some details in section III.

II. THE CONTINUUM MODEL

Common to all approaches using the continuum model is the classical assumption that the solute molecule lies inside a cavity embedded in the solvent medium. The medium is assumed to be a continuous polarizable dielectric which is polarized by the solute lying in the cavity and in turn reacts on it by a "reaction field" (Onsager, 1936). Assuming the solute to be represented by a set of fixed point charges in the cavity, and the cavity to be a sphere of radius a, Kirkwood (1934, 1938) has obtained the explicit expression of the reaction potential $V(r_i)$ at any point inside the sphere, by solving the classical equations of electrostatics inside and outside the sphere, satisfying the appropriate boundary conditions. Thus :

$$[1] \qquad V(r_i) = \sum_j \left(\frac{e_j}{a}\right) \sum_{l=o}^{\infty} \frac{(l+1)(1-\varepsilon)}{\varepsilon(l+1)+l} \left(\frac{\vec{r}_i \vec{r}_j}{a^2}\right)^l P_l(\cos\Theta_{ji})$$

where $\vec{r}_i$ defines the position of the ith charge and Θ_{ji} the angle between $\vec{r}_j$ and $\vec{r}_i$.

The energy of interaction, that is the electrostatic part of the solvation energy, between the solute and the bulk dielectric can then be expressed as

$$[2] \qquad U = \frac{1}{2} \sum_i e_i \, V(r_i)$$

The insertion of this classical model into quantum-mechanical calculation has been attempted by Hylton et al. (1974) in the following way : they use, instead of the classical point charge distribution e_i, the continuous quantum mechanical charge distribution $\rho(r)$ of the solute molecule and thereby convert the classical expression of U into a quantum mechanical operator :

$$U = \frac{1}{2} \int \rho(r)\ V(r)\ dr \qquad [3]$$

which may be added to the usual free molecule hamiltonian to yield the hamiltonian of the solute molecule in the presence of the bulk solvent :

$$H = H^{o} + U \qquad [4]$$

U may be written as a sum of multipole terms :

$$U = \sum_{l} U_{l} \qquad [5]$$

Thus

$$U_{o} = -\frac{1}{2}\,\frac{Q^{2}}{a}\left(1 - \frac{1}{\varepsilon}\right) \qquad [6]$$

is the Born-charging term for a charged molecule of global charge Q, and similarly :

$$U_{1} = -\frac{1}{2}\left[\frac{2(\varepsilon-1)}{2\varepsilon+1}\,\frac{\mu^{2}}{a^{3}}\right] \qquad [7]$$

(where μ is a dipole operator), is the analog of Onsager's dipolar energy.

The utilization of a density function for the solute instead of the point charges of the classical model requires the use of a constraint to force the electrons to remain inside the cavity without introducing discontinuities in the wave functions. For this purpose, a "penalty" function C is inserted in the hamiltonian (4) :

$$C = \sum_{i} \left(\frac{r_i}{a}\right)^{n} \qquad [8]$$

the value of n being choosen so as to insure that the corresponding spurious added energy term remains small relative to the solvation energy.

The eigenvalues and eigenfunctions of the total operator

$$H = H_o + U + C \quad [9]$$

are computed in a self-consistent manner. If the molecular orbitals are expressed as a linear combination of floating spherical gaussians (FSGO's,Frost (1967)) the expression of all necessary integrals may be obtained in closed form (Hylton McCreery et al., 1976).

Aside from the choice of n in the penalty function, the scheme involves the choice of the number of terms, l_{max}, in the expansion of the potential and of two arbitrary constants, ε and, foremost, the cavity radius a. The number N of FSGO's may also in principle be varied. A study of this particular choice together with that of the effect of varying l_{max} and n has been done for the solvation of the helium atom in CCl_4, with fixed values of ε and a. (Hylton et al., 1974). Further computations on the solvation of the series methane to n-butane in n-hexane have adopted $l_{max} = 3$ and n > 10. In the same study (Hylton Mc Creery, et al., 1976) it was shown that there is unfortunately a strong dependence of the calculated solvent effect on the choice of a : for ethane in hexane the solvation energy varies from -49.24 to -6.88 kcal/mole when a goes from $d + r_H$ to $d + 2.5\ r_H$ (d being the distance from the center of mass of the solute to the farthest hydrogen and r_H the Van der Waals radius of H).

The uniform utilization of a spherical cavity has not been discussed in this context but is another obvious inconvenience of the model which is not easy to remedy to in practical computations.

In all computations with this model to date the bulk dielectric constant of the solvent has been adopted, a feature which has obvious shortcomings in that it does not take into account the strong inhomogeneities which may occur, particularly in the inner solvation shell due to the direct solute-solvent interactions at the microscopic level (Pullman, A., 1973). Two possibilities have been considered to alleviate this difficulty. The most straightforward is to include the first solvation shell(s) inside the cavity together with the solute molecule. A model computation used a monohydrated formamide system in this fashion (Burch et al., 1976) but it is clear that the significance of the results will be assessed only when at least all the molecules of the first solvation layer are included. A complete cluster of 4 water molecules around a solvated electron has been embedded in a recent continuum SCF computation by Newton (1975) (vide infra).

Another possibility of introducing the inhomogeneity of the solvent has been recently proposed by Beveridge and Schnuelle (1975), consisting of the introduction of successive concentric shells of dielectric continua, each with a different ad hoc value of the dielectric constant. The appropriate Kirkwood potential for a discrete charge distribution has been derived in the case of two such concentric shells by the above authors. No explicit SCF computation is available as yet with this scheme.

Apart from Hylton et al.'s transcription of the continuum model, a similarly-based formulation of an UHF scheme appropriate for the treatment of the solvated electron has been developed by Newton (1973-1975) and used to discuss the trapping of electrons in NH_3 and in H_2O. The potential used is the expression of the reversible work to charge up the dielectric, and a cluster of at least the first solvation layer is inserted inside the cavity.

Expressions for the appropriate integrals as well as a thorough discussion of the approach is given in the 1975 paper.

Another recent attempt at a transcription of the classical continuum model into a quantum mechanical framework due to Tapia and Goscinski (1975) is limited to the reaction field term inserted into the SCF equations (SCRF theory). Using both the quantum-mechanical average, with respect to the solvent wave function, of the classical expression of the dipole-dipole solute-solvent interaction, and its statistical average over the solution configurations, they express the effective field acting on the solute molecule in a cavity as

$$\vec{G} = \bar{\bar{g}} . \vec{M} \qquad [10]$$

here $\vec{M}$ is the dipole moment in the presence of the solvent and $\bar{\bar{g}}$ is a tensor depending both on the static dielectric constant and on the solute's geometrical parameters (in practice, $\bar{\bar{g}}$ is reduced to the scalar $\frac{2}{a^3}\frac{\varepsilon-1}{2\varepsilon+1}$).

The SCRF equations including the reaction field must be solved iteratively : the SCF equations are first solved for the isolated solute; the results are used to build the effective operator matrix for $g \neq o$ and an iterative procedure is carried out till consistency. The procedure has been implemented in a CNDO framework only, in sample computations on solvated CO, H_2CO and $(CH_3)_2CO$ for different values of g. Typical results are an appreciable variation of the dipole moments under the influence of the solvent (up to 40%), and an enhancement of the effective polarizabilities. The solvation energy is appreciably sensitive to the use of the modified dipole in (10) instead of the initial one, and to the deformation effects implied in the use of an iterated wave function.

Another recent utilization of the SCRF CNDO scheme by Tapia *et al.* (1975) supplemented by a second order perturbation computation of the dispersion energy was made on a water dimer : the proton potential curve, binding energy, dipole moment and polarizability were computed for different values of g.

Altogether these different attempts at treating the solvent effect at the macroscopic level do not seem to have been applied in a sufficiently large and systematic manner to the study of actual chemical problems to enable a clear-cut appreciation of their validity. Until now more efforts have been spent by the authors on elaborating the different schemes than on their practical utilization. In view of the nature of the hypotheses implied in all these approaches and of the technical approximations used in the model studies it is necessary to wait for a deeper investigation before being able to estimate their real potentialities.

III. THE SUPERMOLECULE APPROACH TO SOLVATION.

A. *The strategy*

It has become customary in the recent past to call supermolecule or supersystem, an entity made of two (or more) molecules in interaction, when they are treated as a single system by the methods of quantum chemistry, computing the global wave function of the whole set of electrons of the entire system in the field of all the nuclei fixed in the appropriate mutual relative arrangement. The supermolecule approach to the problem of solvation consists in fact of three steps which are carried out in succession :

1°) In the first step one spans in every possible way the hypersurface of *interaction of the substrate with one single solvent molecule*, approaching it from various directions, turning it around the solute, and turning it also about local axis, every point on the surface corresponding to one supermolecule computation. It is clear that the better the method utilized, the more reliable the results are : this requires at least the use of the self-consistent field molecular orbital method in an *ab initio* fashion, in order to avoid the parameter problem of the semi-empirical schemes and, perhaps still more, their known and unknown artefacts.

Such a search yields a number of informations which constitute the basic characteristics of the solute-solvent interaction at the microscopic level : the *minima* on the energy hypersurface indicate the most stable positions for one single solvent molecule, the value of the binding energy, the distance of equilibrium and relative orientations of the solute and solvent, but moreover, when the spanning is done in sufficient details, one obtains the *lability characteristics* of the interaction which tell how much it is possible to depart from the most stable positions without too much altering the binding. This knowledge is fundamental in order to understand the fluidity of the solvation phenomenon. From this point of view, it is convenient to distinguish the *in situ*, the *ad situm* and the *extra situm* labilities. The first one tells us how much the solvent molecule, once fixed at a given binding site, may rotate about its local axis. The *ad situm* lability describes the possibility of moving the solvent molecule in the neighbourhood of the most stable position and away from it. Finally the *extra situm* lability provides informations on the energy characteristics of the possible movement of the solvent when it happens to be in contact with the solute but in a region far from a stable site. Examples may be seen on figs. 1 and 2 which are taken from the study on monohydration of formamide (Alagona *et al.*, 1973) : the five in-plane minima on the energy hypersurface are represented by the water molecules drawn in thick lines in fig. 1. Examples of *in situ*, *ad situm* and *extra situm* lability are indicated by A, B and C respectively. Other examples of *in situ* lability are shown on fig. 2 for the site indicated in the figure. Two kinds of curves occur according to the rotation one considers : whereas the out-of-plane rotation of the H-bonded proton, as well as its in-plane rotations are rather unfavorable, the rotation around the H-bond axis is relatively easy up to a relatively wide angle. This U-shape-type curve is quite characteristic of the *in situ* lability for rotation about the hydrogen bond axis for water acting as a proton donor, at least in the absence of very bulky groups hindering the rotation (Pullman, 1975). Worth noting is the fact that the *in situ* lability may be obtained to a very good approximation by computing only the electrostatic component of the SCF binding energy (Dreyfus and Pullman, 1970; Bonaccorsi *et al.*, 1971; Alagona *et al.*, 1972, 1973).

Apart from the above-mentioned characteristics of the microscopic interactions, which are concerned with the energies and the positions, the first part of the supermolecule exploration yields a large amount of information through the usual sub-products of quantum mechanical molecular computations, namely the Mulliken populations or better the electron density distribution, the dipole moments, the molecular potential, the orbital energies, etc.. These quantities properly compared to the corresponding ones for the isolated solute and solvent molecules can bring a precious contribution to the understanding of the *manifestations* of solvation which one obtains by various physico-chemical techniques (i.r., Raman, u.v., N.M.R., ESCA, UPS, etc.).

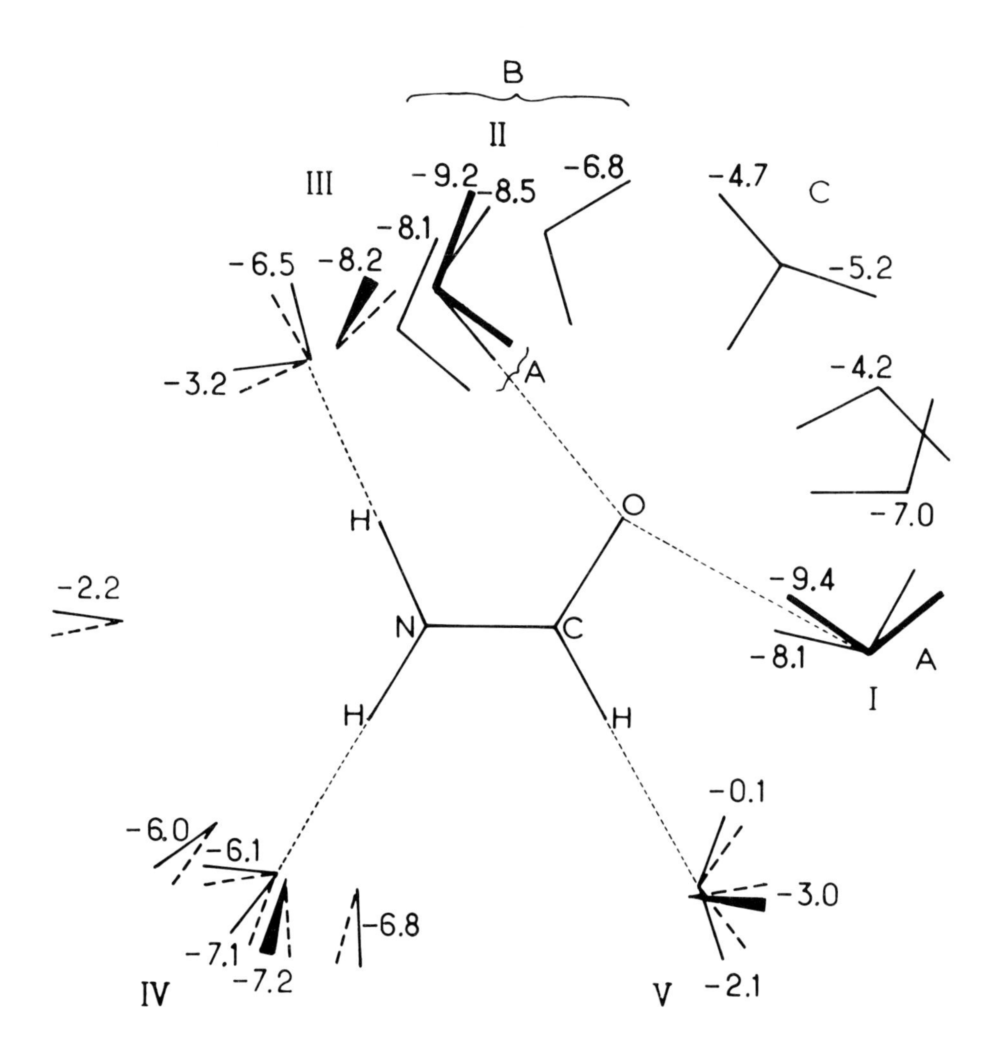

Fig. 1.- In-plane hydration of formamide (ΔE : kcal/mole)

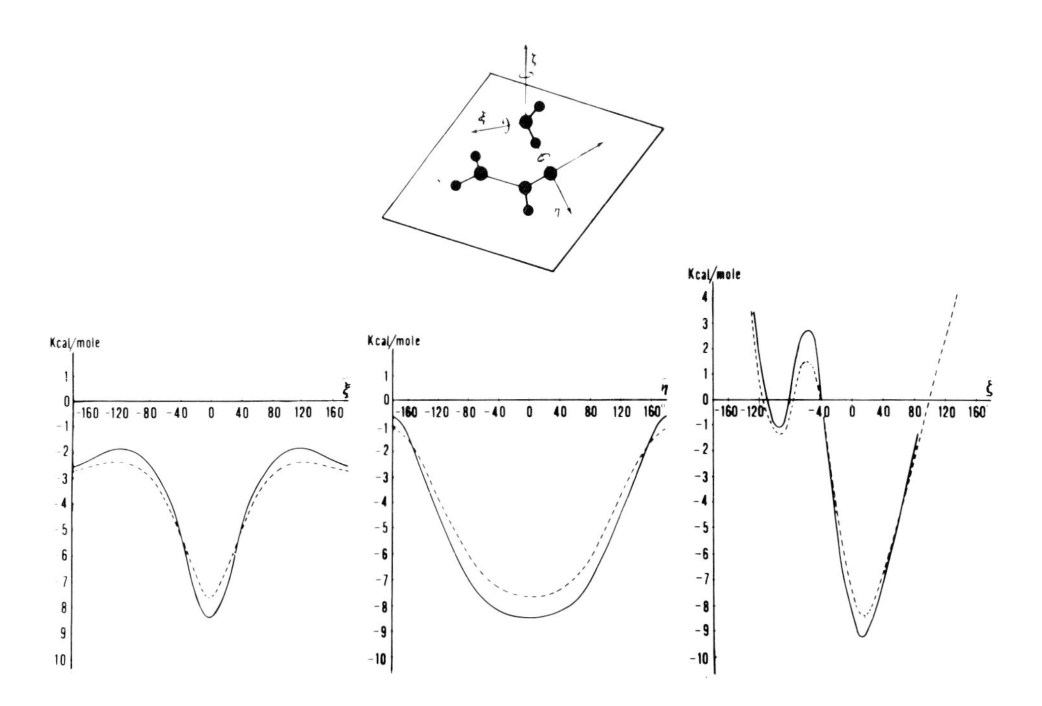

Fig. 2.- *In situ* lability of water in hydration site II of formamide.

2°) However important and accurate are the informations obtained in the above-described study, they are not sufficient to have a proper image of the solvation, and a second step is required : it consists of introducing within the same method, more solvent molecules so as to constitute at least the first solvation shell around the solute. Using as a guide the information gained in the first part of the study, the solvent molecules are added in a stepwise fashion, reoptimizing the positions at every step. (This is not an additive procedure, but the results yield a good deal of information on the amount of additivity present in every case). The kind of stepwise addition of solvent molecules may further be used for building a second or a third layer, whenever one suspects that the structuration (or "polarization" in the wide sense) of the solvent by the substrate may go beyond the first shell.

3°) It is only where a fair image of the microscopic interactions around the solute has been obtained that the last step of the supermolecule model may be carried out, namely the computation of the properties of the solute in the presence of the solvent : for this purpose, the molecules of solvent are set about the solute in the most important positions determined previously, and a supersystem computation is performed of the properties of the ensemble.

The three steps of this kind of study have been carried out in our laboratory on a large number of molecules in a systematic fashion. Examples are given below. Other groups have carried out partial studies using one step or the other. Numerous computations belong to step 1 at least in a partial way (Kollman et al., Morokuma et al., Del Bene et al.). Most of these studies have been limited to the search of the minima on the energy hypersurface and some of the lability characteristics. Step 2 has rarely been carried out to completion by other groups. Notable exceptions are the recent study of the first hydration shell of formamide (Hinton, 1976) and the early work on OH^- and H_3O hydration by Newton and Ehrenson (1971) to which should be added the computations on the hydration of water itself (for a review, see Schuster, 1974). Finally, step 3 has been accomplished directly by Cremaschi et al. (1972, 1973, 1975a,b) mostly in the CNDO approximation, using arbitrarily choosen positions for setting the water molecules around the solutes, for computing the effect of solvation on reaction activation energies, rotation barriers and ultra-violet transitions. Iwata and Morokuma (1973, 1974) have used the three steps in a study of the n π^* and π π^* transitions of the carbonyl group. (See also Del Bene et al., 1975).

Before examining some specific cases, two general remarks must be made concerning the supermolecule approach :

i) Since its emphasis is primarily on the microscopic interactions occuring at the solute-solvent interface and between the polarized solvent molecules, the model is particularly adapted to the study of polar solvents and of their interactions with polar molecules. Its very character makes it particularly apt to deal with problems of hydration, where the well-known potential ability of the water molecule to act as a double proton-donor and /or as a double proton-acceptor often raises the problem of the existence of successive shells.

ii) The second remark concerns the accuracy of the results. It is clear that the amount of computation needed to carry out steps 1 and 2 at the SCF *ab initio* level is large, even for relatively small solute molecules. This precludes the use of very extended basis sets and may throw doubt on the validity of the conclusions. We shall first examine on an example what kind of accuracy is to be expected in the representation of solvation.

B. Accuracy of the model : The solvation of ammonium ions.

An experimental mass-spectrometer study of the competitive solvation of the ammonium ion by water and ammonia vapors (Hogg and Kebarle 1965; Hogg *et al*. 1966; Searles and Kebarle, 1968) has shown that NH_3 molecules are taken up preferentially to water, up to an addition of four molecules in the first solvation shell. This was unexpected at that time and it was suggested, on the basis of the experimental data, that the arrangement of the solvent molecules around the ion might be of type I (inclusion between N^+H directions) instead of II (hydrogen bonding along N^+H bonds), although no definite choice could be made. (See fig. 3).

The problem provides a good test of the ability of the computations to account for the observed preferences of the ion for binding ammonia or water and to possibly throw some light on the type of binding.

The monosolvation of NH_4^+ by NH_3 and H_2O respectively was thus studied (Pullman and Armbruster, 1974) as described in part IIIA, using both an STO 3G and a 4-31 G basis set. The results obtained with the STO 3G basis are summarized in Table I for three directions of approach : one along the NH bond, one bissecting two NH bonds in their plane, and one bissecting the angle made by three NH bonds. Rotation of water or NH_3 about the direction of approach was allowed in each case so as to obtain the most stable arrangement.

The data given in the table allow two clear-cut conclusions : (i) For both H_2O and NH_3, the most stable addition compound is, by far, the one which involves fixation along an NH bond, in strong preference to a bisecting position. The difference is sufficiently large to make very unlikely an inversion of this result upon refinements of the computation. (ii) The affinity of NH_4^+ for NH_3 is definitely larger than its affinity for water, in agreement with the afore-mentioned experimental conclusions.

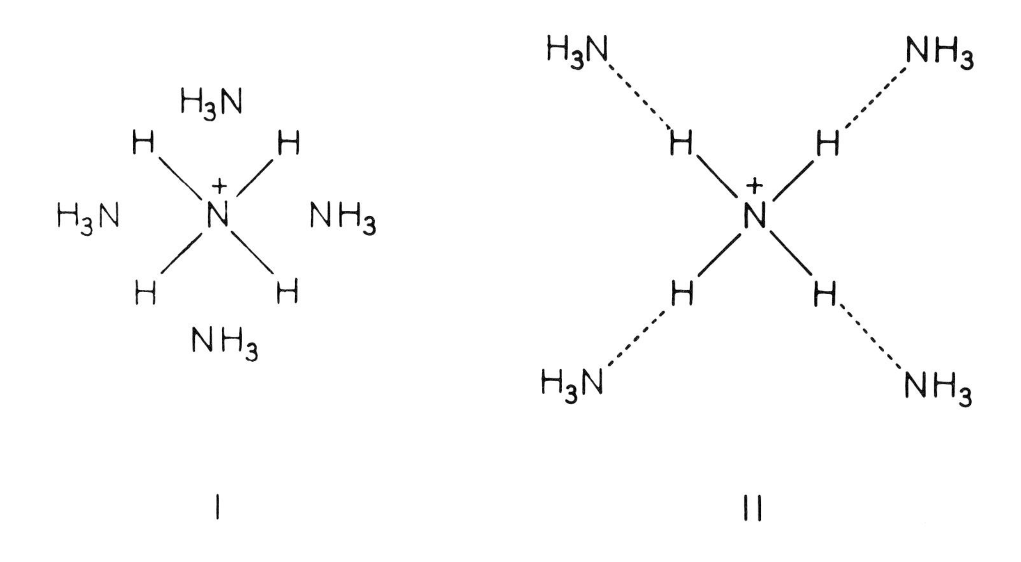

Fig. 3 - Two considered structures for the NH_4^+ $(NH_3)_4$ - adduct.

Another quite general result of the computation is the considerable in situ lability of the position of the solvent molecule about its axis : The variation in the stabilization energy upon rotation of NH_3 or H_2O is practically negligible. This is certainely an important property which relates to the fluidity of the solvation. Note in this connection that the conclusion i) above indicates a relatively limited ad situm lability which shows that in this respect NH_4^+ ought not be assimilated to the ordinary much more spherical alkali ions of similar radius.

Concerning the numerical values of the stabilization energies and the closeness of approach, they are overestimated due to the use of the STO 3G basis set. This has been observed both in the case of ordinary hydrogen bonds (Kollman and Allen, 1972), and in the case of ionic binding to monovalent cations (Perricaudet and Pullman, 1973). The situation here is of an intermediary type consisting of a hydrogen bond to a cation. Examination of the amount of electron transfer and decomposition of the binding ener-

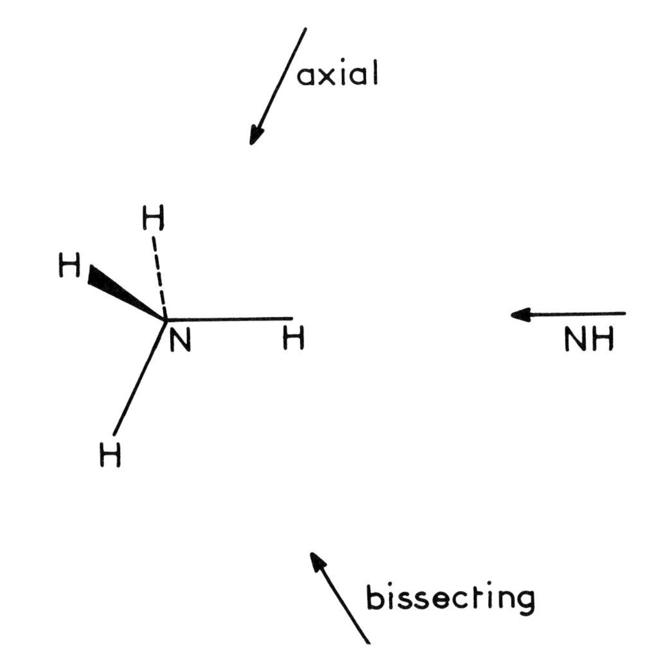

Fig. 4 Three main directions of approach of NH_3 or H_2O towards NH_4^+

Table I

Values of ΔE (kcal/mole) and R(Å) (STO 3G)

Solvent		NH	Axial	Bissecting	Exp.
NH_3	ΔE	42.2	16.2	14.3	24.8
	R_{NN}	2.5	2.8	2.75	
H_2O	ΔE	37.3	16.0	—	17.3
	R_{NO}	2.4	2.6	—	

gy have shown (Pullman et al., 1976) that the STO 3G basis favors the existence of a spurious charge transfer contribution due to the utilization by each ligand of the basis set of its partner to improve its own representation inside the supermolecule. Use of a 4-31 G basis improves the results by a factor of about 0.75 and increases the distances by about 0.25 Å, without changing the other conclusions above.

Under these conditions, the next question concerns the possible accuracy of the next step in the supermolecule approach, namely the study of polysolvation. Here again the ammonium ion provides us with experimental information coming from more recent gas-phase ion-cluster equilibria measurements (Payzant et al., 1973) yielding the enthalpies of stepwise addition, $\Delta H_{n-1,n}$, of NH_3 and H_2O to NH_4^+ from n=1 to 5. The evolution of the experimental values with increasing n is shown in fig. 5 together with the evolution of the corresponding quantities computed in the STO 3G basis according to the procedure described in IIIA namely with stepwise reoptimization (Pullman and Armbruster, 1975).

Two essential features appear upon examination of the data : (i) The evolution of the computed numbers indicates that "with increasing number of ligands, the NH_4^+ interactions with water molecules become progressively closer in magnitude to those with ammonia. A crossover occurs in the curves such that the $|\Delta H_{4,5}|$ value for water becomes larger than that for ammonia". The resemblance to the evolution of the experimental values is such that we have been able to use within quotation marks for commenting the theoretical results the very words utilized by Payzant et al. (1973) for commenting the experimental data.
(ii) The numerical overestimation of the binding energies mentioned above, considerable for n=1, decreases strongly upon progressive solvation for both solvents as shown by the stepwise decrease of the spacing between the experimental and theoretical curves.

These results have clearly far-reaching implications : not only do they confirm the possibility of obtaining satisfactory information of a qualitative nature in working with a small basis set, but they also indicate that the error made by computing solvation properties for this kind of ions using an STO 3G basis is much less important for computations involving complete or near-complete solvation than one might have inferred on the basis of the results concerning fixation of a single solvent molecule.

A similar decrease in the exaggeration of the binding energies by the STO 3G basis upon stepwise addition of water molecules to the system $(OH^-)H_2O$ has been obtained recently (B. Pullman et al., 1975) in a comparison (see fig. 6) with the corresponding values computed with a 4-31 G basis by Newton and Ehrenson (1971). Both observations indicate that the reason for this situation

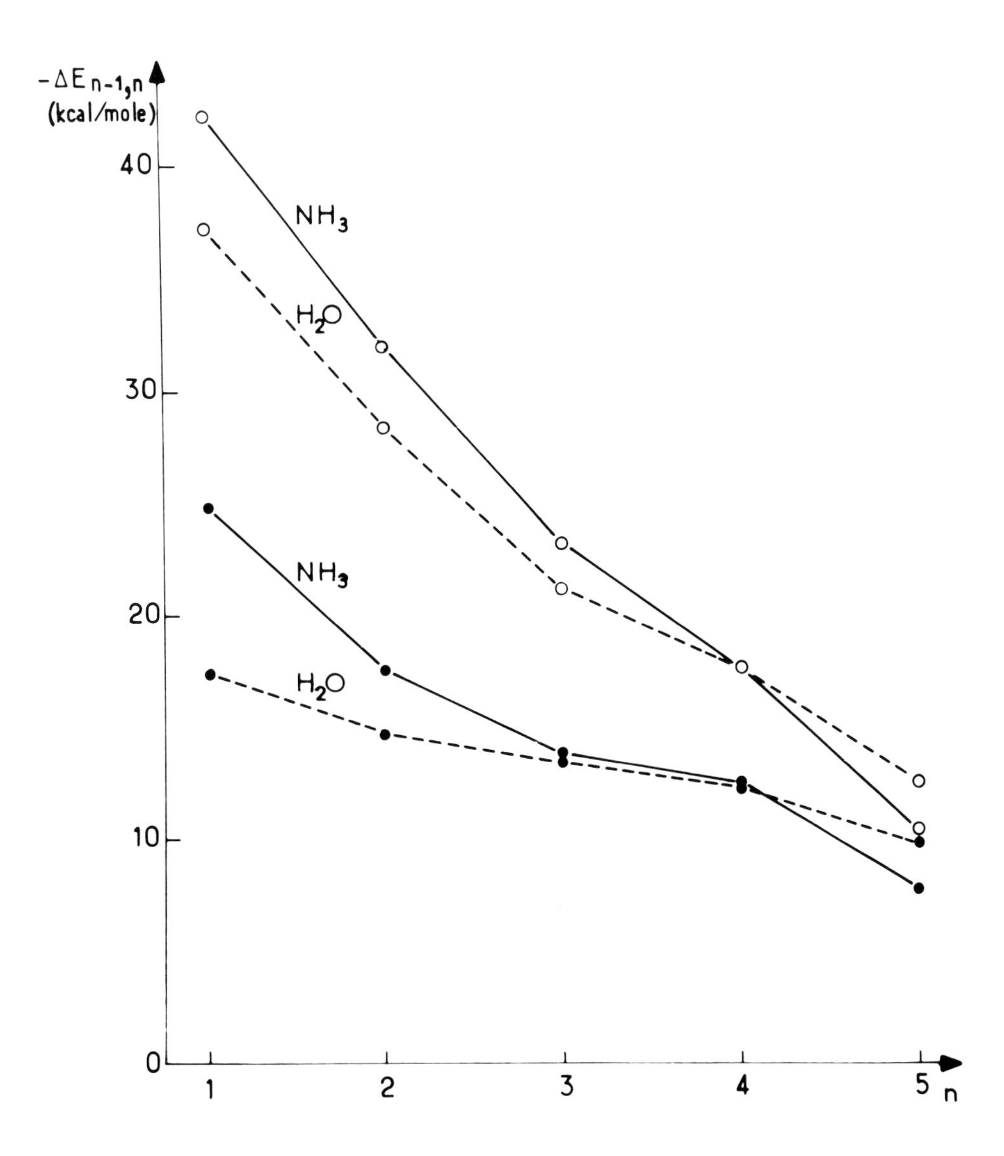

Fig. 5. - Theoretical (two upper curves) and experimental (two lower curves) variation of $\Delta E_{n-1,n}$ form n = 1 to 5 for stepwise addition of NH_3 and H_2O to NH_4^+

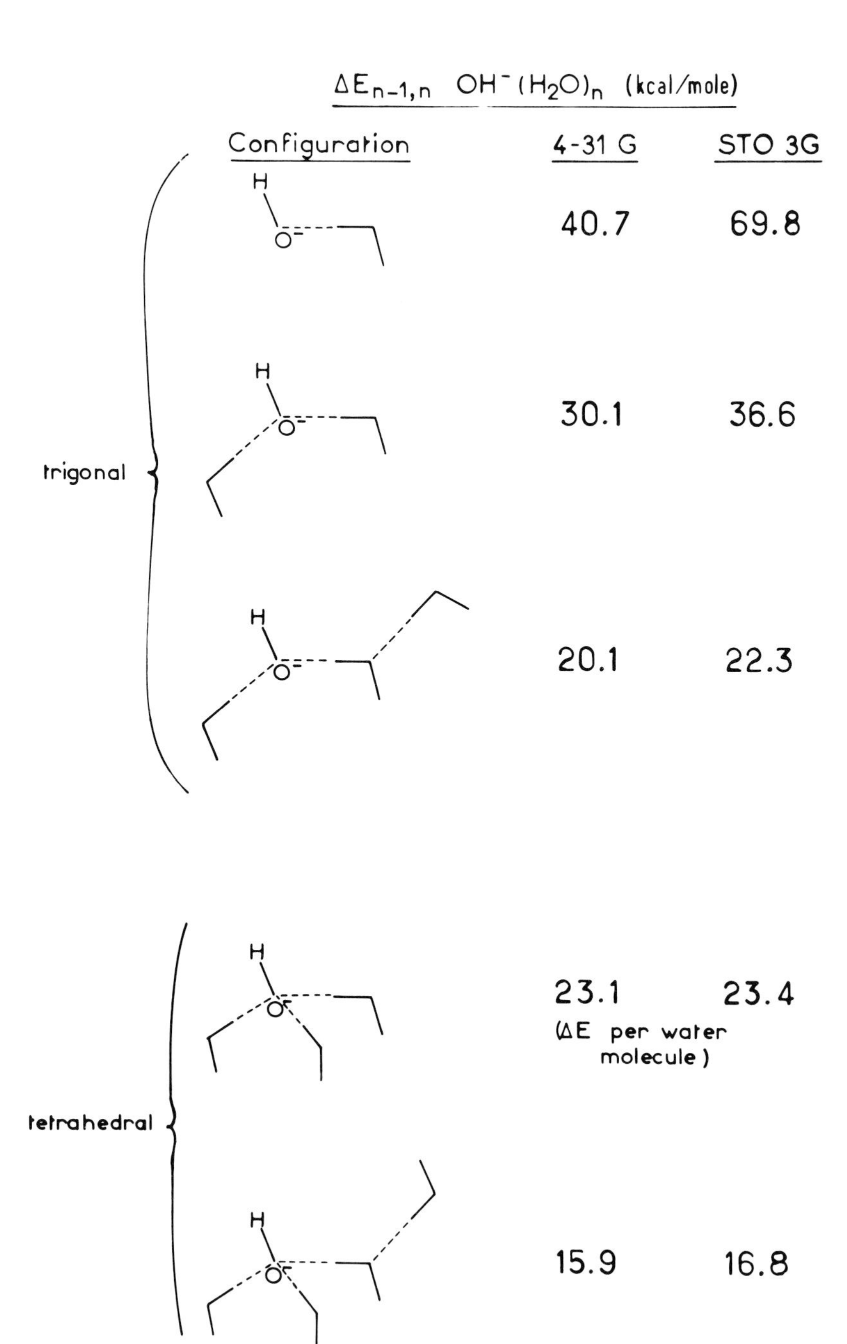

Fig. 6 - Computed values of $\Delta E_{n-1,n}$ for stepwise addition of water to OH^-.

resides most likely in the decrease of the spurious basis-set extension stabilization, upon addition of more and more molecules of ligand to the central ion.

Worth noting is the fact that one may still go one step further towards an understanding of the reasons for the crossing-over of the curves for NH_3 and H_2O when enough molecules are added to the central ion : in the first solvation shell, the ligands are directly attached to the NH_4^+ ion, but in the outer shell they attach to molecules of NH_3 or H_2O which, although partially polarized, are more and more similar to "normal" NH_3 or H_2O when n increases. In the limit, the respective $\Delta E_{n-1,n}$ must converge to the binding energies of the NH_3 and H_2O dimers. These energies being -3.7 and -5.6 kcal/mole (linear dimer) respectively using the STO 3G basis, a crossing of the curves is to be expected.

Furthermore, the underlying features which command the order $NH_3 > H_2O$ in the binding to NH_4^+ and the opposite order $NH_3 < H_2O$ in the dimer formation can be rationalized as follows : as underlined by Payzant et al. (1973), NH_3 has a much larger proton affinity than H_2O, thus its attraction for the highly charged hydrogen of NH_4^+ should be the largest; indeed the (exact) computation of the Coulomb attractive component of the binding energy using the procedure of Dreyfus and Pullman (1970) yields values (table 2) in the order expected. At the other extreme, the same computation for the dimers (table 2) indicates a reversal in the order of the pure Coulomb attraction in parallel to the reversal in the order of the binding energies. In this case, the hydrogen to be bound is appreciably less positive in NH_3, (+0.160e), than in H_2O (+0.183 e) so that in the dimers, the large intrinsic attraction of NH_3 exerts itself on a small positive charge whereas the smaller intrinsic attraction of H_2O acts on a larger positive charge.

Table 2.
The Coulomb components (E_C) of the binding energies, E_{SCF}(kcal/mole), to NH_4^+ and in the homodimers.

	NH_4^+		Dimer	
	$-\Delta E_C$	$-\Delta E_{SCF}$	$-\Delta E_C$	$-\Delta E_{SCF}$
NH_3	46.4	42.3	4.8	3.7
H_2O	38.5	37.2	6.8	5.6

The conclusion of this section as to the accuracy to expect should thus be a temperate optimistic view : although absolute numbers are not to be expected from an STO 3G study, the qualitative image of the solvation even up to such fine details as the competition between NH_3 and H_2O, the reversal of the order observed after n=4, and the asymptotic behavior, is correct.

We proceed in the next section to describe a case where the three steps of the supermolecule approach have been carried out in details.

C. A complete example : hydration of the phosphate group

The phosphodiester linkage is a potential *locus* of hydration both in the nucleic acids, in the polar heads of phospholipids and in many other fundamental biological components, owing to its anionic character. We have studied in details the characteristics of its hydration on the model of the dimethylphosphate anion DMP^-. This anion is represented in fig. 7, in its *gg* conformation, (where both methyl groups are in a *gauche* conformation with respect to rotation about the PO single bonds) which is the intrinsically preferred one, as computed by *ab initio* STO 3G (Newton, 1973) STO 3G + *d* orbitals on phosphorus (B. Pullman *et al.* 1975 and PCILO (Perahia *et al.*, 1974).

The terminal methyl groups have been fixed in a staggered arrangement. For easy recognition, the anionic oxygens carry the odd and the ester oxygens the even numbers. (Note that DMP^- is a *mono* anion, but the negative charge is shared equivalent by O_1 and O_3 due to the symmetry).

This conformer possesses a two-fold symmetry axis bissecting the O_1PO_3 angle so that a 180° rotation about it brings the atoms O_1, P, O_4, C_4 in coincidence with O_3, P, O_2, C_2. Consequently, the planes O_1PO_4 and O_3PO_2 are equivalent and the same is true for the planes O_1PO_2 and O_3PO_4.

The geometry of both the ion and the water molecule were kept constant.

The computations were carried out *ab initio* by the SCF LCAO procedure with an STO 3G basis set, using the program Gauss 70 and its extension made in our laboratory to accomodate 105 contracted functions. The detailed results concerning the energies and distances have been given in the original papers and we do not come back here on the details, neither on the well-established fact that the energies of binding are too large by roughly 5/3 to 2 for single hydration, and the distances too short, the error per water molecule decreasing upon increasing solvation (*vide supra*).

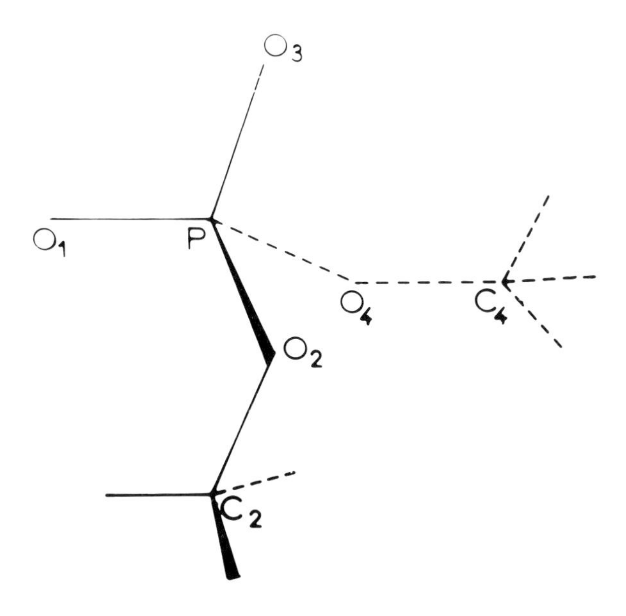

Fig. 7. Dimethyl phosphate (DMP^-) in the gg conformation.

1) Monohydration

The results concerning the possibilities of hydration by one water molecule are best visualized by first examining the situation in the various OPO-planes of the phosphate. (Fig. 8a,b,c). The plane O_1PO_3 contains the site corresponding to the minimum minimorum (-18.6 kcal/mole) on the energy hypersurface : this occurs in two equivalent positions when water is bound to one of the anionic oxygens, O_1 or O_3, by one H-bond, at the exterior of the OPO angle (positions E_{13} and E_{31}). But there are also, in that plane,other positions nearly as good in energy (E'_{31} and E'_{13}), and 4 kcal/mole below (E''_{13} and E''_{31}). Moreover, there is also a bridged adduct B_{13}, only 1.8 kcal/mole less stable than E_{31} comparable to that found in the case of the formate ion (Port and Pullman, 1974). However, in contrast to this last case, there are here a number of other OPO-planes as show fig. 8b and 8c : each of these planes contains one external binding site E (to the anionic oxygen) similar to E_{13} and E_{31}, and another site E' obtained from E by a 180° rotation about the PO axis. Moreover, a favorable bridged position B appears also in the equivalent planes O_1PO_4 and O_3PO_2, with an inclination of the water molecule towards the anionic oxygen, a situation reflecting the distribution of the electrostatic potential around the phosphate ion (Berthod and Pullman,

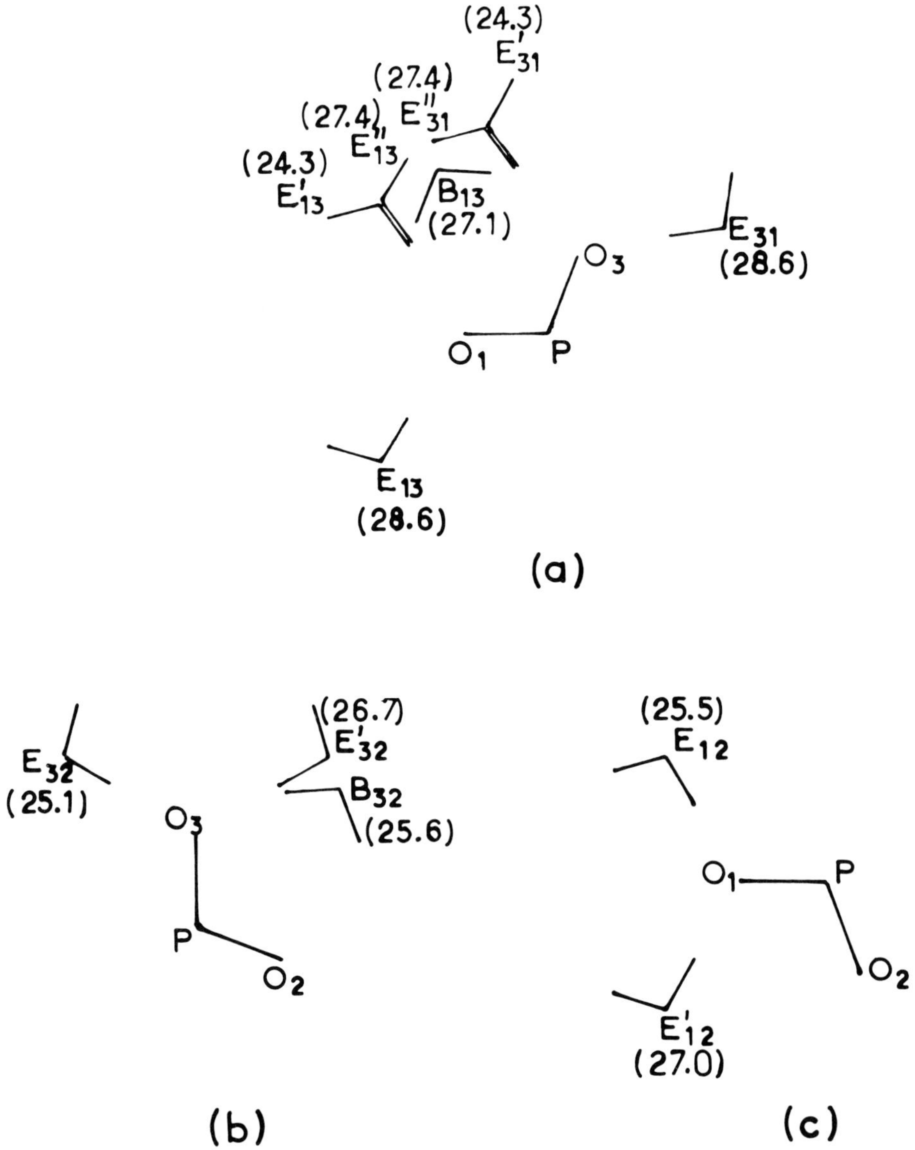

(24.3)
E'_{31}
(27.4)
(27.4) E''_{31}
E''_{13}
(24.3)
E'_{13}
B_{13}
(27.1)
O_3
E_{31}
(28.6)
O_1 P
E_{13}
(28.6)
(a)

E_{32}
(25.1)
(26.7)
E'_{32}
B_{32}
(25.6)
O_3
P
O_2
(25.5)
E_{12}
O_1 P
O_2
E'_{12}
(27.0)
(b) (c)

Fig. 8 - The preferred hydration sites in the three sorts of OPO-planes in DMP^-.

1975). The corresponding bridge positions in the planes O_1PO_2 and O_3PO_4 are less favorable, which is another reflection of the potential distribution in the planes.

On the whole, the distribution of the favorable hydration sites in the neighbourhood of the anionic oxygens points to the existence of a whole zone of attraction for water about these two atoms : this is confirmed by studying the variations in the binding energy obtained by rotating the plane of the water molecule about the PO_1 axis, from its most stable position E_{13}, in such a way that $O_1...HO$ describes a cone about this axis. (See fig. 9). The circular zone of very large attraction for water corresponds to the existence of a circular zone of intrinsic attraction for a proton (Berthod and Pullman, 1975), the slight differences in the two distributions in these zones reflecting the obvious differences between the proton and a water molecule, and also the second-order delocalization effects.

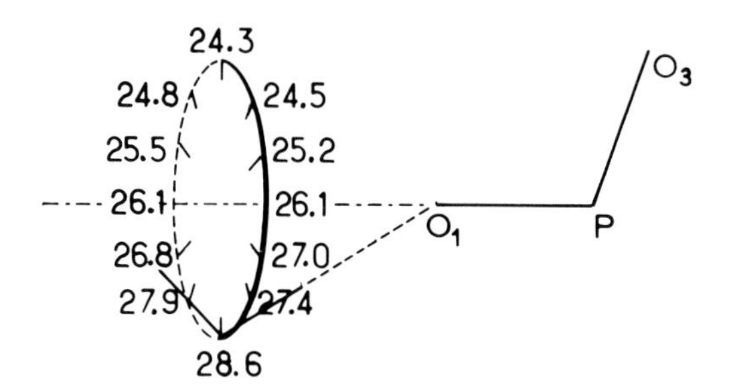

Fig. 9. Variation of $-\Delta E$ (kcal/mole) for rotation of water out of position E_{13}, about PO_1 axis.

Aside from the lability of the binding observed in the above-mentioned rotation, one may note the considerable in situ and ad situm labilities manifested in positions B_{13}, E'_{13}, E''_{13}, E'_{31}, E''_{31} in fig. 8.

The finding of such a large number of nearly equivalent positions for one molecule of water around the phosphate oxygens is a strong indication that there will be many possible schemes for polyhydration.

2) The first solvation shell.

The results obtained for simultaneous fixation of a number of water molecules to the phosphate ion are given in Table III. It is indeed observed that many possibilities appear. Up to six water molecules may be accomodated in the first hydration shell. Intermediate hydration involving two to five water molecules may of course take place, a number of energetically nearly equivalent possibilities occuring for a given number of water molecules (3 for n = 2 to 4, 2 for n = 5). The average energy of water attachment is approximately -23 kcal/mole for n = 2 and 3 and decreases when n increases further, down to -17.3 for n = 6.

Table III - Polyhydration of DMP^- (STO 3G) with n water molecules in the first shell.

n	occupied positions	$-\Delta E_{tot/n}$ (kcal/mole)
2	E_{13} E_{31}	27.1
	B_{14} E'_{12}	23.6
	B_{14} B_{12}	20.4
3	B_{13} B_{32} B_{14}	23.0
	B_{13} E'_{32} E'_{14}	23.5
	B_{13} E_{31} E_{13}	24.7
4	E_{13} E_{31} E_{14} E_{32}	21.9
	E'_{12} E'_{34} E'_{14} E'_{32}	21.7
	E_{14} E_{32} E'_{14} E'_{32}	21.3
5	B_{13} B_{32} B_{14} E'_{12} E'_{34}	20.3
	B_{13} E'_{32} E'_{14} E'_{12} E'_{34}	20.3
6	E_{31} E_{32} E_{34} E_{12} E_{13} E_{14}	17.3

Examples of these polyhydrates are given in figs. 10 and 11.

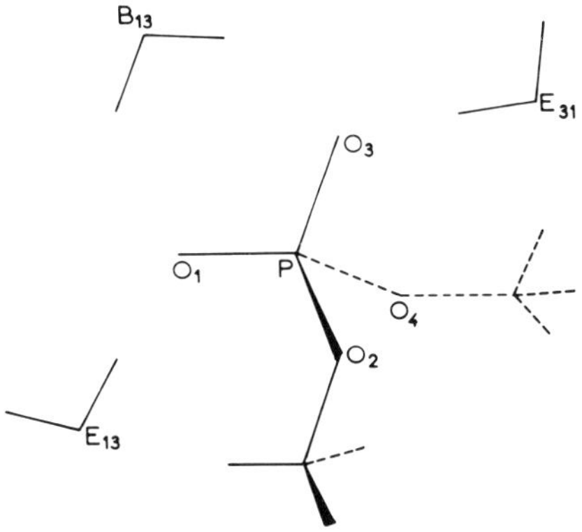

Fig. 10 - The preferred trihydrate of DMP^-.

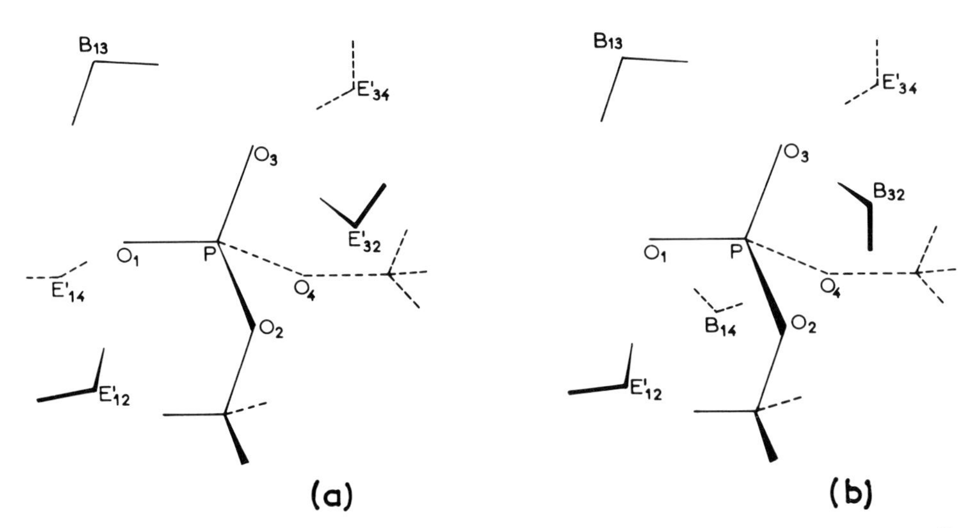

Fig. 11 - Two energetically equivalent pentahydrates of DMP^-.

3) The second solvation shell and beyond.

In the case of an ionic substrate like DMP^- the question arises of the possibility of the building-up of further solvation shells with water molecules hydrogen-bonded specifically to those of the first layer, and if the answer is positive, how far does the structuration go ?

From a theoretical point of view, the problem may be approached in two ways : the simplest approach is to consider a monohydrate of DMP^- and try to attach a second molecule of water to the water already bound, the procedure being then extended to the attachment of further water molecules representing further shells.

The results for successive stepwise linear addition (tricoordination) of water molecules to the principal sites of the first hydration shell, computed in this approximation are indicated in Table IV. It is found in this model that as an average the interaction energies of the water molecules of the second hydration shell represent about 60% of the interaction energies of the water molecules of the first shell and that the energies of interaction of the water molecules of the third hydration shell represent in this approximation about 35-40% of the energies of hydration of the first hydration shell, and still surpass appreciably the energy of normal water-water interaction which, in the most favorable dimer is -6.4 kcal/mole (with the same STO 3G basis).

Thus a reasoning based on the model relating to single water molecule placed on each hydration shell would conclude to a structuration of water by DMP^- going at least up to a third and perhaps to a fourth layer.

Table IV.- Hydration energy ΔE of $DMP^- \ldots n(H_2O)$ (one H_2O molecule in each shell, linearly bound) (kcal/mole)

Site of location of the first water molecule	1st shell	2nd shell (a)	3rd shell (a)
B_{13}	-27.1	-17.6	-12.1
E_{13}	-28.6	-14.0	-10.0
B_{32}	-25.6	-14.3	
E'_{32}	-26.7	-13.4	
E'_{34}	-27.0	-11.0	

(a) increment in ΔE.

Table V - Hydration energies of polyhydrated DMP^- (kcal/mole)

Number of water molecules	Location	Hydration energy per water of 2nd shell	Hydration energy per water of 3rd shell
6	Three in the first shell : B_{13}, E_{13}, E_{31} three linearly in the second shell	-12.9	
3	One in the first shell : B_{13} two tetrahedrally in the second shell	-15.4	
7	One in the first shell : B_{13} two tetrahedrally in the second shell four tetrahedrally in the third shell		-8.1
	six in the first shell one linearly in the second shell	-11.9	

Another model, however, leading to a more precise evaluation of the values of the mean energies of interaction characteristic of the different shells may be expected from computations on polyhydration, involving simultaneously a larger amount of water molecules. Three typical results are presented in Table V. The fundamental indication is that polyhydration diminishes the mean energy of interaction of the water molecules of each hydration shell: thus, in the hexahydrate of table V, the mean value of the individual hydration energies of the three waters of the 1st shell changes from -28.1 to -24.7 kcal/mole in the trihydrate. That of the second shell drops from -15.2 to -12.9. It may be inferred from this situation that the individual values of the hydration energies of the water molecules of the 3rd shell which were of the order of -10 kcal/mole, will similarly become reduced in a polyhydration scheme to values not much greater than the energy of dimerization of water itself (-6.4 kcal/mole in the STO 3G basis). It seems therefore relatively improbable that organized "bound" water layers may extend around DMP^- beyond the 2nd shell, although residual organized such fragments may persist in the vicinity of some particularly favorable hydration sites.

Table V presents also two results of polyhydration involving tetrahedral coordination of the water molecules of the second and third hydration shells. The value of -15.4 kcal/mole obtained as the main value of the energy of interaction for the two water molecules fixed in the second hydration shell on a water molecule in position B_{13} of the first shell may be compared with the value of -17.6 kcal/mole found in Table IV for a single water molecule fixed trigonally in the second shell at the same site. The decrease of the mean energy of interaction for the water molecule fixed tetrahedrally in the third hydration shell (third line of Table V) is worth stressing.

Finally, the last line in Table V gives the result of one larger computation involving a complete first hydration shell of six water molecules plus one water molecule of the second shell added to the first one at site E_{13}. The energy of interaction of the water molecule of the second shell is then equal to -11.9 kcal/mole, a value significantly smaller than that of -14 kcal/mole found when two water molecules were considered alone at the E_{13} site. The completion of the first hydration shell has thus the effect of reducing the binding energy of the water molecules of the second hydration shell and this effect extends doubtlessly to further shells. This result strengthens the conclusion that organized "bound" water is most probably essentially limited to the two first hydration layers around DMP^-.

This conclusion was reached on the basis of energy data. There are other ways to look at the problem : one is to consider the perturbation brought about in the electronic structure of the water molecules bound to the phosphate group and its possible transmission to further water layers. This can be visualized by the distribution of electronic charges obtained by a Mulliken population analysis. It must of course be kept in mind that the charges defined there do not correspond to an "observable" in the quantum-mechanical sense, and represent only a conventional convenient partition of the exact electron-density distribution into atom-centered fractions. Moreover, the numerical values of the charges depend on the basis set used in the computations : in the present case the absence of *d* functions on the phosphorus atom results in a strong exaggeration of the numerical values of the net charges on the phosphate ion, but the relative values as well as their behavior under a given perturbation are sufficiently independent of the basis set for providing useful information in a problem such as the present one. (See D. Perahia *et al.*, 1975, for a discussion of the effect of the *d* orbitals).

It is also known that the utilization of an STO 3G basis set leads to an overestimation of the charge transfer (vide supra). While these restrictions have to be kept in mind in evaluating the *quantitative* significance of the results, they are of secondary importance in a comparative, qualitative study.

Fig. 12 presents the distribution of the net electronic charges in free DMP^- (a), in H_2O (b) and in the first hydration shell of the two monoadducts B_{13} (c) and E_{31} (d) at their equilibrium O...H distances : it is seen that the structure of the "bound" water is significantly perturbed with respect to that of free water : its oxygen atom has become appreciably more negative and its hydrogen(s) engaged in the hydrogen bond(s) with DMP^- have become slightly but significantly more positive. Altogether the binding of water to DMP^- produces a partial transfer of electrons from the latter to the former. The bound water carries thus an excess of electronic density from 0.120 to 0.142 e which may be compared to the case of water dimers or polymers in which there is already a small charge transfer from the proton acceptor to the proton donor molecule : for the water dimer the corresponding value (STO 3G) is 0.034 e. This enhancement of the charge transfer is a result of the anionic nature of the solute which transmits a small partial anionic character to the bound water molecule of the first layer, thereby enhancing its attraction towards other water molecules.

Polyhydration in the first shell indicates that the presence of a number of water molecules <u>reduces</u> in each of them the amount of charge transfer found for a single molecule. This is illustrated in fig. 13 which presents the distribution of the electronic charges in the trihydrate $B_{13} - E'_{14} - E'_{32}$ of DMP^- (all water molecules in the first hydration shell).

When solvation is extended to the second hydration shell the situation may appear in two different ways according to the two different models adopted in building the successive layers : in fig. 14 there is one water molecule in the first shell and one in the second. In fig. 15, the first shell is completed to 6 molecules of water, with one water molecule in the second shell. Let us denote them 1/1, and 6/1 respectively.

The comparison of fig. 14 (1/1) with fig. 12 would suggest that the overall transfer of charge from DMP^- to the water molecules takes place predominantly at the benefit of the terminal molecule : thus the excess of charge of the water molecule of the first hydration shell is only 0.041 e while that of the water molecule of the second hydration shell amounts to 0.094 e. This last number is nevertheless smaller than the 0.120 e transferred to the single water molecule of fig. 12, indicating a decreasing ability of the water molecules of the second shell for further binding. This is equally shown by the decrease in negative charge on the oxygen of the terminal water (-0.462 instead of -0.508 e in fig. 12).

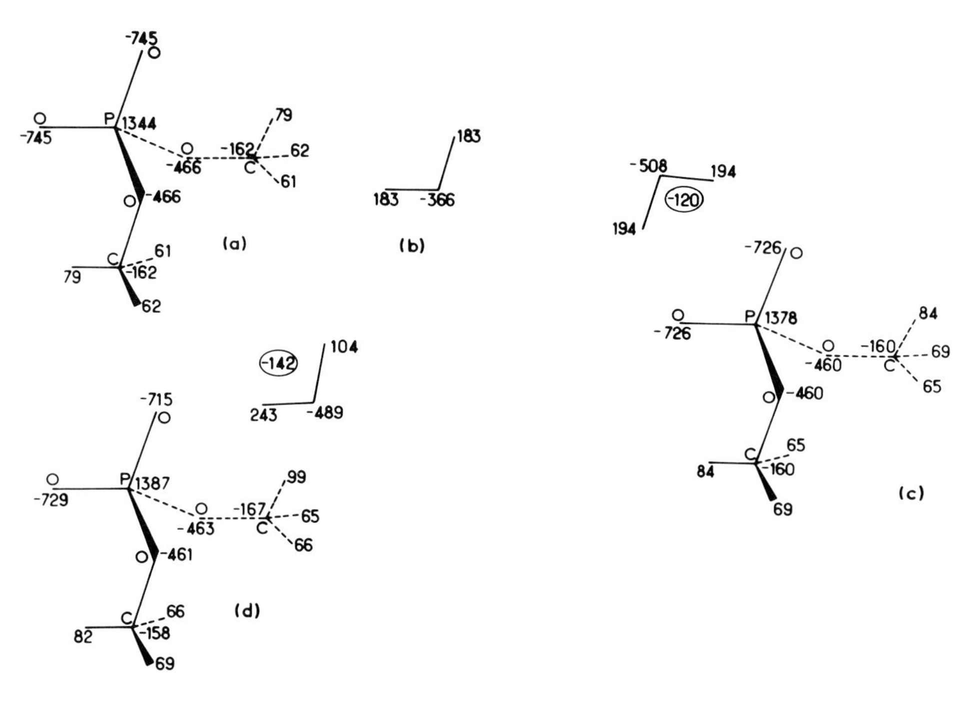

Fig. 12 - Net charges in DMP^- (a), H_2O (b) and two monohydrates (c) - (d).

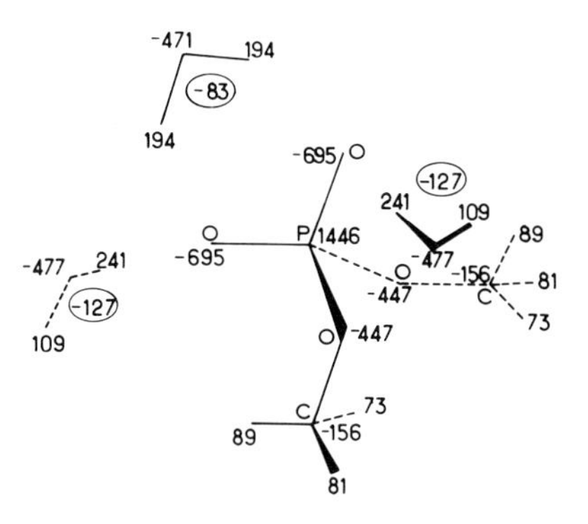

Fig. 13 - Net charges in a trihydrate of DMP^-

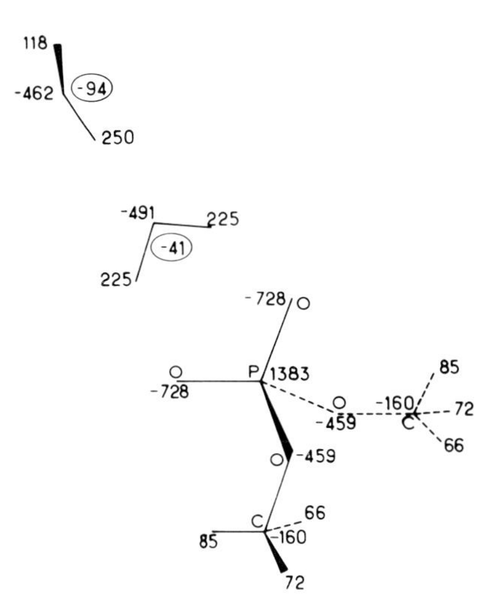

Fig. 14 - Net charges in the dihydrate 1/1 of DMP$^-$.

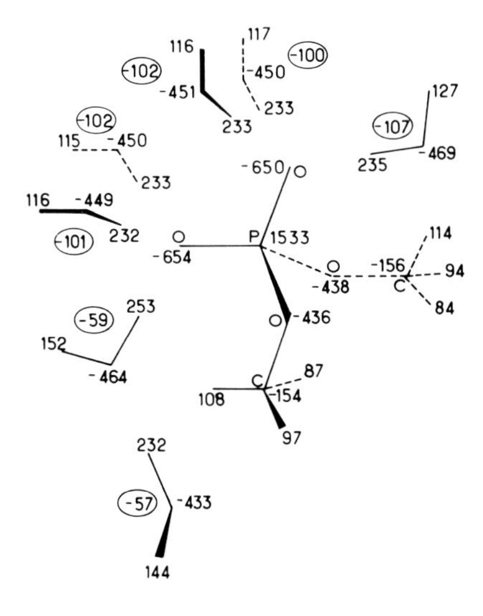

Fig. 15 - Net charges in the 6/1 heptahydrate of DMP$^-$.

A more correct and significant picture may be observed in the heptahydrate (6/1) of fig. 15. It appears here that the transfer of electronic charges leads to an excess of approximately 0.1 e at each water molecule of the first hydration shell, with the exception of the molecule in site E_{13} (bound to the water of the second shell), where the excess is smaller. Furthermore, on the water molecule of the second shell, the charge transfer amounts only to 0.057 e. Similarly to the situation in case 1/1, the negative charge on the oxygen atom of the water molecule of the second shell has decreased to -0.433 e.

It may thus be estimated that in a reality which involves the completion of rather large hydration shells, the polarization of the water molecules due to the effect of the perturbing anionic solute will decrease in successive solvation layers. The decrease may be estimated as relatively rapid and it seems probable that the water molecules of the third hydration shell around DMP^- will be only very slightly perturbed if at all. Compared with the 1/1 data of fig. 14, this result shows again the necessity of taking into account the totality or at least large parts of the first hydration shell in drawing conclusions about the characteristics of the second layer, in agreement with the previous same conclusion based on energy computations.

Another interesting way to consider the possible structuration of water around the solute is to consider the modifications of the molecular electrostatic potential of the bound water molecules with respect to that of isolated water or of water bound to itself : an example of the situation is given below (for more details, see Berthod and Pullman, 1976) Figs. 16 and 17 give the molecular potential (STO 3G) of a water dimer, and that of the hexahydrate of DMP^- in the O_1PO_3 plane. Their comparison shows the strong enhancement of the attractive character for a proton of the DMP^--bound water with respect to that of the molecules of the bulk solution. The perturbation of the potential of the molecules of the second shell have been studied (loc cit) in the same fashion.

All these results point to the conclusion based on energy computations that "bound" perturbed water around DMP^- is essentially restricted to the first two hydration shells. Experimental information about the hydration of the phosphate group comes mostly from studies of phospholipids. It is relatively abundant (see references in B. Pullman et al. (1975)) and although it does not lead to a unique scheme and does not fix precisely the preferred sites of hydration, it indicates a number of "bound" water molecules which altogether is comparable with that suggested by the theoretical studies : depending upon the experimental conditions and techniques utilized the number of water molecules in the primary hydration shell (most strongly bound) at the polar head of

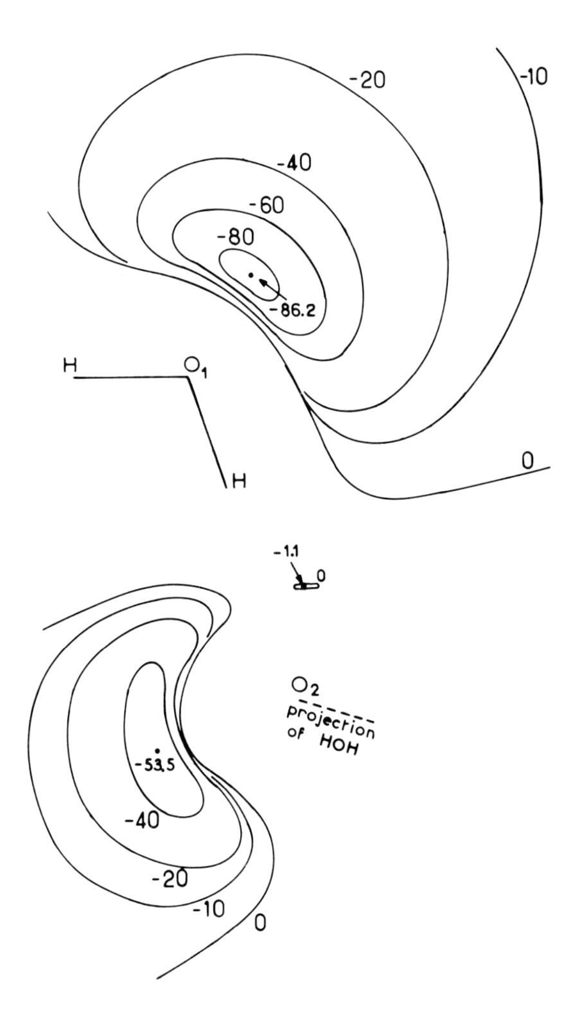

Fig. 16- Electrostatic potential of the most stable water dimer.

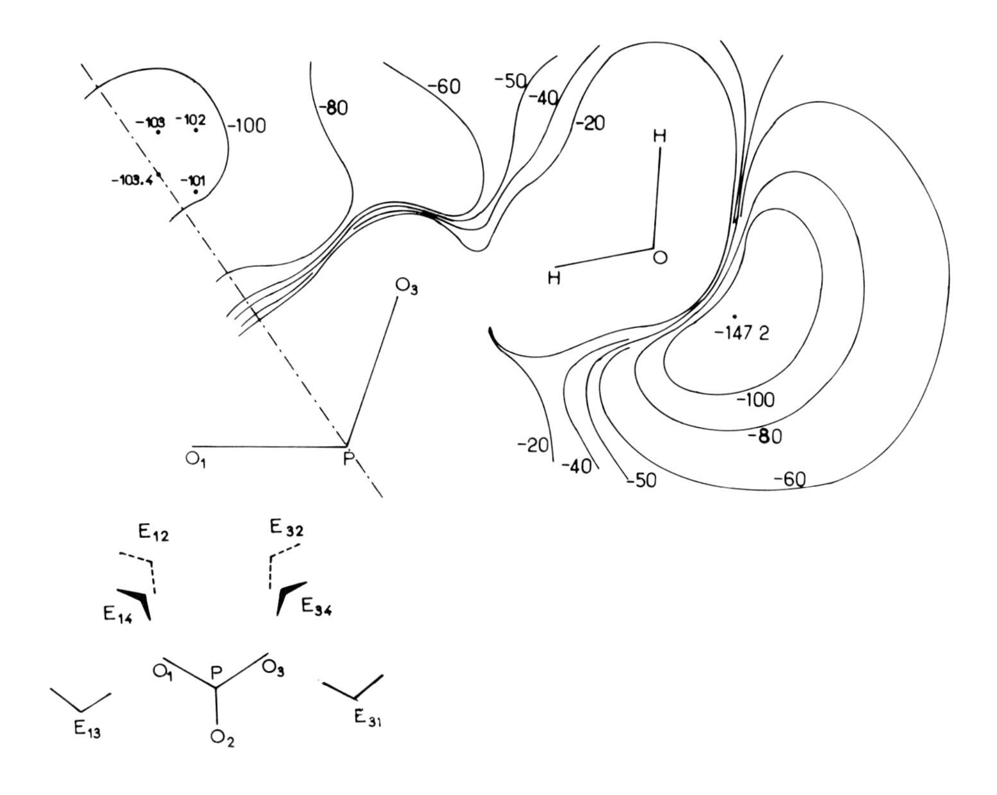

Fig. 17 - Electrostatic potential in the O_1PO_3 plane of the hexahydrate of DMP^- (structure at bottom left).

phosphatidylcholine (which was more abundantly studied than phosphatidylethanolamine) varies from 2 to 6. As, in these molecules no strong hydrogen bond is expected around the cationic head, this number may be considered as relevant essentially to the phosphate group. A secondary shell of 4-8 molecules of water, less strongly bound that the previous ones, is sometimes distinguished. The preferential fixation of 4-6 water molecules on the phosphates of the nucleic acids has similarly been proposed as a result of infra-red studies (Falk *et al*.1963; Hartman *et al*., 1973).

4) Effect of hydration on the properties of the phosphate group.

The last step of the supermolecule approach to solvation is the utilization of the previously gained information for the evaluation of the solvent effect. One example concerning the phosphate group will be given : it is concerned with the effect of solvation on the conformational stability. As mentionned earlier the *gg* conformation about the P-O bonds is the one predicted by computation as the most stable for the free molecules of the type of dimethylphosphate or disugar phosphate or dinucleoside monophosphate or the polar head of phospholipids. It is also the one observed predominantly for this type of compound in X-ray crystallographic studies, where the values of the corresponding torsion angles spread generally between 60°-90° or 270°-300° (for a review see Pullman B. and Saran, 1976). The problem may, however, be raised as concerns the situation in solution for which less precise knowledge is available. As it is practically impossible, because of the complexity of the situation, to construct conformational energy *maps* for the hydrated species which would indicate the conformational possibilities as a function of the continuous rotation about the two P-O bonds, for different degrees of hydration, we have computed the essential features of the hydration scheme of the *gt* (or *tg*) and *tt* forms and compared them with those of the *gg* form. In this way conclusions may be drawn at least about the effect of hydration on the relative stabilization or destabilization of these three typical forms.

For the *gt* form a complete study of the monohydration scheme with optimization for each position was done as for the *gg* case. The hydration sites being very little different from those of the *gg* conformer, the same symbols were kept. By extension, the same positions were also kept for the *tt* form.

Table VI summarizes the essential results obtained. Its upper part indicates the values of the energies of binding for the attachment of one water molecule to the principal sites of hydration of the *gg* and *gt* conformers. The column (*gt*-*gg*) indicates the energy difference between the two forms in the free dimethylphosphate anion and in its different monohydrates. It is seen that according to the site of the monohydration this difference, which is 3.4 kcal/mole in favor of the *gg* form in the free molecule, oscillates between the values of 1-3.9 kcal/mole in the monohydrates. Depending upon its position, a water molecule may thus have a stabilizing or a destabilizing effect, although always a small one, upon the *gt* form with respect to the *gg* one.

Table VI - Hydration of the *gg*, *gt* and *tt* forms of DMP^-

n	Symbol (a)	Symbol (b)	Energy (kcal/mole) *gg*	*gt* (a)	*gt* (b)	*gt-gg* (a)	*gt-gg* (b)	*tt*	*tt-gg*
0			0	3.4		3.4		8.0	8.0
1	B_{13}		-27.1	-27.6		2.9			
	E_{13}	E_{31}	-28.6	-29.6	-28.1	2.4	3.9		
	B_{32}	B_{14}	-25.6	-25.6		3.3			
	E'_{32}	E'_{14}	-26.7	-26.7	-28.1	3.4	2.0		
	E_{32}	E_{14}	-25.1	-25.9	-26.4	2.6	2.1		
	E_{12}	E_{34}	-25.5	-26.8	-25.7	2.1	3.2		
	E'_{12}	E'_{34}	-27.0	-27.6	-27.3	2.8	3.1		
2	$E_{13}E_{31}$		-27.1	-27.4		2.9			
3	$B_{13}E_{13}E_{31}$		-24.7	-25.0		2.6			
4	$E_{13}E_{14}E_{31}E_{32}$		-21.9	-22.2		2.2			
5	See Fig. 11a		-20.3	-20.7		1.1			
6	See Fig. 17		-17.3	-17.9		-0.01		-18.3	1.8

(a) and (b) are equivalent in *gg*. For n=0, energies are given with respect to *gg*. Values in columns *gg*, *gt*, *tt* are $\Delta E_{tot}/n$; *gt-gg* and *tt-gg* is the difference in energy between the two conformers hydrated as indicated.

A more regular evolution of the situation occurs in the polyhydrates, as seen in the lower part of the Table. Both in the *gg* and in the *gt* form the binding energy per water molecule decreases with increasing hydration. The essential result, however, is that increasing hydration brings also the energies of the two conformers continuously closer to each other, to the point that when the first hydration shell is completed with its six water molecules, the two forms are practically equienergetical. For a lesser water content, the *gg* form predominates but its excess of stability is less than it is in the free molecule.

Only one computation was carried out for the *tt* form and it is indicated in the last column of Table VI.In view of the preceding considerations it is quite illustrative. In the free molecule the ab initio computations indicate that the tt form is 8 kcal /mole less stable than the gg form. Considering the energy of the tt form hydrated with six water molecules it appears that this form is then only 1.8 kcal/mole less stable than the gg form. It seems probable that the tt form with intermediate degrees of hydration will be relatively still less stable with respect to the other two forms.

The principal conclusion of this comparison is thus that polyhydration is destabilizing the gg form relative to the gt and tt forms. The gg form still remains, however, the most stable one for all degrees of hydration with the exception of hydration with six water molecules when the gt form becomes energetically equivalent to it. The tt form always remains the least stable one.

It may be pertinent to relate this theoretical result on the increase of the importance of the elongated forms of the phosphate group upon hydration to some recent observations on the conformation of purine and pyrimidine nucleotides in solution as studied by NMR spectroscopy in the light of theoretical computations of the relevant P-C spin-spin coupling constants in model systems. With the standard numbering system utilized in the study of mono- and polynucleotides (Figure 18) it is the $^2J_{PC_5'}$ spin-spin coupling which may be related to the torsion angle about the P-$O_{5'}$ bond. In all non-cyclic mono and polynucleotides studied, the experimental values of this coupling are 4-5 Hz (Smith et al. 1973). Theoretical computations on this coupling constant as a function of the torsion angle about the P-O bond (Giessner-Prettre and Pullman, 1974) on the model of ethyl phosphate : CH_3-CH_2-O-PO_3^- have indicated that the experimental coupling constants are compatible only with values of the torsion angles about the PO bonds ranging between 120° and 150° (or 210° and 240°, NMR not permitting the distinction between the two cases). This result suggests strongly that the torsion angles about the P-O bonds may have values in solution (the experimental conditions of NMR measurement) which are different from those measured in crystal phase or that there is an equilibrium in solution between several conformers corresponding to different values of these torsion angles, the average conformation having a somewhat larger value for this angle than has been observed in crystals in most cases. These results bear obviously a close relationship to the above described results indicating that hydration may promote a possible elongation of the structure of the phosphate group.

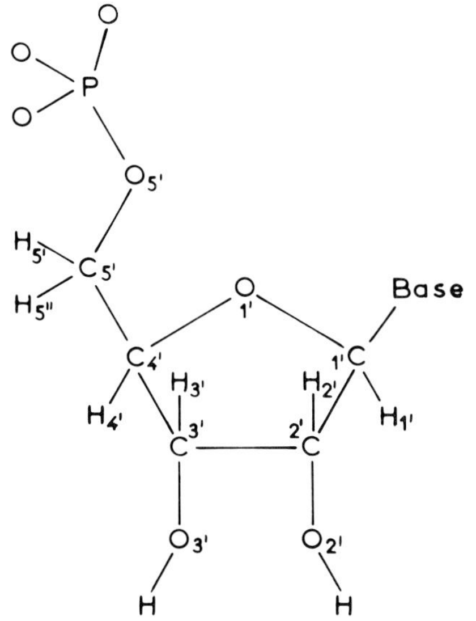

Fig. 18 - Atom-numbering in 5'- β nucleotides.

D. Other applications.

We have given in what precedes a full description of the supermolecule approach to the solvation of the phosphate group carried out entirely in an *ab initio* fashion. As mentionned earlier, whereas the determination of the hydration scheme proper requires the use of *ab initio* methods — at least until a large enough catalog of the hydration of a wide variety of compounds is established — the last part of the study may be performed by semi-empirical methods. A large number of conformational studies has been carried out in our laboratory (for details see Pullman A. and Pullman B., 1975), using the PCILO method (Diner *et al*, 1969).

Table VII indicates the essential compounds which have been studied in this fashion as well as their most stable conformation in vacuum and in water.

Applications other than to conformational problems have been carried out recently in the field of tautomeric equilibria, in the nucleic acid bases (Kwiatkowski, 1975) and in uracil anions (Kwiatkowski and B. Pullman, 1976).

Table VII.
Influence of hydration on the conformation of pharmacological compounds.

Substance	Method	most stable conformation	
		in vacuum	in water
GABA $^{-}OOC-(CH_2)_3-\overset{+}{N}H_3$	PCILO Ab initio	Folded Folded	Various extended
Histamine monocation $C_3N_2H_3-(CH_2)_2-\overset{+}{N}H_3$	PCILO Ab initio	gauche gauche	gauche/trans
Histamine dication $C_3\overset{+}{N}_2H_4-(CH_2)_2-\overset{+}{N}H_3$	PCILO Ab initio	trans trans	gauche/trans
Antihistamine drugs $(R_1)(R_2)X-CH_2-CH_2-\overset{+}{N}H(Me)_2$	PCILO	gauche	gauche/trans
Acetylcholine $CH_3OOC-(CH_2)_2-\overset{+}{N}(CH_3)_3$	PCILO Ab initio	gauche gauche	gauche gauche (add)
Phenethylamines $(Ph)_s(CHR)\ CH_2\ \overset{+}{N}H_3$	PCILO Ab initio	gauche/trans gauche	trans trans
Polar head of phospholipids $POO^{-}(CH_2)_2\ \overset{+}{N}R_3$	PCILO Ab initio	Folded Folded	Extended
Serotonine $(C_8N\ HOH)-(CH_2)_2-\overset{+}{N}H_3$	PCILO	gauche	gauche/trans
Bufotenine $(C_8NHOH)-(CH_2)_2-\overset{+}{N}(CH_3)_3$	PCILO	gauche	gauche

REFERENCES

Alagona, G., R. Cimiraglia, E. Scrocco and J. Tomasi (1972) Theor. Chim. Acta, 25, 103.
Alagona, G., A. Pullman, E. Scrocco and J. Tomasi (1973) Int. Journal of Peptide and Protein Research, 5, 251.
Berthod, H. and A. Pullman (1975) Chem. Phys. Letters, 32, 233.
Berthod, H. and A. Pullman (1976) (in preparation).
Beveridge, D.L., M.M. Kelly and R.J. Radna (1974) J. Amer. Chem. Soc. 96, 3769.
Beveridge, D.L. and G.W. Schnuelle (1975) J. Phys. Chem. 79,2562.
Bonaccorsi, R., C. Petrongolo, E. Scrocco and J. Tomasi (1971) Theor. Chim. Acta, 20, 331.
Brown, R.D. and C.A. Coulson (1958) in "Le Calcul des Fonctions d'Ondes Moléculaires" C.N.R.S. Paris, p. 313.
Burch, J.L., K.S. Raghuveer and R.E. Christoffersen (1976) in "Environmental Effects on Molecular Structure and Properties" 8th Jerusalem Symposium, B. Pullman (Ed.) D. Reidel Publishing Company, Dordrecht, Holland, p. 17.
Coulson, C.A. (1960) Proc. Roy. Soc. A, 255, 69.
Cremaschi, P., A. Gamba and M. Simonetta (1972) Theor. Chim. Acta, 25, 237.
Cremaschi, P., A. Gamba and M. Simonetta (1973) Theor. Chim. Acta, 31, 155.
Cremaschi, P. and M. Simonetta (1975a) Theor. Chim. Acta, 37, 31.
Cremaschi, P. and M. Simonetta (1975b) J. of Mol. Structure, 29,39.
Del Bene, J.E. (1975) J. Chem. Phys. 63, 4666.
Diner, S., J.P. Malrieu, F. Jordan and M. Gilbert (1969) Theor. Chim. Acta, 15, 100.
Dreyfus, M. and A. Pullman (1970) Theor. Chim. Acta, 19, 20.
Falk, M., K.A. Hartman, Jr. and R.C. Lord (1963) J. Amer. Chem. Soc. 85, 387.
Frost, A.A. (1967) J. Chem. Phys. 47, 3707.
Giessner-Prettre, C. and B. Pullman (1974) J. Theor. Biol., 48, 425.
Hartman, K.A., R.C. Lord and G.J. Thomas (1973) in "Physical Chemical Properties of Nucleic Acids", J. Duchesne (Ed.) Academic Press, New York, vol. 2, p.1.
Hinton, J. (1967) Communication at this congress and submitted to J. Amer. Chem. Soc.
Hogg, A.M., R.M. Haynes and P. Kebarle (1966) J. Amer. Chem. Soc. 88, 28.
Hogg, A.M. and P. Kebarle (1965) J. Chem. Phys. 43, 449.
Hopfinger, A.J. (1973) in "Conformational Properties of Macromolecules, Acad. Press, New York.
Hylton, J., R.E. Christoffersen and G.G. Hall (1974) Chem. Phys. Letters, 26, 501.
Hylton Mc Creery, J., J., R.E. Christoffersen and G.G. Hall (1975) in press,preprint.
Iwata, S. and K. Morokuma (1973) J. Amer. Chem. Soc. 95, 7563.

Iwata, S. and K. Morokuma (1974) J. Amer. Chem. Soc., 97, 966.
Jortner, J. (1962) Mol. Phys. 5, 257.
Jortner J. and C.A. Coulson (1961) Mol. Phys. 4, 451.
Kirkwood, J.G. (1935) J. Chem. Phys., 2, 351.
Kirkwood, J.G. (1939) J. Chem. Phys. 7, 911.
Kirkwood, J.G. and F.H. Westheimer (1938) J. Chem. Phys., 6, 506.
Kollman, P.A. and L.C. Allen (1972) Chem. Rev. 72, 283.
Kwiatkowski, S. (1975) Studia Biophysica 46, 79.
Kwiatkowski, S. and B. Pullman (1976) Theor. Chim. Acta, in press.
Newton, M.C. (1973) J. Chem. Phys. 58, 5833.
Newton, M.C. (1975) J. Phys. Chem. 79, 2795.
Newton, M.D. and S. Ehrenson (1971) J. Amer. Chem. Soc., 93, 4971.
Newton, M.D. (1973) J. Amer. Chem. Soc. 95, 256.
Onsager, L. (1936) J. Amer. Chem. Soc., 58, 1486.
Payzant, J.D., A.J. Cunningham and P. Kebarle (1973) Can. J. Chem. 51, 3242.
Perahia, D., A. Pullman and H. Berthod (1975) Theor. Chim. Acta 40, 47.
Perahia, D., B. Pullman and A. Saran (1974) Biochim. Biophys. Acta, 340, 299.
Perricaudet, M. and A. Pullman (1973) FEBS Letters, 34, 222.
Port, G.N.J. and A. Pullman (1974) Int. J. of Quant. Chem. Quant. Biol. Symp. 1, 21.
Pullman, A. and A.M. Armbruster (1974) Int. J. of Quant. Chem. S8, 169.
Pullman, A., H. Berthod and N. Gresh (1976) Int. J. of Quant. Chem. S10, in press.
Pullman, A. and A.M. Armbruster (1975) Chem. Phys. Letters, 36, 558.
Pullman, A. and B. Pullman (1975) Quarterly Reviews of Biophysics, 7, 505.
Pullman, B., A. Pullman, H. Berthod and N. Gresh (1975) Theor. Chim. Acta, 40, 93.
Pullman, B. and A. Saran (1976) in "Progress in Nucleic Acid Research and Molecular Biology", in press.
Schuster, P. (1974) in "Water as Liquid and Solvent" W.A.P. Lück (Ed.) (Weinheim).
Searles, S.K. and P. Kebarle (1968) J. Phys. Chem. 72, 742.
Smith, I.C.P., H.H. Mantsch, R.D. Lapper, R. Deslauriers and T. Scheich, (1973) in "Conformation of Biological Molecules and Polymers", Bergmann, E.D. and B. Pullman (Eds.), 5th Jerusalem Symposium on Quantum Chemistry and Biochemistry, New York, Academic Press, p. 381.
Tapia, O. and O. Goscinski (1975) Molecular Physics.
Tapia, O., E. Poulain and F. Sussman (1975) Chem. Phys. Letters, 33, 65.

QUANTUM PHARMACOLOGY[1a]

Ralph E. Christoffersen[1b] and Raffaella Pavani Angeli[1c]

Department of Chemistry, University of Kansas,
Lawrence, Kansas 66045 USA

INTRODUCTION

When the formulation of computational quantum mechanics for closed shell molecular systems using Hartree-Fock theory was first given[2,3], there was perhaps a vague hope that, along with the development of efficient computers, a new kind of "analytical tool" might be developed for use on problems of chemical interest. However, few persons would have been brazen enough to suggest that, at an International Congress on Quantum Chemistry just 25 years later, a discussion concerning the current status of "Quantum Pharmacology" would take place. As the following discussion will hopefully indicate, very rapid progress has occurred, both in the development and application of techniques in this area in recent years, and further advances can be expected.

Of particular relevance to these developments are the advances that have and are taking place in the general area of "experimental molecular biology". These developments, including pharmacologic receptor isolation and characterization, as well as the development of a variety of new experimental and theoretical probes, provide the potential for development of truly "rational drug design" programs, that indicate clearly the conceptual, practical, and pragmatic potential for application of theoretical and other techniques to problems in biology.

In the following sections, several aspects of these developments will be assessed. Included is a discussion of the relationship of quantum mechanical studies to the overall problem of drug design, a review of techniques and areas of recent application, and some indications of current studies and trends.

B. Pullman and R. Parr (eds.), The New World of Quantum Chemistry, 189–210.

CURRENT STATUS: METHODOLOGY AND APPLICATIONS

To place this discussion within perspective, it is of interest first to review some of the major factors that are known to influence overall drug action. These include:

1. Absorption
2. Excretion
3. Catabolism
4. Binding to Plasma Protein
5. Penetration of the Blood-Brain Barrier
6. Affinity for the Receptor
7. Intrinsic Activity

Unfortunately, each of these factors (and perhaps others) may have a profound effect upon the overall action of a drug, and each must be quantified if a truly "rational" program of drug design is to be attained. As to the application of quantum mechanics to these areas, it is in the areas of receptor affinity and intrinsic activity (usually classified together as drug-receptor interactions) where the primary contributions may be expected.

In order to apply quantum mechanical techniques to the study of these interactions, it is highly advantageous to have detailed information about the chemical nature of both the drug and its receptor. While such characterizations of drugs can frequently be carried out relatively routinely, analogous studies of receptors are not always possible. Hence, before discussing the kinds of quantum mechanical techniques that are available and the kinds of applications that have occurred, it is appropriate first to speak briefly of the state-of-the-art in pharmacologic receptor isolation and characterization.

Advances in Experimental Molecular Biology

Unfortunately, the "rule" (rather than the "exception") in the case of receptor isolation and characterization is that the detailed chemical structure is not usually known. However, recent developments in this area indicate that major advances can be expected in the near future.

For example, the lactose repressor protein (that can be considered to be the "drug" in this context) selectively chooses one out of six million nucleotide sequences in the E coli genome (the "receptor" in this context), and binds it to prevent expression of the genes for lactose metabolism in pancreatic DNase. In 1973, two reports appeared concerning this system[4,5] that give an indication of the substantial progress that has been made in experimental molecular biology, and portends things to come. The first of these[4] was the identification of both the chemical nature and sequence of amino acids in the lac repressor protein. As seen in

Table I. Amino Acid Composition of the Lac-Repressor Protein and Nucleotide Sequence in the Lac Operator.

AMINO-ACID COMPOSITION OF THE LAC-REPRESSOR

AMINO ACID	NO. OF RESIDUES	AMINO ACID	NO. OF RESIDUES
LYSINE	11	GLYCINE	22
HISTIDINE	7	ALANINE	44
ARGININE	19	CYSTEINE	3
ASPARTIC ACID	15	VALINE	33
ASPARAGINE	11	METHIONINE	9
THREONINE	18	ISOLEUCINE	17
SERINE	30	LEUCINE	40
GLUTAMIC ACID	13	TYROSINE	8
GLUTAMINE	28	PHENYLALANINE	4
PROLINE	13	TRYPTOPHAN	2

TOTAL NUMBER OF RESIDUES = 347

NUCLEIC ACID SEQUENCE OF THE LAC OPERATOR

5' --- T G G [A A T T G T] G [A] G [C] G [G] A [T] A [A C A A T T] 3'

3' --- A C C [T T A A C A] C [T] C [G] C [C] T [A] T [T G T T A A] 5'

Table I, 347 residues are involved, with 20 different kinds of amino acids present. In the following paper[5], the nucleotide sequence of the region (called the "lac operator" region) of the particular DNA with which the lac repressor interacts was determined. As also seen in Table I, the operator region is a double stranded segment of DNA that is approximately 27 base-pairs long. In addition, it has a palendromic symmetry associated with it, as indicated by the regions inside the boxes, that involves a 2-fold symmetry with respect to a single turn of the DNA double helix. These were important contributions, for they outlined the chemical identities of both the "drug" (i.e., the repressor protein) and its "receptor" (i.e., the operator region of the particular DNA of interest).

In addition to these chemical characterizations, additional geometric data have also become available. For example, microcrystals of the lac repressor protein have been prepared and examined using electron microscopic and powder X-ray diffraction techniques.[6] From these studies, a new model for the repressor-operator interaction was proposed that is consistent with both the asymmetric shape of the repressor protein and the 2-fold palendromic symmetry found in the operator. As illustrated in Figure 1, the proposed model envisages the repressor acting as a tetramer, and binding to the operator region of DNA with its long axis aligned with the long axis of DNA.

Figure 1. Proposed lac repressor-operator complex.

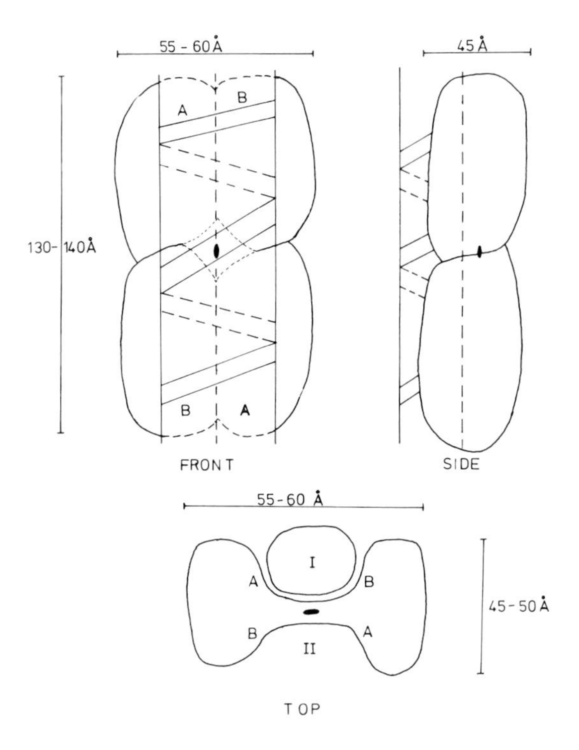

Detailed studies such as indicated above are not limited only to the lac repressor-operator system. For example, substantially larger operator regions in the λ-phage DNA have been isolated and sequenced.[7] In addition, other kinds of receptors, including enzymatic[8,9] and others[10], are being characterized at the molecular level.

Studies such as these have important implications, since they indicate that drug-receptor interactions can now be probed in detail at molecular resolution, at least for some cases. The importance of developments such as these to quantum mechanical and other kinds of studies of drug binding and intrinsic activity is also clear, since it implies that, at least for some systems, characterization and development of structure-activity relationships need no longer be based solely on information concerning the drug alone.

Quantum Mechanical Studies of Drug Design

Turning next to the application of quantum mechanical techniques to systems of pharmacological interest, it is appropriate first to identify the techniques in use. These include elementary Hückel theory[11] (HMO), extended Hückel theory[12] (EHT), iterated extended Hückel theory[13,14] (IEHT), empirical potentials[15], CNDO[16], INDO[17], PCILO[18], X-α[19], MINDO[20], the *ab initio* molecular fragment approach[21], and other *ab initio* techniques[22,23]. In addition, a variety of techniques have been developed for investigation of the effect of solvent on molecules[24-40], that are discussed in detail in the preceding article.

As to the application of these techniques to drug design, a comprehensive review through the end of calendar year 1974 has recently been given.[41] In this review, all articles that used techniques more sophisticated than elementary Hückel theory were sought. This gives rise to a period of approximately 10 years of applications, after the introduction of extended Hückel theory in 1963-64.[12] During this period, a total of 130 papers in 22 different journals or books was found, covering the 27 different areas listed in Table II.

It should be noted that the number of studies in the various areas is far from uniform. For example, the literature during this period is dominated primarily by studies on agents that affect nerve function, with particular emphasis on adrenergics,

Table II. Areas of Application of Quantum Mechanics in Drug Design.

Adrenergics	Anticoagulants
Analgetics	Vasopressor Agents
Anesthetics	Steroid Hormones
Antiarrhythmic	Prostaglandins
Anticonvulsants	Polypeptide, Protein, and other Hormones
Antidepressants	Thyroid Agents
Antiparkinsonian	Hypoglycemic Agents
Antipsychotic	Antibiotics
Cholinergic and anticholinergic	Antiallergenic Agents
Emetics and antiemetics	Carbohydrates
Hallucinogens	Vitamins
Hypnotics	Insecticides
Hypotensive and Hypertensive Agents	Antiinflammatory
	Antisecretory

analgesics, cholinergics, antipsychotics, hallucinogens, and tryptaminergics. In addition, with the exception of the studies using the PCILO techniques, virtually all studies were done within the framework of Hartree-Fock (i.e., molecular orbital) theory. Thus, the main information of interest that could be gleaned from these studies was related to ground state equilibrium properties, such as conformational analyses, charge distribution, molecular orbital structure, etc. In addition, most studies during this period were done on isolated molecules, in which both solvent effects and interactions with a receptor were neglected.

As an example of the kind of information that was obtained in studies during this period and how it can be used in drug design, it is useful to consider the class of molecules known as hallucinogens. In this case, only the drugs themselves were studied, because of the absence of detailed data concerning the hallucinogenic receptor(s). However, some 14 different quantum mechanical studies[41] of molecules in this class took place through 1974, including studies of geometric structure, electronic structure, and the effect of solvent on the allowed conformations of the molecules.

To illustrate the nature of the geometric information that was obtained, the PCILO studies on serotonin[42,43] provide a good example. In Figure 2, the serotonin molecule (5-hydroxy-tryptamine) in its cationic form is depicted, along with conformational energy maps for the free molecule and the molecule in its "solvated" form

Figure 2. Conformational Energy Maps for Isolated and "Solvated" Cationic Serotonin. Contour levels are given in units of kcal/mole.

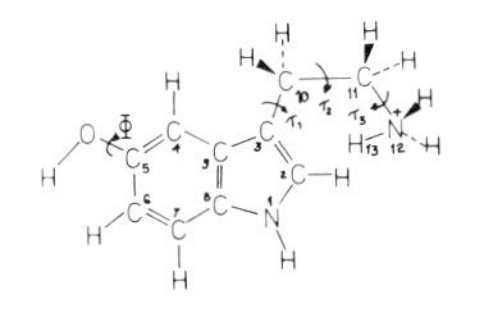

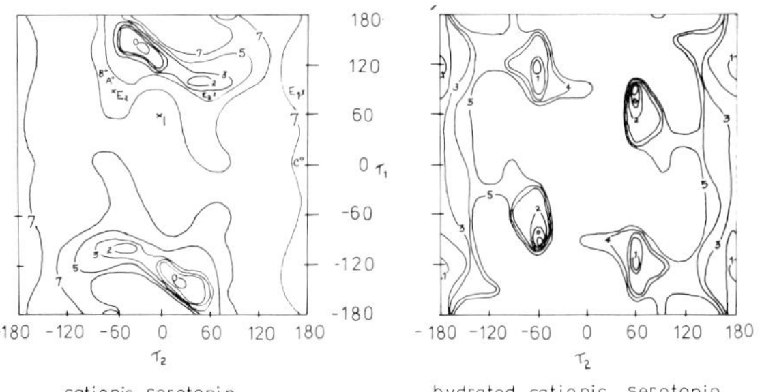

with three water molecules placed around the cationic ammonium moiety. The particular angles of interest in these studies were τ_1 and τ_2, two of the dihedral angles that describe the conformation of the side chain.

For the free molecule, it is seen from Figure 2 that the minimum energy is one in which $\tau_1 \cong 140°$ and $\tau_2 \cong -20°$, with a secondary minimum 2 kcal/mole above the global minimum at $\tau_1 \cong 100°$ and $\tau_2 \cong 60°$. Minima calculated via other procedures are also shown (E_1, E_2, X_1)[44,45] as well as the values for these dihedral angles found in several different crystallographic studies (A, B, C).[46,47] Clearly, not all methods provide equivalent results.

In contrast to the <u>gauche</u> form that predominates the free molecule conformational energy map, the corresponding map done in the presence of three water molecules gives rise to near equivalence of both the <u>gauche</u> and <u>trans</u> forms, which is consistent with NMR results in aqueous solution.[48] Thus, while there remain some problems with methodology, examples such as these indicate that a reasonably detailed description of important geometric features of molecules of interest can be obtained, both theoretically and experimentally.

To illustrate the use to which electronic structural information has been made, it is useful to consider INDO studies[49] on a series of eighteen molecules related to serotonin. In these investigations, various measures of electronic structural features were calculated from the SCF wavefunctions, and a systematic investigation was carried out to ascertain which possible parameters might correlate statistically with <u>in vitro</u> data concerning the ability of the molecules to contract rat fundus strips (relative to serotonin). As a result, the following equation was found to provide the best correlation with the available data:

$$\log (1/c) = 18.09 f_1 - 74.44 q_1 + 1.182 \pi_7 - 13.061 ,$$

with a correlation coefficient of 0.962, where f_1 is the frontier density on atom 1, q_1 is the net total charge on atom 1, and π_7 is the hydrophobic substituent constant for C_7. While it is not possible with certainty to deduce receptor characteristics and/or mechanism of action from such a statistical relationship, it is certainly a very useful relationship from the point of view of possible directions for synthesis of new agents with high efficacy. Thus, even when the nature of the receptor is not known, it is possible in some cases to deduce quite useful statistical relationships. When it is considered that perhaps one in ten to fifteen thousand molecules may actually prove to be a successful drug, use of relationships such as these to increase the cost-effectiveness of synthesis programs (even in the absence of detailed receptor information) has obvious advantages and desirability.

Before leaving the question of what kinds of studies have and can be carried out, it is useful to review the work that has appeared in 1975 in this area. Using quantum mechanical techniques at a level of sophistication greater than elementary Hückel theory, there have been 24 studies[50-73] that relate reasonably directly to problems of drug design. Many other studies on simpler model systems have also occurred, but have been omitted for brevity. The studies included here have been on systems that include adrenergics[51,64], analgesics[53,54,60,62,63,66], antibiotics[58] anti-cancer[61], anti-allergy[59,67], cholinergics[52,55], enzymes and enzyme-substrate interactions[69,73], DNA[57,70,71], and membranes[50,56,65,68,72]. Of these studies, those on analgesics and enzyme-substrate interactions are of interest to describe more fully, due to the relatively large number of studies in the analgesic area during this period, and the indication of possible new directions for computational studies in the future as illustrated by the enzyme-substrate interaction studies.

In the area of analgesics, the molecules that were studied are depicted in Figure 3. Three different techniques were used for most of the studies, <u>i.e.</u>, PCILO, INDO, and <u>ab initio</u> (STO-3G), and the protonated forms of these molecules were most commonly used. The PCILO and INDO studies were used to explore conformational degrees of freedom, while the <u>ab initio</u> studies were designed primarily to extract electronic structural features of interest.

In morphine, nalorphine, and N-phenethylmorphine studies[54], conformational degrees of freedom in the vicinity of the quaternary nitrogen were examined. For low-energy conformers, electronic structural features for both the protonated and unprotonated forms were obtained. Two types of low-energy conformers were found that were suggested to be related to agonist and antagonist behavior. In <u>ab initio</u> studies of morphine and nalorphine[62], the molecular orbital structure and charge distribution were examined, and compared with photoelectron spectra. Among other things, it was found that the charge distribution around the quaternary nitrogen was remarkably insensitive to substitution at that nitrogen.

In the case of meperidine, prodine, and desmethylprodine[53], PCILO studies investigating the dependence of energy on τ_1 and τ_2 were carried out. The results were consistent with X-ray data, but also indicated additional low-energy conformers. In general, phenyl equatorial conformers were favored over phenyl axial conformers, and these data were used to suggest that 4-phenylpiperidine narcotics act in an equatorial conformation at the same cationic receptor site as the rigid opiates.

In other studies[60], ESCA spectra was correlated with electron density calculations based on CNDO/2 studies for morphine, nalorphine, oxymorphone, naloxone, levorphanol, levallorphan, methadone,

Figure 3. Analgesic Agents Studied.

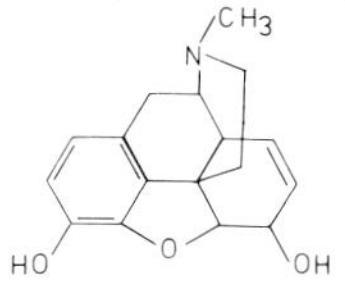

Morphine

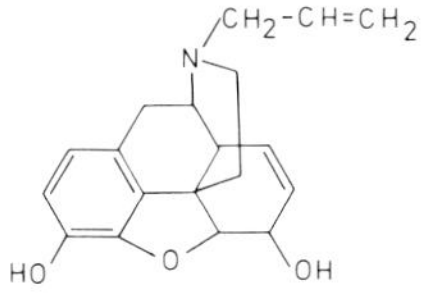

Nalorphine

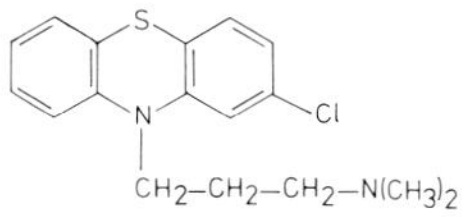

Chlorpromazine

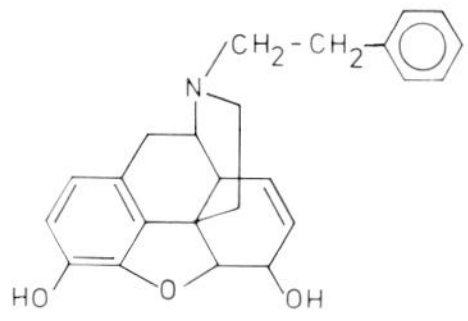

N-phenethylmorphine

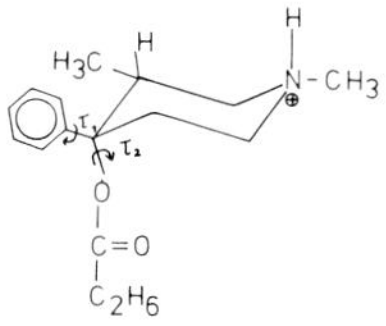

Prodine

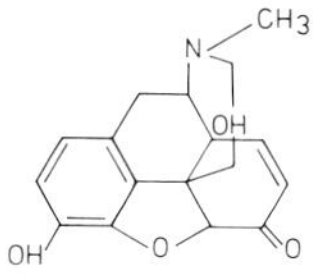

Hydromorphone

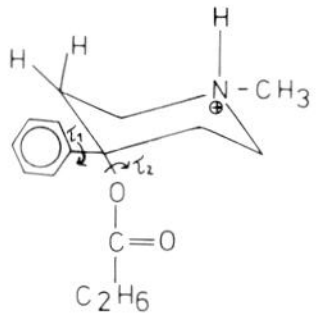

Desmethylprodine

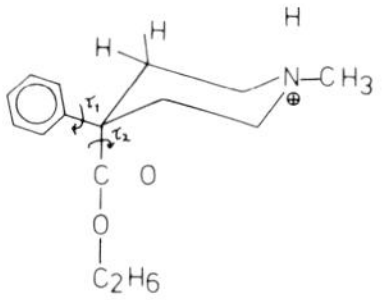

Meperidine

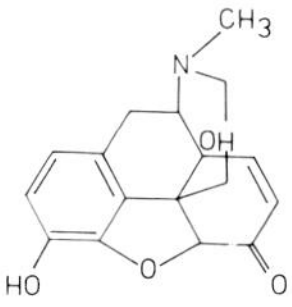

Oxymorphone

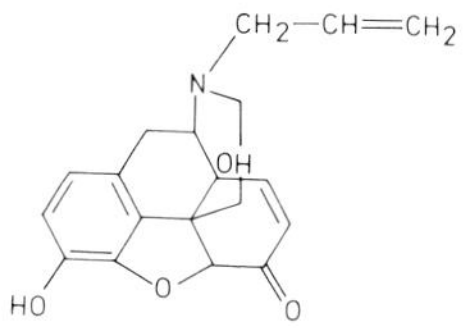

Naloxone

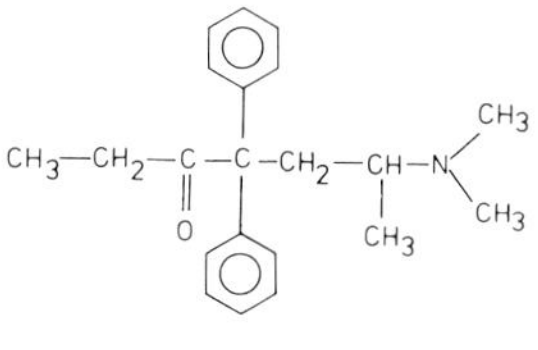

Methadone

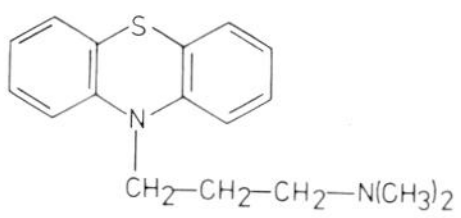

Promazine

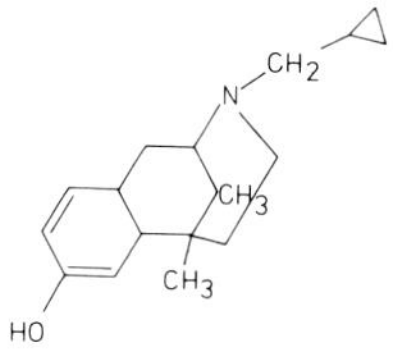

Cyclazocine

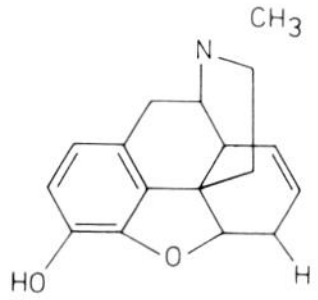

Levorphanol

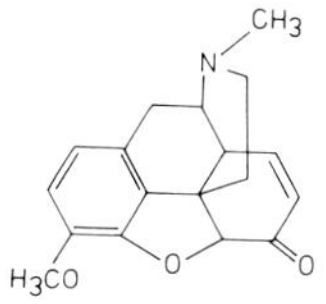

Hydrocodone

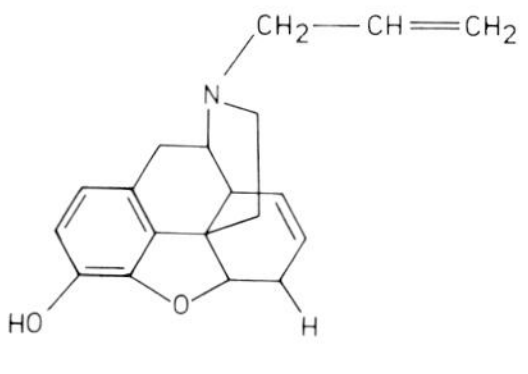

Levallorphan

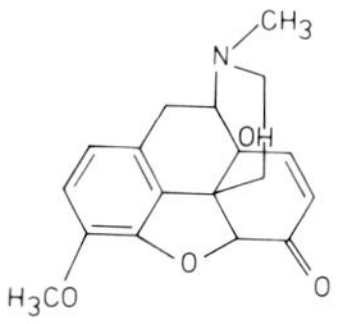

Oxycodone

and cyclazocine. Measured pK_a values were also related to calculated densities, and the degree of localization of positive charge around the nitrogen nucleus and its possible relationship to narcotic activity was discussed. Trends in pK_a values were also studied in oxymorphone, hydromorphone, oxycodone, and hydrocodone, using the PCILO technique, including a discussion of the possible effects of the solvent.[66]

While studies such as these cannot be expected, by themselves, to characterize all of the conformational degrees of freedom or electronic structural features of possible interest, they do illustrate several important points. First, relatively extensive conformational studies on relatively large drugs are now reasonably straightforward to carry out. Next, for the specific case of narcotic agonists and antagonists, a reasonably detailed understanding of both geometric and electronic structural features of the isolated drugs is emerging. This is particular significance in light of the progress being made on isolation and characterization of the opiate receptor[74-76], and the recent identification of enkephalin as a structurally unrelated molecule that apparently also acts at the analgesic receptor.[77] One can expect very interesting insights from studies of interactions among these moieties in the near future.

The other type of quantum mechanical study that is of interest to note from publications in 1975 is one that may be indicative of the kinds of studies that may be possible in a variety of areas in drug research in the not-too-distant future. The particular study of interest in this regard is a study of the charge relay system and tetrahedral intermediate in the acylation of serine proteinases[69]. In these studies, approximate molecular orbital calculations using the PRDDO technique[78] were carried out on a model of the active site of serine proteinases (e.g., chymotrypsin, trypsin), which is the "receptor" in this case, interacting with a model substrate, which is the "drug" in this case. The particular system studied included the active site residues Ser-195, His-57, and Asp-102, that were modeled by methanol, imidazole, and formic acid, respectively. The substrate was modeled by formamide, and the entire model system is depicted in Figure 4.

The basic question of interest involves the role (or lack thereof) of the Asp-102 and His-57 moieties in the reaction of Ser-195 with the substrate to form a tetrahedral intermediate. To investigate this, the first step was to optimize the position of the four groups relative to each other. One of the main features of interest that results is the placement of $H^{\delta 1}$ so that its distance from the two oxygen atoms of the formate anion differ only by 0.1Å, which is in agreement with experimental findings[79] of a bifurcated hydrogen bond between His-57 and Asp-102. The other point of interest is that the calculated Mulliken overlap

Figure 4. Model of the Charge Relay System in Chymotrypsin.

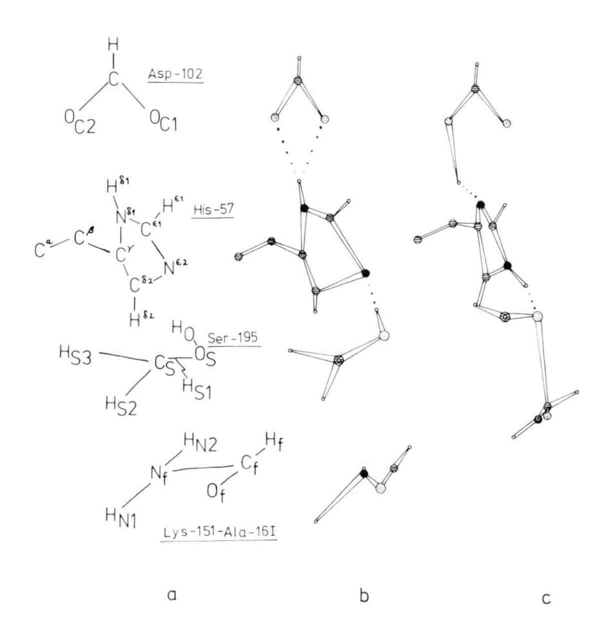

population for the $H^{\delta1}$ — Oa_2 pair is nearly four times that of the $H^{\delta1}$ — Oa_1 pair. This latter observation led to transfer of $H^{\delta1}$ to Oa_2 in the subsequent study of the charge transfer step.

Investigation of the charge transfer step, that results ultimately in the activation of serine as a nucleophile, indicated that a transfer of H_O from Ser-195 to His-57 accompanied by simultaneous transfer of $H^{\delta1}$ to Asp-102 was substantially superior (energetically) to transfer of the $H^{\delta1}$ proton to His-102 first, followed by transfer of H_O to $N^{\epsilon2}$. It is also of interest that, during this process, the imidazole ring becomes partially negatively charged, and the methoxy group of Ser-195 becomes a powerful nucleophile (i.e., contains substantial negative charge) by the end of the charge transfer step. Thus, the calculations support the contention that Asp-102 is the ultimate proton acceptor in the charge relay system, and provide a rationale as to why protonation of Asp-102 at pH $<\sim 7$ destroys enzymatic activity.

Subsequent attack of the methoxide ion on formamide to form a tetrahedral intermediate was also investigated, which indicates transfer of negative charge from Asp-102 to the carbonyl oxygen of formamide, accompanied by a weakening of the C_f — O_f and the C_f — N_f bonds. In addition, the degree of concertedness between the charge relay step and the formation of a tetrahedral intermediate was investigated. Substantial energetic advantage was found for the nonconcerted pathway, with attack of the seryl oxygen on the peptide linkage not beginning until after the proton transfer has nearly been completed.

Of course, many uncertainties necessarily remain as a result of these initial studies, such as incomplete variation of geometric degrees of freedom and the suitability of investigation of such reaction surfaces without inclusion of correlation effects. In spite of these limitations, these studies indicate that an important new kind of study, i.e., of the actual "drug-receptor" complex, is now possible at least in some cases.

CURRENT ACTIVITIES AND FUTURE DIRECTIONS

To provide an additional indication of the direction that can be expected in the future, a description of long term studies that have been begun in our laboratory in anti-cancer drug design will be described.

While it is not yet possible to assign the entire sequence of bases in a particular DNA, the sequencing and characterization of important regions of DNAs is rapidly becoming possible. In light of such an observation, and with the expectation that characterization of regions of DNAs that are relevant to viral oncogenesis amy take place in the not-too-distant future, a long-term investigation of interactions of drugs with DNA components has been started. The ultimate aim of such studies is to elucidate the nature and magnitude of the interactions that occur so that, when binding to a specific sequence is desired, synthetic strategies will be available to allow synthesis of drugs that will bind selectively to the specific sequence of interest.

As a first step in such a long term study, an investigation of the individual bases and several drugs is being carried out, so that both methodology and properties of individual molecules can be established.

The particular molecules studied include adenine, thymine, guanine, cytosine, uracil, and ethidium. The last molecule can be considered to be an example of a drug, and the first five are components of the receptor. These molecules are depicted in Figure 5, and the geometric data used in the calculations were obtained from X-ray crystallographic studies[80-82].

The quantum mechanical method used to characterize the electronic features of these drugs is the molecular fragment procedure[21], an *ab initio* procedure in which basis orbitals for large molecular

Figure 5. Depiction and Atomic Numbering of Nucleic Acid Components and Ethidium.

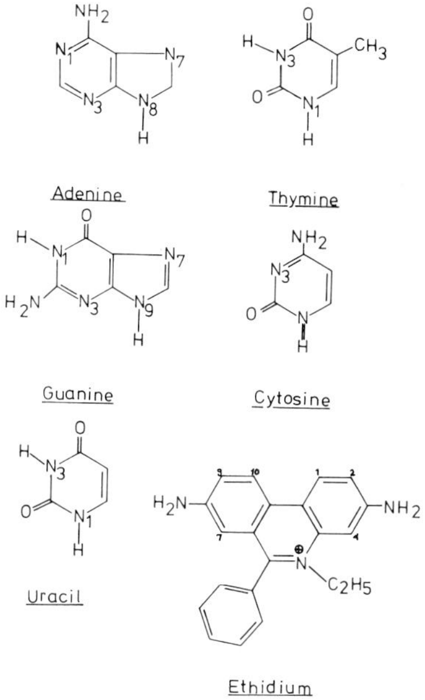

Adenine

Thymine

Guanine

Cytosine

Uracil

Ethidium

calculations are obtained by examination of smaller molecular fragments. The basis functions are floating spherical Gaussian orbitals (FSGO), defined as

$$G_i(\underset{\sim}{r}) = (2/\pi\rho_i^2)^{3/4} \exp\{-(\underset{\sim}{r} - \underset{\sim}{R}_i)^2/\rho_i^2\} ,$$

where ρ_i is the "orbital radius" of G_i, and $\underset{\sim}{R}_i$ is the location of the FSGO, relative to some arbitrary origin. The fragment parameters used in these studies have been described earlier.[83]

Among the calculated electronic structural features of interest for these molecules is the molecular orbital structure, and several of the high-lying filled and low-lying unfilled molecular orbital energies for these molecules are given in Figures 6 and 7. These quantities are of particular interest in this case, since they allow not only estimation of a variety of physical and chemical features of interest, but also allow for an assessment

Figure 6. Energy of High-Lying Filled and Low-Lying Unfilled Molecular Orbitals in Adenine, Thymine, Uracil, Guanine and Cytosine.

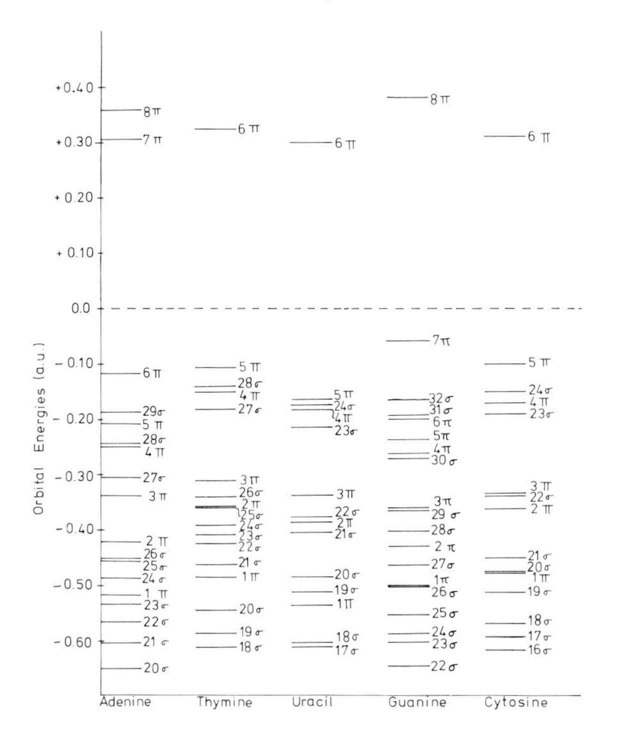

Figure 7. Energy of High-Lying Filled and Low-Lying Unfilled Molecular Orbitals in Ethidium.

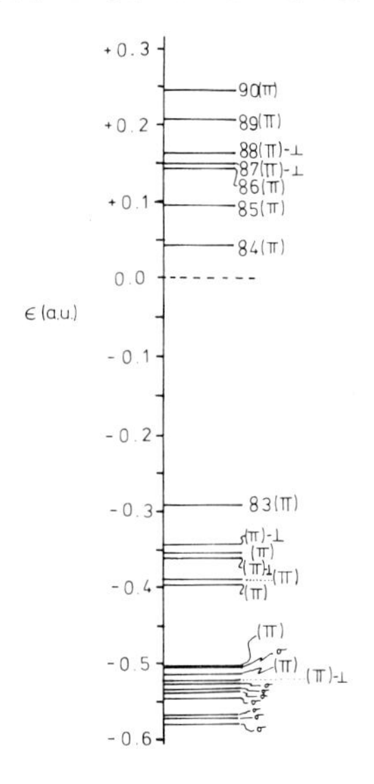

of the methodology to be used. This latter point is of substantial importance since, for systems of the size to be encountered in studies such as these, only relatively small basis sets are computationally feasible to use.

Such a calibration is possible in this case for adenine, thymine, guanine, and cytosine, since other *ab initio* studies on them have been reported[84-88]. One of the most interesting comparisons that can be made to achieve the desired calibration is that of the relative ordering and spacing of molecular orbital energies that results from using different basis sets. For the particular case of interest, a plot of the filled valence molecular orbital energies calculated using the molecular fragment basis versus the corresponding energies calculated by Clementi, *et al.*,[84] is given in Figure 8. The obvious conclusion from such a graph is that the orbital energies of the two basis sets are linearly related. In other words, not only is the relative order of molecular orbitals predicted to be the same, but a linear mapping relates the orbital energies of one basis to the corresponding orbital energies of the

Figure 8. A Plot of Filled Valence Molecular Orbital Energies for Adenine Obtained Using the Molecular Fragment Technique and a Reference Calculation.

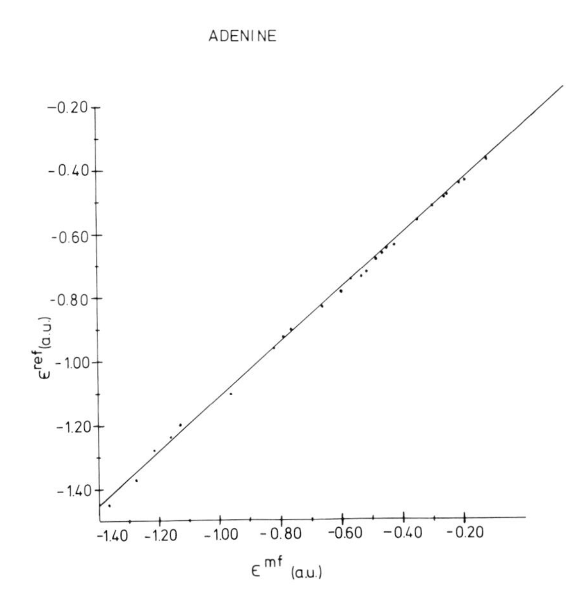

Table III. Linear Regression Equations Obtained for Orbital Energies in Adenine, Thymine, Guanine, and Cytosine.

	Equation	Standard Deviation	Correlation Coefficient	Number of Points
Adenine:				
	$\epsilon^{ref} = 0.8521 \cdot \epsilon^{mf} - 0.2676$	0.0072	0.9992	25
Thymine:				
	$\epsilon^{ref} = 0.8426 \cdot \epsilon^{mf} - 0.3089$	0.0112	0.9981	24
Guanine:				
	$\epsilon^{ref} = 0.8429 \cdot \epsilon^{mf} - 0.2923$	0.0081	0.9988	28
Cytosine:				
	$\epsilon^{ref} = 0.8472 \cdot \epsilon^{mf} - 0.2881$	0.0084	0.9990	21

other basis. Expressed more quantitatively, Table III contains data obtained from a linear regression analysis on the filled valence orbitals of the two basis sets. As the data indicate, a very good linear relationship is obtained, with standard deviations typically below 1% and correlation coefficients greater than 0.99. In addition, the slopes of the lines are all remarkably similar, as illustrated in Figure 9. Thus, it appears that not only is the relative balance of the basis sets the same for

Figure 9. Orbital Energy Linearity Relationships Obtained for Adenine, Guanine, Cytosine, and Thymine.

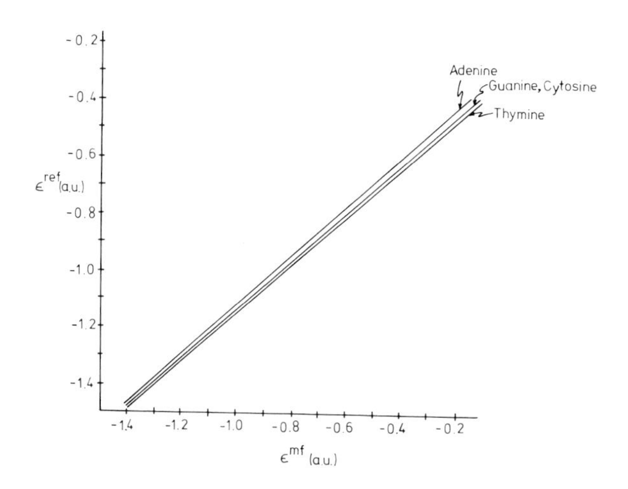

a given molecule, but also among all four molecules. This is very encouraging, for it indicates that, even though small basis sets are required for computational feasibility when drug-receptor interactions are considered, the relative balance of the basis is expected to be the same as when a substantially larger basis is used.

Much additional analysis is currently underway on these molecules and their interactions, using _ab initio_ calculations, point charge models, and a variety of other techniques. However, it would appear that, not only in the anti-cancer area but in many others as well, substantial new kinds of insight into drug-receptor interactions with a concomitant impact on and impetus to drug design can be expected in the relatively near future. Thus, not only statistically-based structure-activity relationships using drug molecules alone can be expected, but true structure-function relationships may be possible through examination of the actual drug-receptor interactions.

ACKNOWLEDGEMENTS

The authors would like to express their appreciation to the University of Kansas for partial support of the computing time required for this work. Also, the collaboration and help from G. M. Maggiora, G. G. Hall, B. V. Cheney, and D. Spangler on several aspects of this work is deeply appreciated. Finally, support from the Istituto Ricerche G. Donegani, Montedison for Raffaella Pavani Angeli while at the University of Kansas is gratefully acknowledged.

REFERENCES

1. a) Supported in part by a grant from the Upjohn Company, Kalamazoo, Michigan 49001. b) Author to whom correspondence should be directed. c) Permanent address: Istituto Ricerche G. Donegani, Montedison, Novara, Italy.

2. C. C. J. Roothan, Rev. Mod. Phys., 23, 69 (1951).

3. G. G. Hall, Proc. Roy. Soc., A205, 541 (1951).

4. K. Beyreuther, K. Adler, N. Geisler, and A. Klemm, Proc. Nat. Acad. Sci. (US), 70, 3576 (1973).

5. W. Gilbert and A. Maxam, Proc. Nat. Acad. Sci. (US), 70, 3581 (1973).

6. T. A. Seitz, T. J. Richmond, D. Wise, and D. Engelman, Proc. Nat. Acad. Sci. (US), 71, 593 (1974).

7. V. Pirrotta, Nature, 254, 114 (1975). See also A. Walz and V. Pirrotta, Nature, 254, 118 (1975).

8. T. A. Seitz, R. Henderson, and D. M. Blow, J. Mol. Biol., 46, 337 (1969).

9. W. N. Lipscomb, J. A. Hartsuck, G. N. Reeke, Jr., F. A. Quiocho, P. H. Bethge, M. L. Ludwig, T. A. Seitz, H. Muirhead, and J. C. Coppola, Brookhaven Symposia in Biology, 21, 24 (1968).

10. L. I. Lowney, K. Schulz, P. J. Lowrey, and A. Goldstein, Science, 183, 749 (1974).

11. See, for example, "Quantum Biochemistry", by B. Pullman and A. Pullman, Interscience, N. Y., 1963.

12. L. L. Lohr and W. N. Lipscomb, J. Chem. Phys., 38, 1607 (1963). See also R. Hoffman, J. Chem. Phys., 39, 1397; 40, 2047, 2474, 2480, 2745 (1964).

13. R. Rein, N. Fukudo, H. Win, G. E. Clarke, and F. Harris, J. Chem. Phys., 45, 4743 (1966).

14. D. G. Carrol, A. T. Armstrong, and S. P. McGlynn, J. Chem. Phys., 44, 1865 (1966). See also M. Zerner and M. Gouterman, Theoret. chim. Acta, 4, 44 (1966).

15. See, for example, R. F. McGuire, F. A. Momany, and H. A. Scheraga, J. Phys. Chim., 76, 375 (1972).

16. J. A. Pople, D. P. Santry, and G. A. Segal, J. Chem. Phys., 43S, 129 (1965); J. A. Pople and G. A. Segal, J. Chem. Phys., 43S, 136 (1965); 44, 3289 (1966).

17. J. A. Pople, D. L. Beveridge, and P. A. Dobosh, J. Chem. Phys., 47, 2026 (1967). See also "Approximate Molecular Orbital Theory", by J. A. Pople and D. L. Beveridge, McGraw-Hill, N. Y., 1970.

18. S. Diner, J. P. Malrieu, F. Jordan, and M. Gilbert, Theoret. chim. Acta, 15, 100 (1969).

19. J. C. Slater and K. H. Johnson, Phys. Rev., B5, 844 (1972).

20. M. J. S. Dewar and E. Haselbach, J. Amer. Chem. Soc., 92, 590 (1970); N. Bodor, M. J. S. Dewar, A. Harget, and E. Haselbach, J. Amer. Chem. Soc., 92, 3854 (1970).

21. R. E. Christoffersen and G. M. Maggiora, Chem. Phys. Letters, 3, 419 (1969). See also D. Spangler, R. McKinney, G. M. Maggiora, L. L. Shipman, and R. E. Christoffersen, Chem. Phys. Letters, 36, 427 (1975) and references contained therein for earlier papers in this series.

22. W. J. Hehre, R. F. Stewart, and J. A. Pople, J. Chem. Phys., 51, 2657 (1969); W. J. Hehre, R. Ditchfield, R. F. Stewart, and J. A. Pople, J. Chem. Phys., 52, 2769 (1970).

23. E. Clementi, J. Chem. Phys., 46, 4731, 4737, 4485 (1967).

24. See, for example, O. Sinanoglu, Adv. Chem. Phys., 12, 283 (1967), and references 25-40.

25. J. G. Kirkwood, J. Chem. Phys., 2, 351 (1934); J. G. Kirkwood and F. H. Westheimer, J. Chem. Phys., 6, 506 (1936).

26. L. Onsager, J. Amer. Chem. Soc., 58, 1486 (1936).

27. B. Linder, J. Chem. Phys., 33, 668 (1960); Adv. Chem. Phys., 12, 225 (1967).

28. O. Chalvet, R. Daudel, and Peradejordi, in Electronic Aspects of Biochemistry, B. Pullman, editor, Acad. Press, N. Y., 1963, p. 283.

29. G. Klopman, Chem. Phys. Letters, 1, 200 (1967).

30. H. A. Sheraga, Chem. Rev., 71, 195 (1971).

31. M. -J. Huron and P. Clavarie, Chem. Phys. Letters, 9, 194 (1971); J. Phys. Chim., 76, 2123 (1972).

32. A. J. Hopfinger, Macromolecules, 4, 731 (1971). See also Conformational Properties of Molecules, Acad. Press, N. Y., 1973.

33. G. Alagona, A. Pullman, E. Scrocco, and J. Tomasi, Int. J. Peptide Protein Res., 5, 251 (1973).

34. M. D. Newton, J. Chem. Phys., 58, 5833 (1973).

35. H. Popkie, H. Kistenmacher, and E. Clementi, J. Chem. Phys., 59, 1325 (1973).

36. S. Yamabe, S. Kato, H. Fujimoto, and K. Fukui, Theoret. chim. Acta, 30, 327 (1973).

37. H. A. Germer, Jr., Theoret. chim. Acta, 34, 145 (1974).

38. D. L. Beveridge, M. M. Kelly, and R. J. Radna, J. Amer. Chem. Soc., 96, 3769 (1974). See also D. L. Beveridge and G. W. Schneulle, J. Phys. Chim., 78, 2064 (1974).

39. K. Morokuma, S. Iwata, and W. A. Lathan, The World of Quantum Chemistry, R. Daudel and B. Pullman, editors, D. Reidel Publ. Co., Dordrecht-Holland, 1974, p. 277.

40. J. Hylton, R. E. Christoffersen, and G. G. Hall, Chem. Phys. Letters, 24, 501 (1974); J. Amer. Chem. Soc., in press.

41. R. E. Christoffersen, in Quantum Mechanics of Molecular Conformations, B. Pullman, editor, J. Wiley and Sons, N. Y., in press.

42. Ph. Courriere, J. -L. Coubeils, and B. Pullman, Comp. Rend., 272D, 1697 (1971).

43. B. Pullman, Ph. Courriere, and H. Berthod, J. Med. Chem., 17, 439 (1974).

44. L. B. Kier, J. Pharm. Sci., 57, 1188 (1968).

45. S. Kang and M. -H. Cho, Theoret. chim. Acta, 22, 176 (1971). See also S. Kang, C. L. Johnson, and J. P. Green, J. Molec. Struct., 15, 453 (1973).

46. I. L. Karle, K. S. Dragovette, and S. A. Brenner, Acta Cryst., 19, 713 (1965).

47. C. E. Bugg and V. Thewalt, Science, 170, 852 (1970). See also V. Thewalt and C. E. Bugg, Acta Cryst., B28, 82 (1972).

48. R. R. Ison, P. Parington, and G. C. K. Roberts, J. Pharm. Pharmacol., 24, 82 (1972).

49. J. P. Green and S. Kang, Molecular Orbital Studies in Chemical Pharmacology, L. B. Kier, editor, Springer, N. Y., 1970, p. 105.

50. T. D. Davis, R. E. Christoffersen, and G. M. Maggiora, J. Amer. Chem. Soc., 97, 1347 (1975).

51. M. Martin, R. Corbo, C. Petrongolo, and J. Tomasi, J. Amer. Chem. Soc., 97, 1338 (1975).

52. B. R. Gelin and R. Karplus, J. Amer. Chem. Soc., 97, 6996 (1975).

53. G. H. Loew and J. R. Jester, J. Med. Chem., 18, 1051 (1975).

54. G. H. Loew and D. S. Berkowitz, J. Med. Chem., 18, 656 (1975).

55. B. Pullman, H. Berthod, and N. Gresh, Comp. Rend., D280, 1741 (1975).

56. B. Pullman, H. Berthod, and N. Gresh, FEBS Letters, 53, 199 (1975).

57. S. Adams, S. Nir, and R. Rein, Int. J. Quantum Chem., 9, 701 (1975).

58. D. B. Boyd, R. B. Hermann, D. E. Presti, and M. M. Marsh, J. Med. Chem., 18, 408 (1975).

59. L. Farnell, W. G. Richards, and C. R. Ganellin, J. Med. Chem., 18, 662 (1975).

60. L. J. Saethre, T. A. Carlson, J. J. Kaufman, and W. S. Koski, Molec. Pharmacol., 11, 492 (1975).

61. Y. Miyaji, H. Ichikawa, and M. Ogata, Chem. Pharm. Bull., 23, 1256 (1975).

62. H. E. Popkie, W. S. Koski, and J. J. Kaufman, Int. J. Quantum Chem. (1975).

63. H. E. Popkie and J. J. Kaufman, Int. J. Quantum Chem. (1975).

64. G. G. Hall, C. J. Miller, and G. W. Schnuelle, J. Theoret. Biol., 53, 475 (1975).

65. B. Pullman, in Jerusalem Symp. on Quantum Chem. and Biochem., 8, 55 (1975).

66. G. Loew, H. Weinstein, and D. Berkowitz, in Jerusalem Symp. on Quantum Chem. and Biochem., 8, 239 (1975).

67. S. P. Gupta, G. Govil, and R. K. Mishra, J. Theoret. Biol., 51, 13 (1975).

68. B. Pullman, Ph. Courriere, and H. Berthod, Molec. Pharmacol., 11, 268 (1975).

69. S. Scheiner, D. A. Kleier, and W. N. Lipscomb, Proc. Nat. Acad. Sci. (US), 72, 2606 (1975).

70. M. Geller and B. Lesyng, Biochim. Biophys. Acta, 407, 407 (1975).

71. I. Kulakowska, M. Geller, B. Lesyng, K. Bolewska, and K. L. Wierzchowski, Biochim. Biophys. Acta, 407, 420 (1975).

72. T. Weller and H. Frischleder, Chem. and Phys. of Lipids, 15, 5 (1975).

73. D. L. Breen, J. Theoret. Biol., 53, 101 (1975).

74. C. B. Pert and S. H. Snyder, Proc. Nat. Acad. Sci. (US), 70, 2243 (1973).

75. H. H. Loh, T. M. Cho, Y. C. Wu, and E. L. Way, Life Sci., 14, 2231 (1974).

76. A. Goldstein, Life. Sci, 14, 614 (1974).

77. L. Iverson, Nature, 258, 567 (1975).

78. T. A. Halgren and W. N. Lipscomb, J. Chem. Phys., 58, 1569 (1973).

79. M. Krieger, L. M. Kay, and R. M. Stroud, J. Mol. Biol., 83, 209 (1974).

80. M. Spencer, Acta Cryst., 12, 59 (1959).

81. R. F. Stewart and L. H. Jenson, Acta Cryst., 23, 1102 (1967).

82. M. Hospital and B. Busetta, Comp. Rend., D268, 1232 (1969).

83. R. E. Christoffersen, D. Spangler, G. M. Maggiora, and G. G. Hall, J. Amer. Chem. Soc., 95, 8526 (1973).

84. E. Clementi, J. M. Andre, M. Cl. Andre, D. Klint, and D. Hahn, Acta Physica Acad. Scient. Hung., 27, 493 (1969).

85. B. Mely and A. Pullman, Theoret. chim. Acta, 13, 278 (1969).

86. A. Pullman, M. Dreyfus, and B. Mely, Theoret. chim. Acta, 16, 85 (1970).

87. L. C. Snyder, R. G. Shulman, and D. B. Neumann, J. Chem. Phys., 53, 256 (1970).

88. E. Clementi, J. Mehl, and W. von Niessen, J. Chem. Phys., 54, 508 (1971).

SYMPOSIUM IV. POTENTIAL SURFACES, TRANSITION STATES, AND INTERMEDIATES IN CHEMICAL AND PHOTOCHEMICAL PROCESSES

Chairman : M. Karplus

Chemistry Department,
Harvard University,
Cambridge, Massachusetts, U.S.A.

CALCULATION OF POTENTIAL SURFACES FOR GROUND AND EXCITED STATES

S.D. Peyerimhoff and R.J. Buenker

Lehrstuhl für Theoretische Chemie
Institut für Physikalische Chemie
Universität Bonn, 53 Bonn, W. Germany

INTRODUCTION

There is a wide variety of information which can be extracted from the knowledge of potential energy surfaces of molecules, and hence the motivations for calculating such surfaces are manifold, arising from a broad area of topics in chemistry and physics. Let us divide the surface for the purpose of our present discussion into several (partially overlapping) parts to obtain a general view of the type of problems which arise in the calculation of such quantities and to see what kind of data pertinent to experiment can be obtained therefrom.

- The region in the neighborhood of the potential minimum is a fairly narrow portion of the total surface and one expects to predict details of the equilibrium structure of the system under study, i.e. bond lengths, bond angles, oftentimes force constants, spectroscopic constants ω, B, α, ωx etc.

- Sections joining the minimum to some dissociation limit can give binding energies, activation energies, transition state geometries and related quantities.

- The multidimensional surface can be used to obtain (together with appropriate trajectory calculations, for example) the entire reaction path for a system.

- Several complete surfaces (oftentimes interacting)

B. Pullman and R. Parr (eds.), The New World of Quantum Chemistry, 213–240.

are needed for the description of complicated reactions; they are also basic for the treatment of details of electronic spectra.

The accuracy (i.e. level of sophistication) required for determining a given potential surface will depend very much on the problem itself. In what follows we want to discuss the success of ab initio calculations in obtaining results for the individual parts of the potential energy surfaces described above; no comparison with data obtained from various semiempirical methods will be attempted, however.

II. GENERAL ASPECTS OF THE CALCULATIONS

In principle there is no essential difference between treating ground state potential surfaces or those of excited state species. The AO basis set has to be adequate in either case, i.e. it must include functions necessary to properly describe the characteristics of the dominant electronic configurations of the state under consideration. For the ground state case the AO basis should consist of the usual valence-type s and p atomic species. For excited states the picture changes only because of the need to describe Rydberg-type species in addition to those of the more familiar (although certainly not more abundant) valence states; investigations for Rydberg states (or mixed valence-Rydberg species) then requires additional long-range (or diffuse) basis functions, if only to satisfactorily represent the excited or upper orbital in such states [1]. Furthermore, polarization functions in the form of either higher spherical harmonics (d,f etc) located at the atomic centers or s and p bond functions placed at the center of the chemical bond are oftentimes required for an upgraded level of treatment of both ground and excited state potential surfaces.

For obvious reasons the method chosen for the theoretical treatment of potential energy surfaces should be a standard treatment which in so far as possible is equivalent for all surfaces and all parts of the surface.

Among all the theoretical methods available for such calculations only a multiconfigurational approach (of one sort or another) qualifies as such a standard treatment adequate for all areas of a given potential surface. Although 10 years ago such calculations were thought to be too costly to be employed routinely, such is no longer the case today; a satisfactory CI treatment is only one to five times and most often no more than three times slower than the corresponding SCF procedure, which has been routinely available for the past 10 or 15 years. A satisfactory CI treatment is thereby one which includes all single-, double- and the most important triple- and higher-excitation species with respect to the leading configuration (or more succinctly all single- and double-excitation species with respect to the set of the most important configurations in a given CI expansion), processed directly or via a combined selection and extrapolation procedure [2,3] . Such a method is essentially equivalent to a full CI in which the energy values depend only on the AO basis set chosen. Hence the MO basis set is immaterial [2,3] and can be chosen according to practical considerations [1-13] (SCF or MCSCF orbital or transformed species such as various NO`s, GVB or localized orbitals etc). And perhaps most importantly all states are accessible by CI treatments (not only those with special spin couplings as in the standard SCF framework) and there is no convergence problem using this approach in contrast to the situation occassionally occurring with SCF routines, particularly when applied to higher-lying excited states.

A single-configurational treatment (from SCF to the Hartree-Fock level) can be quite satisfactory in a number of cases, but such instances are often found (or at least substantiated) only after a multiconfigurational treatment has been carried out; alternatively its adequacy can sometimes safely be assumed on the basis of a comparison with wholly analogous problems for which the SCF method has already been used successfully. Nevertheless there are large classes of potential surface calculations for which SCF results will seriously misrepresent the true experimental situation, as will be discussed in subsequent sections of this paper. Thus it seems quite clear that in the future, as more and more groups develop very practical implementations of multiconfigurational treatments, the trend towards the use of a general routine procedure of this type will continue. As a consequence the arbitrariness of results due to the particular method of treatment undertaken should thereby be greatly reduced and with this development the credibility of such calculated data for persons not specifically working in the field should be enhanced.

III. SIMPLE SURFACES

A. Area around the Potential Minimum

From the time that large-scale quantum mechanical computations first became available, much of their application has been directed toward the determination of this part of the surface in order to obtain equilibrium structural parameters of molecular systems and by now a very large number of bond lengths calculated with various methods is available for comparison with the corresponding experimental data. Rules such as "a general characteristic of the Restricted Hartree-Fock approximation is the prediction of bond distances as much as 0.05 Å too short" [14] have been extracted on the basis of near HF calculations on diatomic molecules and have received a great deal of currency. Such generalizations are not even qualitatively correct unless the secondary configurations next important to the dominant (SCF) species introduce antibonding character into the wavefunction;this latter condition is oftentimes satisfied but by no means in every case, especially not when highly excited states are under consideration. In addition the actual discrepancy between experimental and Hartree-Fock bond lengths will also depend on the steepness of the potential surface and can be much larger than 0.05 Å if the surface is especially flat, in which case the location of the potential minimum becomes especially sensitive to the type of treatment employed.

Furthermore it is known that AO basis sets which are quite restricted are generally deficient in the bonding region and hence are likely to lead to overestimations in equilibrium bond distances (and predict binding energies which are too low). As a result SCF calculations at a level below that of Hartree-Fock can easily be made to yield - in essence by cancellation of errors - the exact equilibrium bond length; but it is clear that such an agreement is rather arbitrary and by no means guarantees that other properties calculated from the same wavefunction will be of acceptable accuracy. Similarly CI calculations at a level significantly below that which we have referred to as "standard" will give generally unreliable results; an example of all these trends is given in Table 1, which contains bond lengths for the first three states of O_2 calculated by various methods. Indeed, the most limited treatment of all gives the "best" bond distance for the $^3\Sigma_g^-$ ground state of O_2;

Table I: Equilibrium distances (in A^{o}) for various states of O_2 obtained from several treatments.

	$^3\Sigma_g^-$	$^1\Delta_g$	$^1\Sigma_g^+$	$(^1\Delta_g-^3\Sigma_g^-)$	$(^1\Sigma_g^+-^3\Sigma_g^-)$
SCF, DZD a)	1.200	1.201	1.202	0.001	0.002
SCF, DZD + p a)	1.169	---	---	---	---
SCF, near HF	1.152	---	---	---	---
CI, full min basis	1.30	1.33	1.34	0.030	0.040
CI(PCMO) b), DZD + P	1.244	1.252	1.262	0.008	0.018
INO, DZD + P	1.238	---	---	---	---
CI(INO), DZD + P	1.234	1.242	1.258	0.008	0.024
CI "standard"	1.215	---	---	---	---
Exptl.	1.207	1.215	1.227	0.008	0.020

a) DZD + P refers to an AO basis of double zeta quality augmented by diffuse and polarization functions. More details can be found in the original work [9].

b) CI(PCMO) refers to the treatment in which the SCF MO`s of the parent configuration (i.e. $^3\Sigma_g^-$ or $^1\Delta_g$ or $^1\Sigma_g^+$) are used in the CI treatment.

the arbitariness of the agreement is already seen by the fact, however, that the same treatment fails to note any significant change in bond distance between the ground and the $^1\Delta_g$ and $^1\Sigma_g^+$ excited states of O_2. Indeed, the correct relative bond length ordering for the three states is only obtained satisfactorily with CI treatments using a better than minimal AO basis set.

Some of the dangers in putting too much faith in single-determinantal results can be seen from Fig. 1 which shows SCF and CI CC-stretching potential curves of the ground $X^2\Sigma^+$ and first excited $A\ ^2\Pi$ states of C_2H. Although the difference in ground state equilibrium bond length between SCF and multiconfigurational treatment is only 0.05 bohr, the calculated CI vertical excitation energy shifts from 0.96 eV to 0.76 eV as a consequence of altering the bond distance by this amount (the difference between SCF and CI transition energies is over 1.0 eV, as will be discussed in Sect. 4).

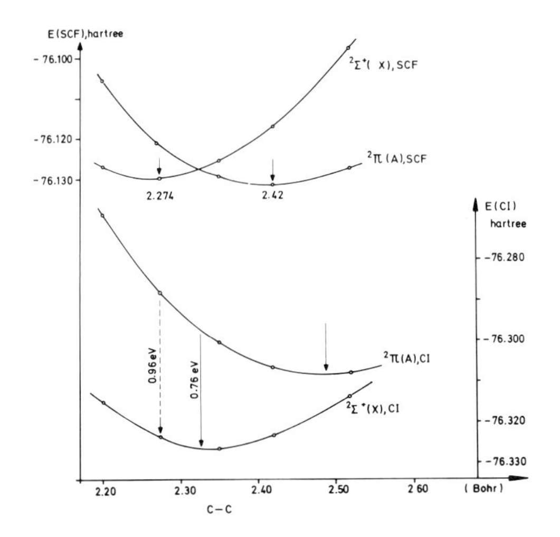

Fig. 1. Comparison of SCF and CI CC-stretch potential curves for the two lowest states of linear C_2H. The excitation energy is also indicated for the case in which the SCF minimum (2.274 bohr) is employed and for that which refers to the CI ground state minimum.

The general situation is wholly analogous for the calculation of bond angles. Other spectroscopic data in the form of force constants, vibrational frequencies, and rotational constants depend even more strongly on the shape of the potential surface around its minimum. From the large number of tables found in the literature the only reasonable conclusion to be drawn is that such quantities are not reliably predicted within the

Hartree-Fock approximation, although results for certain classes of molecules are more accurate than for others [14] . The data of Fig. 2 showing the ethylene

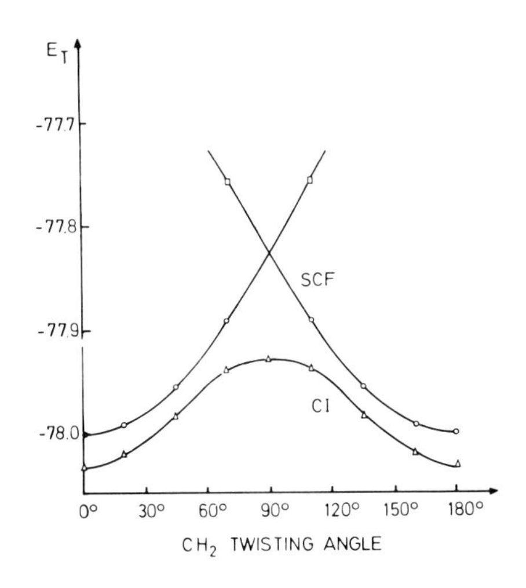

Fig. 2. Calculated SCF and CI twisting potential curves of ethylene (planar 0°, perpendicular 90°).

ground state torsional potential curve in both CI and SCF representations give another striking example of the shortcomings inherent in the SCF method. Even though in both cases the 90° point is correctly predicted to correspond to an energy maximum, it is clear that the overall appearance of the two potential curves is quite different in the two methods, with the CI results yielding by far the more realistic description of this quantity*. Similarly the complete potential curves of O_2 shown in Fig. 3 demonstrate that even though the calculated R value is close to the exact result, the SCF curves themselves are again far away from reality; this finding underscores the fact that agreement in bond length does not itself insure similar accuracy for the entire potential curve.

In summary then experience with a number of systems has indicated that the type of standard CI treatment described above (with an AO basis of at least double-zeta quality plus a moderate number of polarization

---

* The use of complex MO's in this instance succeeds in reproducing the correct qualitative behavior at the potential maximum while retaining the single-determinantal formalism, but this result is nevertheless equivalent to a 2x2 CI in the conventional sense.

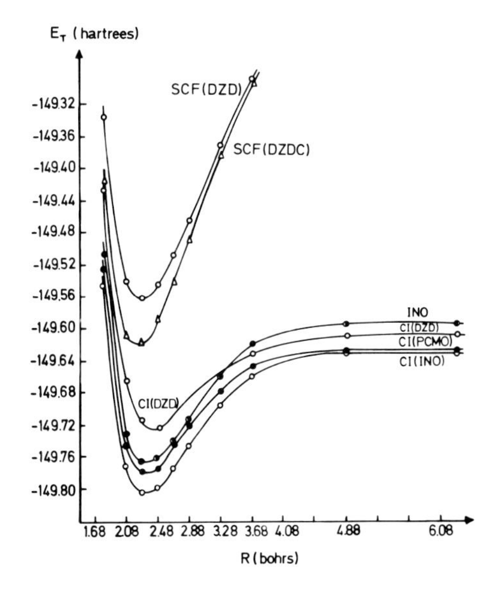

Fig. 3. Potential curves for the $^3\Sigma_g^-$ ground state of O_2 obtained from various treatment (DZD = double zeta AO basis plus diffuse functions, DZDC same with additional polarization species; details about the various CI curves are given in the original reference).

species) can be expected to predict equilibrium bond distances and bond angles to within roughly 0.02 $\mathring{A}$ and 2-3 degrees respectively on a quite general basis as long as the potential surface in the region of the minimum is not unusually flat. In a given case a single-determinantal treatment can obtain somewhat better accuracy than such CI methods, merely because a fundamentally multiconfigurational effect is cancelled out numerically as a result of a compensating error caused by the use of an overly restrictive AO basis.

B. Slices of the Surface toward Dissociation

The two requirements for a satisfactory multiconfigurational treatment mentioned in the beginning, namely the use of a set of main configurations as the basis for constructing the configuration space, and the inclusion (directly or indirectly) of all such generated species (not just a selected sample) in the final secular equation can be illustrated best for calculatinss dealing with the intermediate to long-range portions of the potential surface. The data in Table 2 show the well-known make-up of the dominant part of the wavefunction for the $^3\Sigma_g^-$ state of O_2 at equilibrium and infinity, and clearly demonstrate (as has been pointed out many times before) that the configuration space to be generated must give an equal opportunity to all such strongly interacting species,

Table II: Dominant configurations in a CI expansion of O_2 ($^3\Sigma_g^-$) at equilibrium bond length R_e and for the separated atoms. The square of the approximate coefficient is always given ahead of the configuration.

	R_e		separated atoms
0.94	$3\sigma_g^2\ 1\pi_u^4\ 1\pi_g^2$	1/8	$3\sigma_g^2\ 1\pi_u^4\ 1\pi_g^2$
		1/8	$3\sigma_u^2\ 1\pi_u^4\ 1\pi_g^2$
		1/8	$3\sigma_g^2\ 1\pi_u^2\ 1\pi_g^4$
		1/8	$3\sigma_u^2\ 1\pi_u^2\ 1\pi_g^4$
		1/2	$3\sigma_g\ 3\sigma_u\ 1\pi_u^3\ 1\pi_g^3$

i.e. this space should consist of all single and double excitations relative to a series of the most important configurations, not just to the leading (SCF) term. A second more novel aspect of such calculations can be extracted from Fig. 4. This diagram contains potential curves for three states of O_2 calculated in two ways: first by a truncated CI in which only those configurations which give an energy lowering relative to the dominant part of the wavefunction (main configurations) of more than 20 microhartree (0.6 meV) are included; secondly by a treatment in which no *a priori* selection of the originally generated space is made (these results are obtained using the extrapolation method [3] with essentially no additional work or increase in computer usage). It is obvious that the two figures disagree strongly, and that, despite the extremely small value for the selection threshold in the first instance, it is very unwise to simply neglect all the more weakly-interacting configurations entirely in the final result; only Fig. 4b reproduces the true experimental situation.

Two further remarks are pertinent in this connection. If the same selected set of configurations is maintained for all parts of the potential surface the treatment can easily be prejudiced so that one part of the surface is described much more accurately

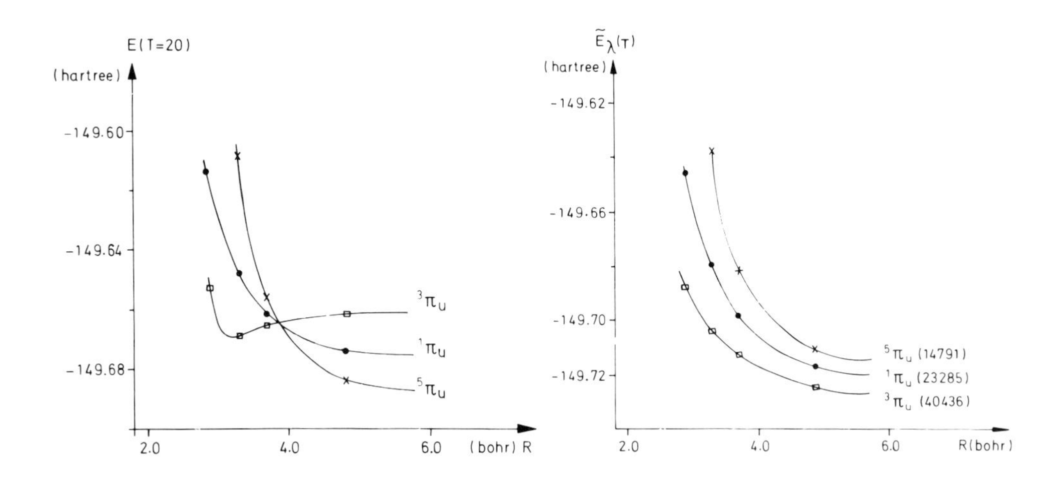

Fig. 4. Potential curves for the $^{5,3,1}\Pi_u$ states of O_2 obtained from a CI calculation at selection threshold T = 20 μhartree (Fig. a) and by extrapolating to zero threshold (Fig. b).

than another, leading, for example, to the prediction of minima which are too shallow, i.e. to the underestimation of dissociation energies [9] . Furthermore, in employing bond functions to describe polarization effects care has to be taken that such species do not simply contribute to the short-range part of the potential curve (i.e. to the representation of the atoms as well as the bond), while having virtually no effect at larger distances (separated atoms), thus distorting the form of the potential curve leading to an artificial increase in the binding character of the system.

Certainly the question to be asked at this point is how accurate a slice of a potential surface up to the dissociation limit can really be calculated with the type of standard CI treatment advocated above. For larger systems this test of accuracy can generally be made only in terms of the dissociation energy, while for quite small systems a comparison with the results of very accurate theoretical methods (for systems like H_2 or HeH^+ , for example, [15-17]) is often possible. Alternatively, new experimental techniques [18-20] (translational spectroscopy) can be used to essentially test each point of a potential curve empirically, and an example employing this procedure will be considered below.

The standard CI treatment (in the sense of the Introduction), developed for molecules of the size of butadiene and larger, is employed for the calculation of the He_2^+ potential curve [21] ; the only deviation from the normal procedure for larger systems is the fact that more polarization functions are employed in this case. Pertinent information regarding the results of these calculations is collected in Table 3. Although the absolute minimum of the curve is predicted to be 0.146 ± 0.005 eV too high (because of AO basis set deficiencies), the shape of the entire curve is given extremely well by the calculations: changes necessary to adjust the calculated curve to fit the experimental data are every where less than 1 meV in the bonding region (1.3 to 3.5 bohr) and 15 meV in the dissociation energy. The relative changes in correlation energy in this case are considerably larger (Table 3), averaging nearly 0.4 eV over the entire bond length range.

It would be a mistake, however, to conclude that SCF methods can never be used effectively to study dissociation processes. In fact there are certain important classes of problems for which the need for multiconfigurational treatments is almost non-existent (unless one wants to achieve a very high level of accuracy), the most obvious of which include certain proton addition reactions as well as hydrogen abstractions from certain radicals, for example. Potential curves for protonation of CO at both the carbon and oxygen sites 22 are shown in Fig. 5, and the corresponding results in Table IV demonstrate that the form of the SCF curves is correct to within roughly 0.5 eV, an error which for amny applications is acceptably small. The composition of the CI wavefunction stays approximately constant as protonation continues, this situation indicating in another way the reliability of the SCF approximation in this instance. Similar results are obtained for HN_2^+ [23] .

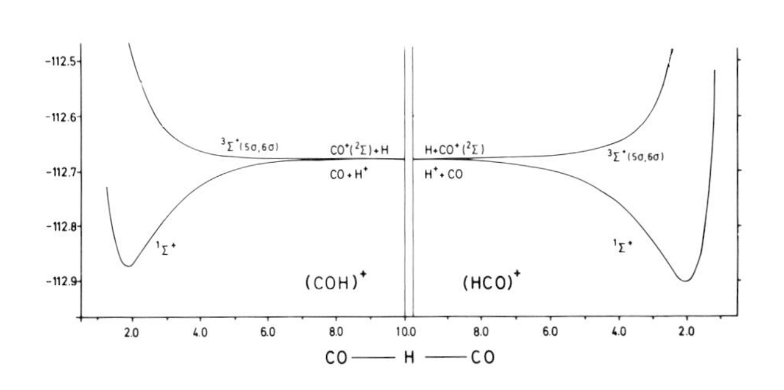

Fig. 5. Potential curves for proton (hydrogen) abstraction in HCO^+ (or COH^+).

Table III: Accuracy test for the He_2^+ ($^1\Sigma_g^+$) potential curve. Given are the energy differences between the experimental curve (from translational spectroscopy) and the calculated values. For comparison the calculated correlation energy is given in the last column. The calculated energy at 2.0626 bohr is 4.98904 hartree [21] .

R (bohr)	ΔE (exptl-calc)[a] (in hartree)	calculated E_{corr}
1.30	0.00001	0.06101
1.40	0.00002	0.06175
1.50	0.00001	0.06255
1.75	0.00000	0.06489
1.90	0.00000	0.06640
1.975	0.00001	0.06712
2.0626	0.00000	0.06795
2.15	0.00002	0.06875
2.25	-0.00001	0.06962
2.50	0.00000	0.07152
2.75	-0.00001	0.07297
3.00	-0.00003	0.07395
3.50	-0.00001	0.07492
4.00	-0.00017	0.07503
5.00	-0.00022	0.07448
5.50	+0.00004	0.07417
7.50	+0.00001	0.07365
10.00	+0.00056	0.07359

[a] All values normalized, so that ΔE = 0 for the minimum. Absolute error probably 0.00535 hartree (≡ 0.146 eV).

The situation is quite similar for the corresponding hydrogen atom abstraction reaction (Fig. 6), except that in this case there is a significant activation barrier to be overcome. The dissociation energy obtained from the SCF treatment [24,25] is (upon proper optimization of the CO bond length and HCO bond angle) 18.8 kcal/mole, in good agreement with the more recent experimental measurements of 17.5 ± 2 and 15.7 ± 1.5 kcal/mole [26,27] , thereby indicating the relatively adequate representation provided by the single-configuration wavefunction in this case. This dissociation is also quite interesting from a symmetry point of view. Consider for example the linear form of the molecule: at small

Table IV: Energy lowerings due to CI at various internuclear distances for HCO^+ ($^1\Sigma_g^+$) and HN_2^+ ($^1\Sigma_g^+$) together with the coefficients for the two leading configurations $5\sigma^2 1\pi^4$ and $5\sigma^2 1\pi^2 2\pi^2$.

	CH (bohr)	ΔE(SCF,CI)(hartree)	c_1/c_2
HCO^+	1.75	0.1719	0.96/-0.18
	2.15	0.1752	0.96/-0.18
	2.35	0.1765	0.96/-0.18
	10.00	0.1577	0.96/-0.14
	NH (bohr)	**ΔE(SCF,CI)(hartree)**	**c_1/c_2**
HN_2^+	1.0	0.1649	0.958/-0.205
	1.6	0.1652	0.958/-0.201
	2.0	0.1700	0.957/-0.198
	2.3	0.1774	--- ---
	2.5	0.1804	0.954/-0.195
	10.0	0.1702	0.958/-0.189

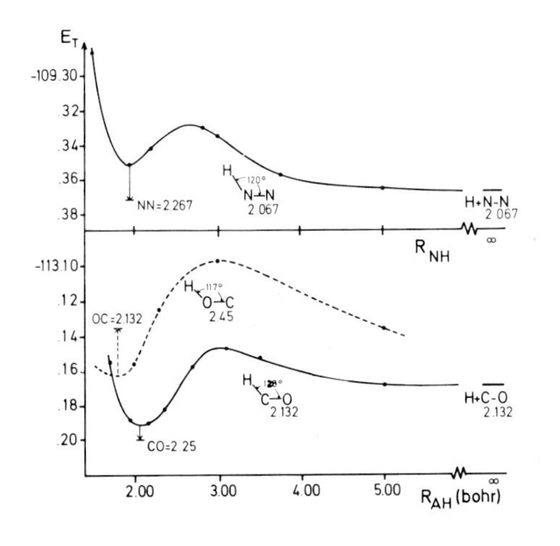

Fig. 6a. Potential curves for hydrogen abstraction in HCO, COH and HN_2. (All distances indicated in the figure are given in bohr.)

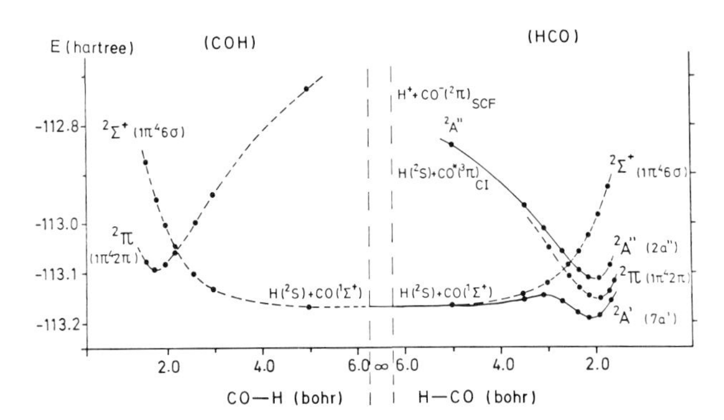

Fig. 6b. Comparison of hydrogen abstraction curves for the linear and bent arrangement of HCO.

internuclear distances the lowest-lying states is $^2\Pi(1\pi^4 2\pi)$, while at large hydrogen separations it is $^2\Sigma^+$ $(1\pi^4 6\sigma)$; since Σ^+ and Π states do not mix, the dissociation for HCO in its linear ground state is symmetry forbidden (Fig. 6b), in the narrowest sense. In actuality the molecule is bent in its lowest state, however, which again can easily be deduced from the qualitative symmetry rules of MO theory [28] (Walsh rules); in this nuclear arrangement the in-plane $^2A'$ component of the $^2\Pi$ state can mix with the A' state corresponding to $^2\Sigma^+$, thereby allowing for a considerably less inihibited dissociation.
The interesting point in addition is that this mixing already appears at the MO level and not only at the CI stage; in other words because of the relatively low symmetry of the bent molecule it is possible for the 7a' MO to be gradually converted from an essentially π-type species at intermediate distances to a pure hydrogen AO at the limit of H-CO dissociation. This example is one of several that might be given which indicates that one-determinantal SCF-type wavefunctions can yield fundamentally better descriptions of a given system if the nuclear symmetry of the problem at hand is quite low.

A similar situation is found in the part of the surface describing the reaction between the amino radical NH_2 and ethylene [29,30] . A relatively good picture of the process can be obtained from SCF calculations, whereas CI is necessary for more quantitative results. For problems of this kind, however, the most critical point seems to be the proper variation of all the possible geometrical variables which change during the attachment or dissociation process. Rather than undertaking a straightforward multidimensional geometry variation, which will generally be prohibitively expensive (by any method), it is always advisable to use chemical intuition to help cut down the scope of the parameter search in attempting to isolate the essential geometrical changes undergone in the course of the reaction.

This part of the disscussion would not be complete without mentioning the work of a number of other people. Special mention should be made, for example, of the many dissociation curves for ground and excited states of CH calculated via CI methods by Lie, Hinze and Liu [5], as well as of the instructive example given by Davidson on the $NO_2 \rightarrow N + O_2$ ground state dissociation [31], and many more. The very elaborate work of Schaefer and Bender and coworkers [7,32] on the linear surface of $F + H_2 \rightarrow FH + H$, or $F + HF \rightarrow FH + F$, for example, and similar studies deserve special mention. Their results at various stages (Table 5) demonstrate once again that SCF treatments might well be capable of giving a good qualitative (or even semi-quantitative) picture, but that a suitable CI treat ment is required to give results of more quantitative reliability on a general basis. This example again illustrates quite clearly that single-determinantal results which are simply assumed as being accurate without making any kind of independent evaluation of the suitability of such treatments in the case at hand (which is sometimes simply neglected because CI programs are not at hand) tend only to confuse the experimentalist, who is often simply not in a position to make such critical judgments for himself.

This section can probably best be summarized in the following way: for the long-range part of the surface multiconfigurational treatments are generally necessary to obtain results of suitable reliability; exceptions, i.e. good results via SCF techniques alone, are possible, however, especially for bonding of protons and H atoms (or sometimes closed shell fragments), especially if a relatively low nuclear symmetry exists. The accuracy which can be obtained with present-day CI treatments depends almost solely on the quality of the AO basis employed; specifically error limits in the order of a few meV, which are generally obtained only for small systems with very elaborate theoretical techniques, can be maintained with practically the same CI methods which are presently employed for larger systems of 30 or more electrons, provided a sufficiently flexible AO basis is chosen.

Table V: Comparison of barrier, exothermicity and saddle point data for the reaction $F + H_2 \rightarrow FH + H$ obtained by various treatments taken from the work of Bender, Schaefer III, O Neil, Pearson [7] .

Treatment	AO Basis	Barrier(kcal)	Exothermicity(kcal)	Saddle Point FH	HH (A°)
SCF	DZ	34.2	-0.58	1.06	0.81
SCF	DZ + P	29.3	13.2	1.18	0.836
CI	DZ	5.7	20.4	1.37	0.81
CI	DZ + P	1.66	34.4	1.54	0.737
Exptl.		1.7	31.2	---	---

C. Complete Multidimensional Surfaces

Complete multidimensional surfaces have been carried out by ab initio methods for only a very few small systems such as H_3^+, H_3, H_4 or HeH_2, for example. This situation will probably prevail for quite some time simply because of practical reasons (computer effort); even though methods become faster and faster the number of points for a chemically interesting complete surface is extremely large, roughly 4^{3N-6} (i.e. assuming an average of four points for codependent optimization of each of the 3N-6 vibrational coordinates of a system). The pertinent question to be answered in this regard, however, is whether knowledge of the entire multidimensional surface is really required to extract the desired chemical information. Given the success of valence bond formulations of structural problems in organic and inorganic chemistry, it seems reasonable to expect that only a few well-chosen slices of the surface for a relatively small number of key vibrational species would need to be investigated in order to allow for a realistic description of the pertinent molecular dynamics; indeed, the basic premise of the Eyring transition-state theory is that a <u>single</u> (weak) vibrational species determines the rate of a chemical reaction. Thus the major effort in this field should probably not be directed toward obtaining all points in an automatic (and of practical necessity technically compromised) way but rather toward determining the most important points of a multidimensional surface in a theoretically reliable manner in order to afford the most efficient computational analysis of the chemical problem at hand.

IV. INTERACTING SURFACES

A. Relative Positions of Potential Curves

The interaction of potential surfaces becomes a key factor whenever electronically excited states are involved. In such cases it is entirely clear that the SCF treatment is not satisfactory, simply because the energy spacing between individual curves is given very unreliably in the single-configurational treatment. One such example was already shown in Sect. III.A for the lowest-lying states of C_2H, for which the SCF is found to be strongly biased toward the excited state species (Fig. 1). A second illustration is contained in Fig. 7, which shows the potential curves for the

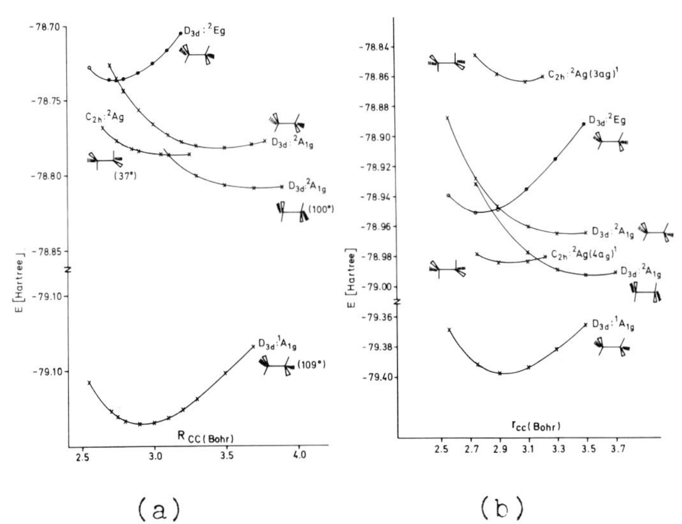

(a) (b)

Fig. 7. Potential curves for the ground state of ethane ($^1A_{1g}$) and for the positive ions of ethane in various geometrical forms obtained from SCF (Fig. a) and CI treatment (Fig. b).

lowest states of the $C_2H_6^+$ ion in various geometries as a function of the CC internuclear distance. It is seen, for example, that in the SCF treatment for umbrella-like D_{3d} conformations the $^2A_{1g}$ minimum energy is 0.045 hartree (29 kcal) lower than that of the ^2E_g species, whereas the corresponding CI result narrows this difference to a value of only 10 kcal; the location of the potential crossing undergone by these two states shifts from 2.75 to 2.92 bohrs in going to the multiconfigurational representation. When the system is allowed to relax in both states the $^2A_{1g}$ species continues to prefer a D_{3d} structure with nearly planar CH_3 groups while one of the ^2E_g components distorts toward a diborane-like structure with reduced symmetry; the difference between the corresponding minimal energies is still quite large in the SCF treatment (Fig.7a) but it is all but eliminated once correlation effects are properly taken into account [33].

The crossing of various excited state potential curves, as illustrated by the O_2 results in Fig. 8, is typical for such interactions: the bound $^3\Sigma_u^-$ curve intersects with the repulsive curves of $^3\Pi_u$, $^1\Pi_u$ and $^5\Pi_u$ symmetry respectively, resulting in the well-known predissociation pattern for the Schumann-Runge band system. These calculated curves are in very good agreement with the small sections of the corresponding curves which are known from experiment. In this connection it should also be mentioned that a large number of excited state calculations have been carried out in the past years by various groups, and that an accuracy of approximately 0.2 eV in the transition energies can be expected regardless of the nature of the state (multiplet structure, Rydberg or valence characteristics), if the CI treatment is carried through with appropriate care. The actual details of the theoretical approach are by no means crucial, which fact was demonstrated quite nicely by calculations

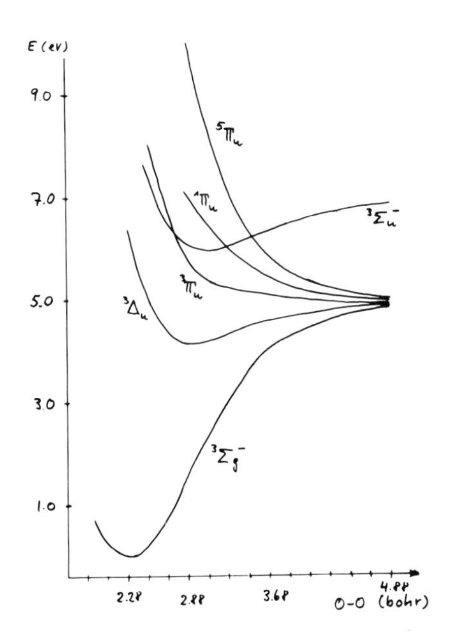

Fig. 8. Calculated CI potential curves for various states of O_2.

on the H_2O spectrum, for example: CI calculations based on SCF MO s using selection and extrapolation techniques[3,34] predict excitation energies to a large number of states by from 0 to 0.2 eV below their corresponding experimental values, while less extensive CI-GVB calculations [35] consistently predict the same states to lie by from 0 to 0.2 eV above the true results; analogous findings are obtained using the vector method [36]. In this application a number of the theoretical predictions predated the experimental work and one discrepancy (for the $^3B_1$ state) was also removed when more accurate measurements were subsequently carried out [37] .

B. Interacting States of Same Symmetry

Interaction of excited states of the same symmetry which lead to avoided crossings are well-known from experiments on diatomic molecules and have been investigated via CI calculations in various cases, including C_2 [38] , CH and the four lowest $^1\Sigma^+$ states of LiF [4](including diabatic curves) as well as in extensive theoretical work on the Li^+Na system in which detailed comparison with inelastic scattering data is possible [39,40].

An example for the interaction of species important for ground state properties is found in considering the angular potential curve of ozone [41,42]. The SCF curves (same symmetry $^1A_1$) of the two interacting states (differing by a double excitation $4b_2^2 \rightarrow 2b_1^2$) in question is shown in Fig. 9a. Depending on the OO distance employed the absolute minimum can be shifted from the open-chain form of the molecule to the cyclic conformer; the latter structure is actually found to be the more stable at the SCF level of treatment when the optimal OO separation is employed at each point, a finding which of course runs contrary to experimental findings. Only after CI is carried out [4] (in this case a rather large INO calculation, i.e. a more restricted procedure than the standard treatment recommended in Sect. II) is the correct behavior for this potential curve forthcoming (Fig. 9b); the CI calculations predict the open ozone triangle to be more stable than the cyclic form by about 16 kcal. Contrary to the previous example with HCO and HN_2, no mixing on the orbital level is possible in the present case ($4b_2$ in-plane and $2b_1$ perpendicular) so that only an actual mixing of configurations can satisfactorily describe the true physical situation.

C. Interacting States of Different Character

The interaction of states of quite different characteristics (mixing of Rydberg and valence states in particular) has obtained relatively little attention until quite recently [43-46] . The general situation is shown schematically in Fig. 10: no mixing is expected if the valence and Rydberg species are well separated energetically (case I, case III), while a strong mixing

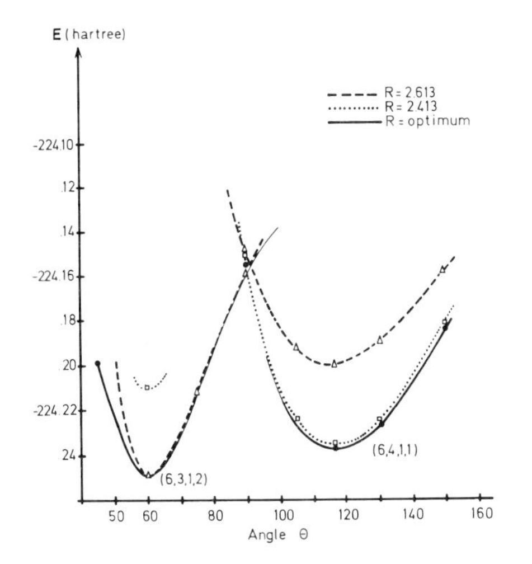

Fig. 9a. Angular potential curves for two states of ozone obtained from SCF treatment for various values of the OO bond distance R. The notation for the configuration (a,b,c,d) refers to the occupation of MO s of a_1, b_2, a_2 and b_1 symmetry.

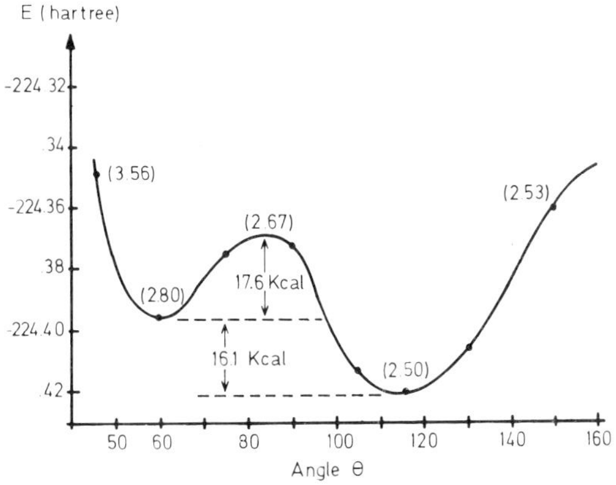

Fig. 9b. Angular potential curve for ozone at optimal R values (indicated in parentheses) obtained from an INO treatment.

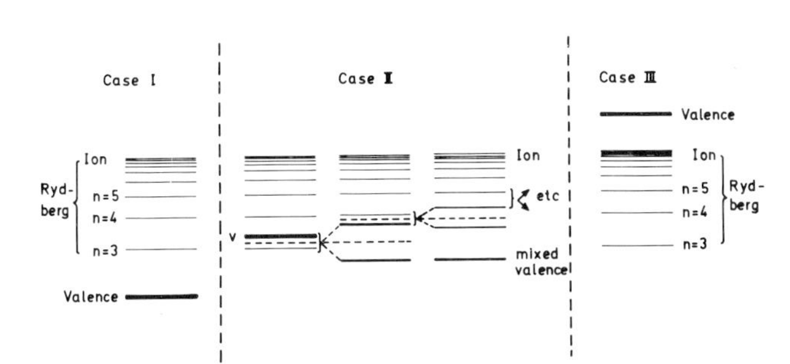

Fig. 10. Schematic diagram showing the interaction between valence and Rydberg states in three different cases.

may occur if the <u>unperturbed</u> valence-shell and Rydberg species of like symmetry are quasi-degenerate, as in case II. Although mixing will generally occur primarily with a single diffuse state, the presence of the other states of the Rydberg series can complicate matters significantly, as indicated in the figure; in such cases the residual valence-shell character not found in the first mixed state can be distributed over a number of higher-lying species of like symmetry, thereby effectively idssipating the strength of the normally localized valence-shell transition over a rather broad range of the electronic spectrum [3,43] . There are many indications that such a set of circumstances is present in the case of the $^1(\pi,\pi^*)$ state of planar ethylene, a center of controversy among experimentalists and theoreticians for much of the past decade [3,47] . In such situations the lowest member of a Rydberg series is not detected as such, since in strongly mixed states experimental techniques tend to emphasize the properties of the valence constituent; for example, Wilkinson's early work on the Rydberg states of ethylene failed to identify a $(\pi,3d\pi)$ species even though results for higher members of the $(\pi,nd\pi)$ series indicated strongly that such a transition should be found in the neighborhood of the N-V (π,π^*) absorption somewhere between 8.0 and 8.5 eV. The degree of mixing clearly depends quite strongly on the relative location of the theoretically unperturbed valence and Rydberg levels and as a result can and often does change with certain types of geometrical variations. Such behavior has also been observed for ethylene, for which the $^1(\pi,\pi^*)$ state is no longer a mixed state at the minimum of its torsional potential curve, i.e. for 90° twisted CH_2 groups.

Another very instructive example showing both the degree of mixing possible between Rydberg and valence states as well as the manner in which such interaction can vary with geometrical changes can be drawn from calculations dealing with the excited states of O_2 (Figs. 11a-c) [48] . The potential curves of the Rydberg states originating from excitation out of the π_g have their minimum at approximately the equilibrium distance of the $^2\Pi_g$ O_2^+ ion (i.e. at smaller values than the equilibrium bond length of O_2 itself) and are essentially parallel to the ionic ground state surface. Thus at small distances the first members of the $^3\Sigma_u^-$, $^3\Delta_u$ (π_g np)states mix with the valence-shell $\pi_u^3\pi_g^3$ states of the same symmetry. The interaction is quite sizeable for the $^3\Sigma_u^-$ state (a splitting of almost 1 eV is calculated between the respective adia batic curves) while it is very small for the $^3\Delta_u$ state of the same electronic configuration; in the latter case the wavefunction is essentially pure valence or pure Rydberg in character at all distances, while both characteristics are present over a rather

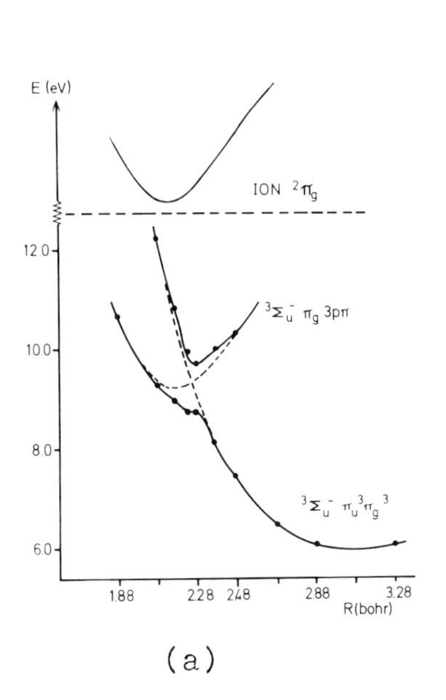

(a)

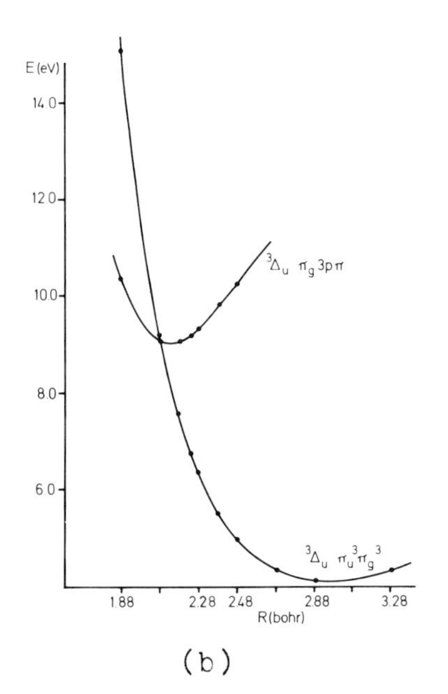

(b)

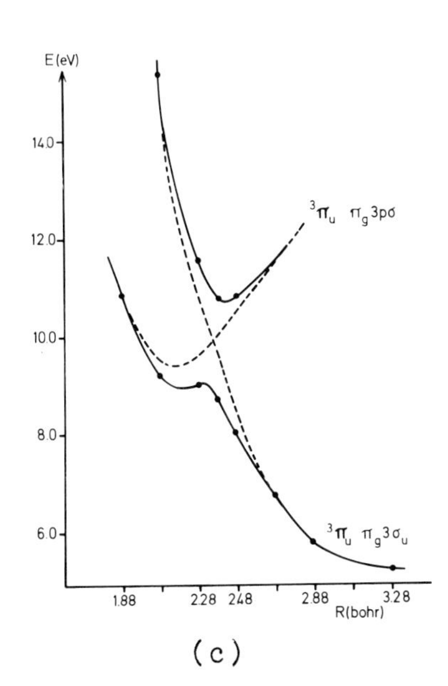

(c)

Fig. 11. Calculated CI potential curves for the lowest two $^3\Sigma_u^-$ (Fig. a), $^3\Delta_u$ (Fig. b) and $^3\Pi_u$ (Fig. c) states of O_2. The energies are taken in eV relative to the $^3\Sigma_g^-$ ground state, distances in bohr). Results for the corresponding pure Rydberg and valence-shell states are also estimated in the figures (dashed lines); the latter curves should be nearly parallel to that of the $^2\Pi_g$ ground state of O_2^+, also included in Fig. a. Since the minimum splitting is quite small for the $^3\Delta_u$ states, the corresponding curves are simply shown to cross in this instance.

larger range of internuclear distances in both $^3\Sigma_u^-$ states (Fig. 11a). The mixing is even more intense for the $^3\Pi_u$ ($\pi_g \rightarrow$ np) and ($\pi_g 3\sigma_u$) states, which differ by only a single excitation (Fig. 11c). Further mixing of the upper potential curve with the higher Rydberg members is also expected, but this aspect of the problem has not been considered explicitly in the O_2 calculations.

D. Further Considerations

The foregoing examples also raise the question as to what extent the adiabatic or diabatic representation of such potential surfaces is meaningful when potential crossings occur. Considerable progress [4,49] has been made in this area in recent times and various methods for calculating the necessary interaction matrix elements are in use, but will not be discussed in this paper. It should be mentioned, however, that attempts to reproduce the experimental differential cross sections for elastic scattering of Li^+-Na at various incident energies on the basis of ab initio potential curves and their calculated interaction have already proven very successful [39] .

For the reproduction or analysis of experimental electronic spectra not only is knowledge of the key potential surfaces required but also the location of the associated vibrational levels within a given absorption system. As long as the various vibrational modes can be separated fairly well it is sufficient to restrict one's attention to related slices of the total potential surface. This procedure has been done in various instances, for example, for CH_2 torsion and CC stretch in C_2H_4, or for the various modes (NN stretch, asymmetric NH stretch, symmetric and antisymmetric trans-bending or NH torsion) in diimide. In some cases, however, vibrations are strongly coupled, and under these circumstances a more general multidimensional analysis corresponding to rather complete surfaces is required. Since a large amount of very accurate data is available from spectroscopic measurements, detailed calculations of electronic spectra appear to offer the most reliable test for the effectiveness of a given theoretical method in representing general potential energy surfaces, a point which explains much of the emphasis which has been directed toward this area in calculations of the past few years.

V. SUMMARY AND OUTLOOK

Applications of <u>ab initio</u> calculations of various types to the determination of potential energy surfaces have led to several general conclusions, in addition to allowing for the solution of a rather large number of specific chemical problems. First, there is now ample evidence that multiconfigurational techniques are required (i.e. over and above methods of SCF type) to obtain a generally reliable description of all portions of a given surface, particularly when excited states are involved. Secondly, as a result of efforts in the area of configuration selection combined with straightforward perturbation and extrapolation techniques it has become possible to carry out multi-configuration treatments practically at the level of a full CI (within a given AO basis) with computational times not greatly exceeding those associated with conventional SCF techniques. Thus it seems that in the future potential energy surface calculations can be continued with even greater confidence, and at the level of accuracy achieved to date they should be a very useful tool in contributing to the solution of problems in chemistry and spectroscopy.

ACKNOWLEDGMENTS

The authors wish to express their gratitude to all their students and coworkers of the past years who have contributed to the theoretical and practical developments in computational methods and to the actual calculations of a number of surfaces; special thanks are due to Drs. P.J. Bruna, W. Butscher, W.E. Kammer, S. Shih and K. Vasudevan. They also want to thank the Deutsche Forschungsgemeinschaft and the Alexander von Humboldt Foundation for continued financial support and the University of Bonn computer center for the services and computer time made available over the past years. Informative discussions on the present subject with various colleagues, especially Drs. I. Shavitt, E.R. Davidson and H. Lefebvre-Brion are also gratefully acknowledged.

REFERENCES

[1] For a discussion see for example S.D. Peyerimhoff and R.J. Buenker, Advances Quantum Chemistry, ed. P.O. Löwdin, Academic Press 1975, Vol. 9, pp. 69

[2] R.J. Buenker and S.D. Peyerimhoff, Theoret. Chim. Acta (Berl.) 35 (1974) 33

[3] R.J. Buenker and S.D. Peyerimhoff, Theoret. Chim. Acta (Berl.) 39 (1975) 217

[4] L.R. Kahn, P.J. Hay and I. Shavitt, J. Chem. Phys. 61 (1974) 3530

[5] G.C. Lie, J. Hinze, B. Liu, J. Chem. Phys. 59 (1973) 1872, 1887

[6] E.R. Davidson, Rev. Mod. Phys. 44 (1972) 451

[7] S.V. O Neil, H.F. Schaefer III and C.F. Bender, Proc. Nat. Acad. Sci. USA 71 (1974) 104

[8] W.J. Hunt, P.J. Hay and W.A. Goddard III, J. Chem. Phys. 57 (1972) 738

[9] S.D. Peyerimhoff and R.J. Buenker, Chem. Phys. Letters 16 (1972) 235

[10] S. Huzinaga and C. Arnau, J. Chem. Phys. 54 (1971) 1948; Phys. Rev. A1 (1970) 1285

[11] R.J. Buenker and S.D. Peyerimhoff, J. Chem. Phys. 53 (1970) 1368

[12] E. Steiner, J. Chem. Phys. 54 (1971) 1114

[13] R.J. Buenker and S.D. Peyerimhoff, Chem. Phys. 9 (1975) 75

[14] H.F. Schaefer III, "The Electronic Structure of Atoms and Molecules", Addison-Wesley Co, (1972) p. 152

[15] W. Kolos and L. Wolniewicz, J. Chem. Phys. 41 (1964) 3663

[16] W. Kolos and J.M. Peek, "New Ab Initio Potential Curve and Quasibound States of HeH^+", Chem. Phys. 1976

[17] L. Wolniewicz, J. Chem. Phys. 43 (1965) 1087

[18] J. Schopman and J. Los, Physica 48 (1970) 190

[19] P.G. Foumier, G. Comtet, R.W. Odom, R. Locht, J.G. Maas, N.P.F.B. von Asselt and J. Los, "A new test of translational spectroscopy: the rotational predissociation of the $X^1\Sigma$ state of HeH^+", preprint

[20] N.P.F.B. von Asselt, J.G. Maas and J. Los, Chem. Phys. 5 (1974) 429 and Chem. Phys. 11 (1975) 253

[21] J.G. Maas, N.P.F.B. von Asselt, P.J.C.M. Nowak, J. Los, S.D. Peyerimhoff and R. J. Buenker, "Ab initio calculation of the $X^2\Sigma_u^+$ state of He_2^+ and adjustments governed by translational spectroscopic measurements", Chem. Phys.

[22] P.J. Bruna, S.D. Peyerimhoff and R.J. Buenker, Chem. Phys. 10 (1975) 323

[23] K. Vasudevan, S.D. Peyerimhoff and R.J. Buenker, Chem. Phys. 5 (1974) 149

[24] S.D. Peyerimhoff and R.J. Buenker, Ber. Bunsengesellschaft 78 (1974) 119

[25] P.J. Bruna, R. J. Buenker and S.D. Peyerimhoff, J. Mol. Structure 32 (1976) 217

[26] P. Warneck, Z. Naturforschg. 26a (1971) 2047

[27] M. A. Haney and J.L. Franklin, Trans. Faraday Soc. 65 (1969) 1794

[28] R.J. Buenker and S.D. Peyerimhoff, Chem. Rev. 74 (1974) 127

[29] S. Shih, R.J. Buenker,S.D. Peyerimhoff and C. Michejda, J. Amer.Chem. Soc.94(1972)7620

[30] R.J. Buenker, in Chemical and Biochemical Reactivity, The Jerusalem Symposia on Quantum Chemistry and Biochemistry VI, Jerusalem 1974

[31] E.R. Davidson, private communication

[32] C.F. Bender, S.V. O Neil, P.K. Pearson and H.F. Schaefer III, Science 176 (1972) 1412; J. Chem. Phys. 56 (1972) 4626

[33] These calculations have been performed primarily by Mr. Richartz and Dr. Bruna of this laboratory

[34] R.J. Buenker and S.D. Peyerimhoff, Chem. Phys. Letters 29 (1974) 253

[35] N.W. Winter, W.A. Goddard III and F. W. Bobrowicz, J. Chem. Phys. 62 (1975) 4325

[36] C.F. Bender, private communication

[37] A. Chutjian, R.I. Hall and S. Trajmar, J. Chem. Phys. 63 (1975) 892

[38] J. Barsuhn, Z. Naturforschg. 27a (1972) 1031

[39] P. Habitz, PhD thesis Bonn 1975

[40] F. v. Busch, Habilitationsschrift Bonn 1975

[41] S. Shih, R.J. Buenker and S.D. Peyerimhoff, Chem. Phys. Letters 28 (1974) 463

[42] P.J. Hay, T.H. Dunning and W.A. Goddard III, J. Chem. Phys. 62 (1975) 3912

[43] R.J. Buenker and S.D. Peyerimhoff, Chem. Phys. Letters 36 (1975) 415

[44] R.S. Mulliken, Intern. J. Quantum Chem. 5S (1971) 83

[45] C. Sandorfy, Lectures at various Conferences

[46] A.E. Douglas in Chemical spectroscopy and photochemistry in the vacuum ultraviolet, eds. C. Sandorfy, P.J. Ausloos and M.B. Robin (Reidel Dordrecht, 1974)

[47] for example: T.H. Dunning, W.J. Hunt, and W.A. Goddard, Chem. Phys. Letters 4 (1969) 147; H. Basch and V. McKoy, J. Chem. Phys. 53 (1970) 1827, J.A. Ryan and J.L.Whitten, Chem. Phys. Letters 15 (1972) 119; R.S. Mulliken, Chem. Phys. Letters 14 (1972) 144, 25 (1974) 305; H. Basch, Chem. Phys. Letters 19 (1973) 323; C.F. Bender, T.H. Dunning, H.F. Schaefer III and W.A. Goddard III, Chem. Phys. Letters 15 (1972) 171

[48] R.J. Buenker and S.D. Peyerimhoff, Chem. Phys. Letters 34 (1975) 225

[49] see for example: M. Desouter-Lecomte, J.C. Leclerc and J.C. Lorquet, Chem. Phys. (1975) for discussion

TRANSITION STATES AND REACTION MECHANISMS IN ORGANIC CHEMISTRY

L. Salem

Laboratoire de Chimie Théorique (ERA n° 549),
Université de Paris-Sud, Centre d'Orsay, 91405 Orsay

I - TRANSITION STATES

Introduction

The conventional description of the chemical reaction as a single equation

$$\text{Reactant(s)} \rightarrow \text{Product(s)}$$

actually encompasses a vast number of microscopic phenomena. These phenomena occur throughout the entire reaction process but it is convenient to distinguish successive stages in which a given phenomenon tends to predominate as in Figure 1. The first stage is vibrational excitation of the reacting molecule - here trans 1,2 dideuterocyclopropane - by multiple collisions with other molecules M in the reaction vessel. The second stage is exchange of acquired vibrational energy between the various modes of the molecule ; overall rotation is certainly involved in this exchange. The third stage is the actual deformation of the molecule along the reaction coordinate - rather than along "useless" vibrational coordinates. Classically speaking, in this third stage the molecule explores, via kinematic trajectories, the potential energy surface.

At a fourth stage the molecule passes through the appropriate col on the surface, or transition "state"[1]. Reaction has then really "occurred", and the three previous phenomena take place again , in reverse order, on the other side of the col. Whereas the first three phenomena together constitute the dynamics of

B. Pullman and R. Parr (eds.), The New World of Quantum Chemistry, 241–269. *All Rights Reserved.*

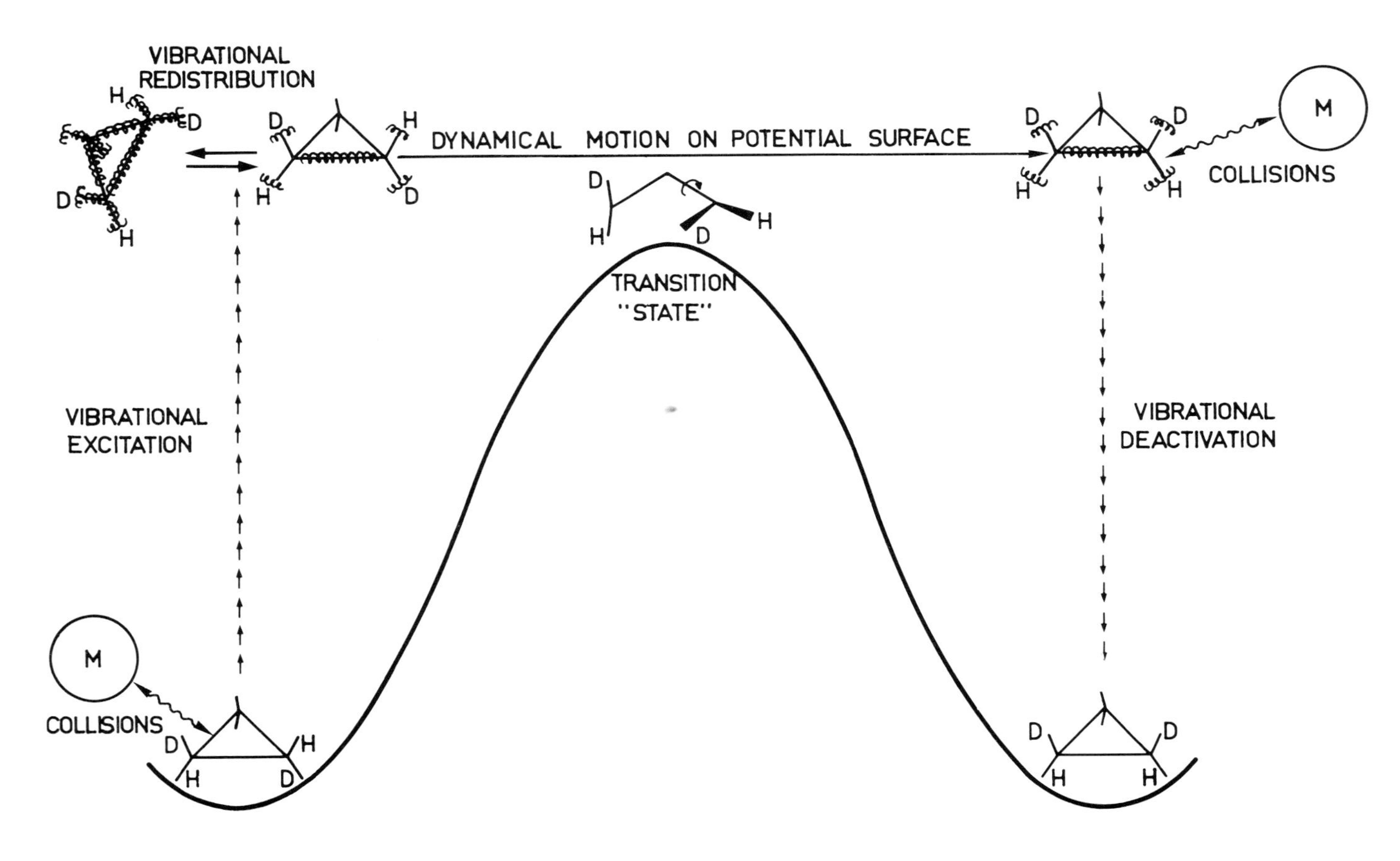

Figure 1 : Schematic description of a unimolecular chemical reaction.

the reaction, the transition "state" belongs to the potential energy surface, and is therefore a static feature of the reaction. In this respect it would have been much more appropriately labelled transition "structure" [2].

Although this meeting has already heard outstanding lectures on dynamics, and although I have been asked to speak only of transition states, I would like to first say some brief words about the three dynamical stages. As an illustration, I will choose precisely the reaction of Figure 1, the geometrical isomerization of cyclopropane :

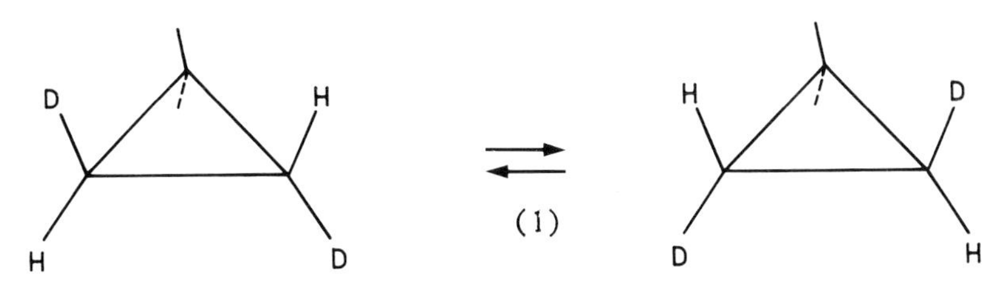

This reaction, which was discovered on a racemic trans di-deutero cyclopropane by Rabinovitch, Schlag and Wiberg[3], has been carried out on a chiral reactant by Berson[4].

Vibrational Excitation

Little is known about this first stage. Although much insight has been gained into the vibrational excitation process of small molecules, the rules governing the vibrational excitation of large polyatomic molecules are still a mystery. However it is possible to develop some qualitative insight with a very crude model[5]. In this model (a) the colliding partner is directed towards the center of gravity of the molecule ; (b) rotational excitation is neglected ; (c) the intermolecular potential is assumed to be a sum of central-force potentials, and (d) the target molecule is assumed to be initially in its ground vibrational state. We then obtain a simple expression for the force acting on a given normal coordinate. These forces are found to have a very specific behavior. For an atom colliding with a symmetric linear triatomic molecule, each force vanishes at some blind angle of attack for which energy transfer will be zero (Figure 2). Similar results are found for attack on an equilateral symmetric triatomic molecule. These calculations give a flicker of hope that some general qualitative rules do exist for the vibrational excitation of polyatomic molecules.

Vibrational Redistribution

Possibly even less is known about the second stage. The manner and the ease with which vibrational energy is exchanged

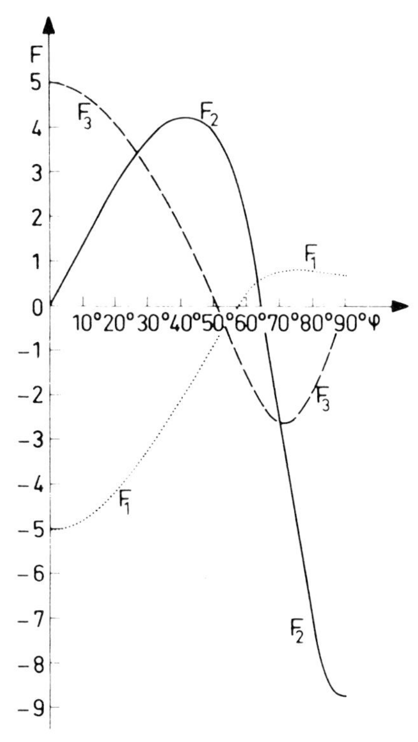

Figure 2 : Variation of the forces acting on the normal coordinates Q_1 (symmetric stretch), Q_2 (bending) and Q_3 (antisymmetric stretch) of a linear triatomic molecule. From ref.5. (Reprinted with permission from J.Amer.Chem.Soc. 97, 3646 (1975). Copyright by the American Chemical Society).

between the "oscillators" composing a molecule is crucial to the validity of unimolecular reaction rate theories such as RRKM theory or Noel Slater's theory. The most interesting qualitative information has been obtained by Gelbart, Freed and Rice[6], and by Nordholm and Rice[7]. The former authors show in particular, in a stochastic model, that if all the molecular energy is put into a single oscillator or bond (which is not allowed to break), the first bond to break is that which is most strongly coupled with the energy-rich bond, even if it is a much stronger bond than others which are weakly coupled to the oscillator.

Geometrical isomerization of Cyclopropane

Before considering the third stage with the classical trajectories, let us turn to the potential energy surface itself. The activation energy for the reaction, and therefore the height of the col, is 64 Kcal/mole, close to the value for isomerization of cyclopropane to propylene. The pioneering work on the nature of the potential surface is due to Hoffmann. In 1968 he demonstrated[8] the existence of an edge-to-edge trimethylene intermediate **I**

with an energy well below that of the two other possible intermediates, edge-to-face trimethylene **II** and face-to-face trimethylene **III**.

Hoffmann suggested that the reaction proceeds through I *via* conrotatory electrocyclic pathway, with synchronous rotation of both terminal methylene groups. Since then it has become clear that there are two distinct reactions : the geometrical isomerization (1) and the optical isomerization

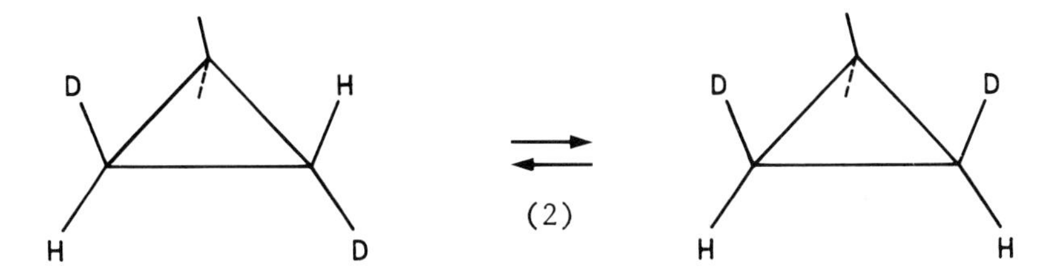

Whereas reaction (2) does require rotation of two groups, this is not necessarily the case for reaction (1).

Our own calculation, which started in 1968, took 3 years and some 600 points, at a time when the calculation of a single geometry lasted 20 minutes of c.p.u. time. An accurate analytical representation of the potential surface for rotation of the terminal CH_2 groups - simplified to eliminate CH_2 pyramidali-

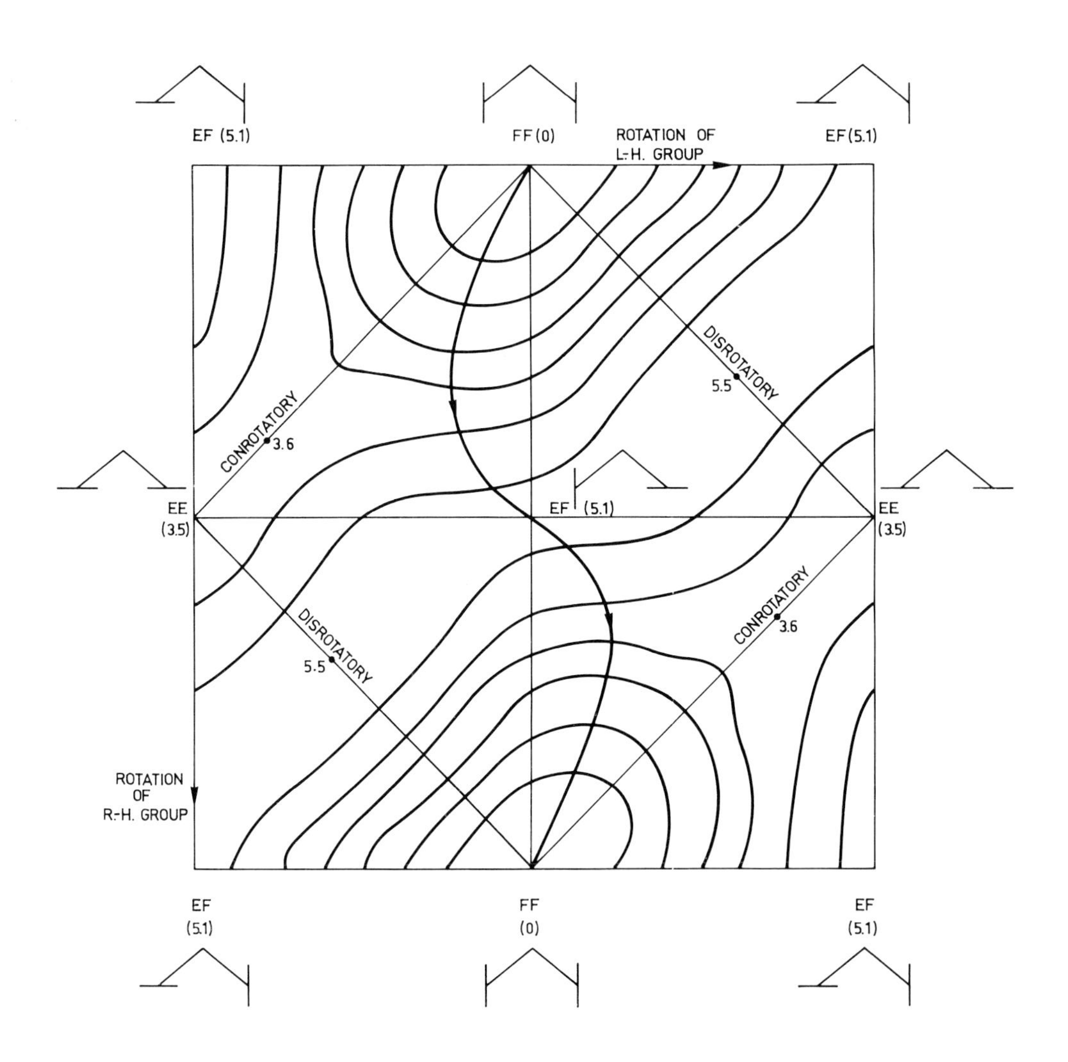

Figure 3 : Potential energy surface for rotation of the terminal CH_2 groups in the trimethylene diradical $\cdot CH_2-CH_2-CH_2\cdot$ Energies, in Kcal/mole, are relative to FF. The energy difference between two equi-potential curves is 0.8 Kcal/mole.

zation - is shown in Figure 3[9]. The conrotatory "valley" discovered by Hoffmann shows up nicely. Starting with a face-to-face diradical (FF, above), geometrical isomerization requires passage to the similar diradical (FF, below) with a single overall methylene rotation. The best static pathway is shown in the Figure and passes through an edge-to-face diradical similar to II. The second methylene group "helps" the motion by first rotating in one direction, and then in the other direction.

The actual optimal static pathway on the surface is slightly more complicated[10], as it involves also some CH_2 pyramidalization and depyramidalization. This path is shown in Figure 4. It is non-synchronous, one hydrogen atom lagging behind the other. In collaboration with Dr. Yves Jean, we have also fully optimized the geometry of the transition "state" ("structure") in the full 21-dimensional hyperspace. Its energy is 1.5 Kcal/mole above that for the best optical isomerization pathway. It was exciting, in June 1971, to "see" the actual transition state after a long effort. Yet our excitement is now moderated by the very modest reliability of our results. The very limited character of the basis set of Slater orbitals used in this first calculation makes the finer details of the surface very uncertain. There is also some error in the relative energies of the intermediates themselves. Berson has shown[4] that, when reactions (1) and (2) are allowed to compete in an optically active trans-dideuterocyclopropane, (2) occurs exclusively. Our calculated energy difference of 0.6 Kcal/mole is therefore too small. Nevertheless it is interesting to observe that the trimethylene diradical is a transition structure, at a col, rather than an intermediate in a secondary minimum[11]. Also our work on trimethylene taught us much about the electronic properties of diradicals (see further). Finally even such a mildly reliable surface is amenable to calculations of dynamical trajectories, i.e. an investigation of phase three of the dynamical reaction process.

Dynamical Trajectories on the Potential Surface

Such calculations have been carried out by Chapuisat and Jean[12] in our laboratory. After considering first the rotational motion of the methylene groups on the surface of Figure 3, they extended their work to include the coupling of these rotations with the motion of ring closure (CCC bending angle 2α). The results for a typical case are shown in Figure 5. The background of each diagram consists of equi-potential energy curves as a function of ring closure (2α) and of concerted methylene rotation (in this particular case an assumed conrotatory motion). Each diagram also shows a trajectory calculated for a given value of the initial energy E°_{ROT} for rotation of the terminal

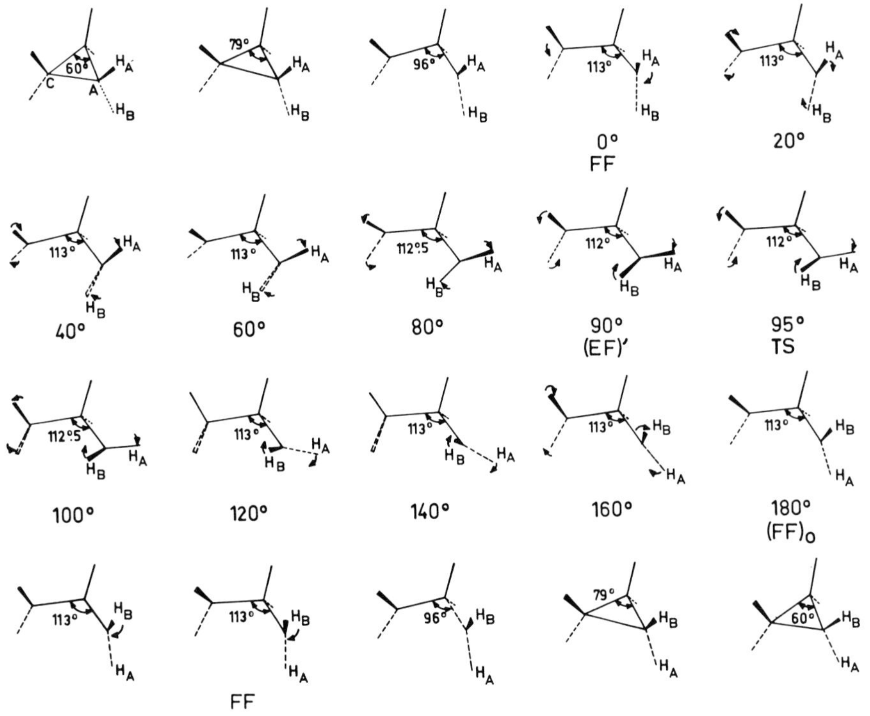

Figure 4 : A nondynamic pathway passing through the transition state. The rotation angle ϕ_A of the principal methylene group is shown below the molecule. Central CCC angles have been rounded off. The notations FF, $(FF)_0$, (EF)', and TS represent, respectively, a face-to-face trimethylene in its stable pyramidalized form, the same with one methylene group trigonal, an edge-to-face diradical in its stable pyramidalized form, and the transition state. From ref. 10 (Reprinted with permission from J.Amer.Chem.Soc. 94, 279 (1972). Copyright by the American Chemical Society).

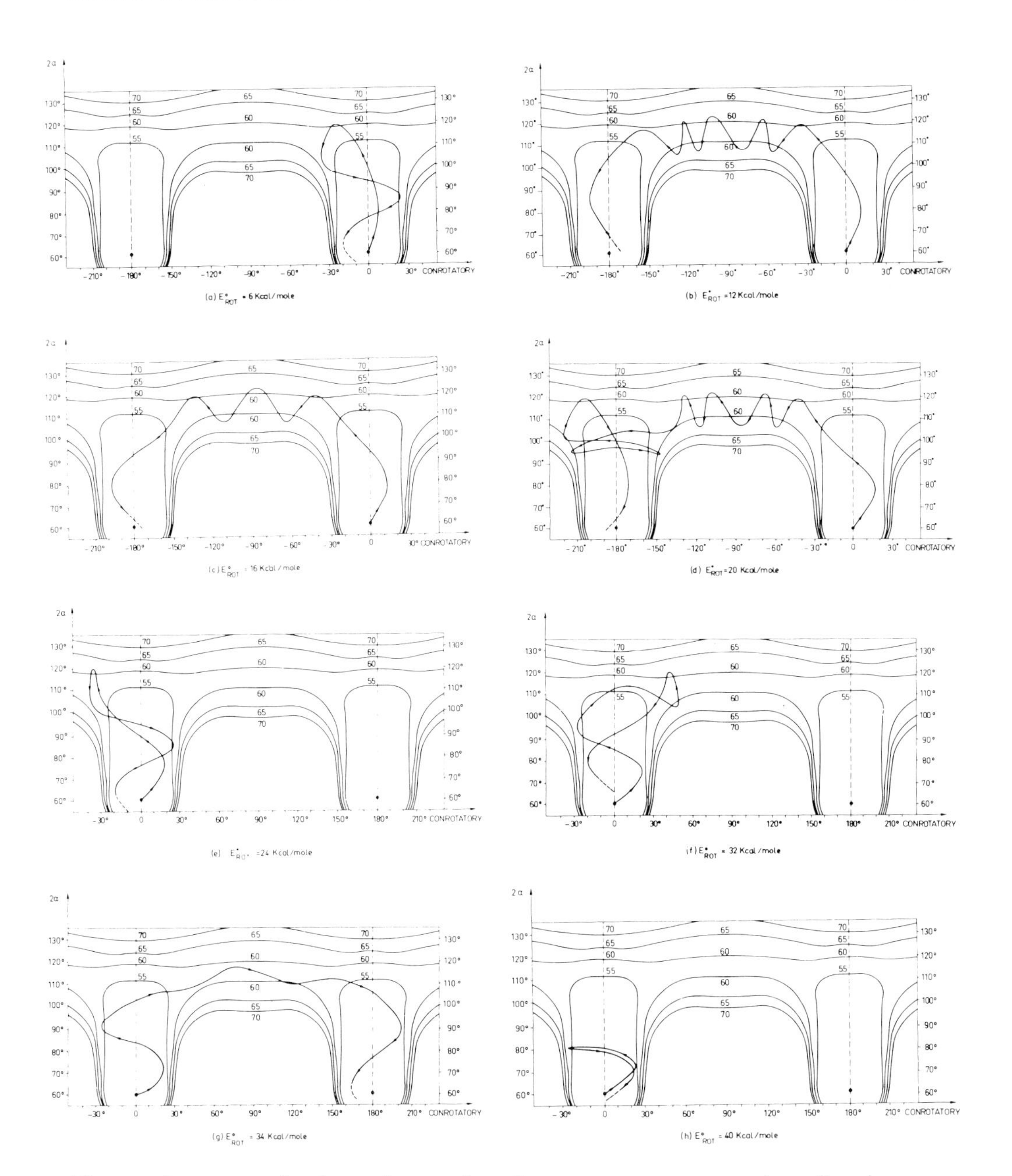

Figure 5 : Dynamical trajectories for conrotatory motion (horizontal coordinate : concerted methylene rotation ; vertical coordinate :ring opening and ring closing). The total energy is held fixed at 1.2 Kcal/mole above the transition state. From ref. 12. (Reprinted with permission from J.Amer.Chem.Soc.97, 6325 (1975). Copyright by the American Chemical Society).

methylene groups. The striking feature is the presence, as the rotor energy increases (the total excess energy being fixed) of energy bands within which trajectories are alternatively non-reactive and reactive. The existence of such "reactive" and "non-reactive" energy bands, which were found independently by Wright in the $H_2 + H_2$ exchange reaction[13], can be explained on the basis of a classical model with a billiard ball rolling up a steep incline with a single exit region at the top[14].

Elimination of HCL from Ethylene Chloride

The dehydrohalogenation of ethyl chloride in the vapour phase is a unimolecular decomposition with a large (60 Kcal/mole) activation energy[15]. A thorough calculation on this reaction by P. Hiberty in our laboratory[16] rules out the possibility of planar anti elimination

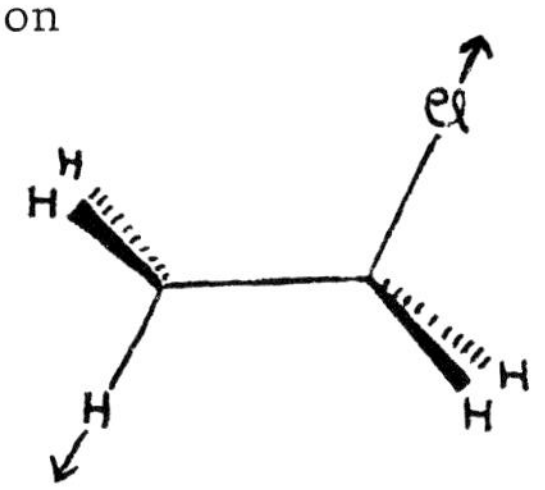

or of non-planar anti elimination :

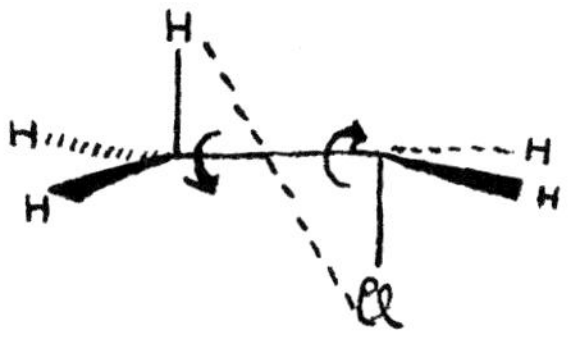

The preferred reaction path is coplanar syn elimination. The corresponding planar four-member transition state appears to be in contradiction with the Woodward-Hoffmann rules, which state that 2s + 2s pericyclic reactions are forbidden. However the conflict is only apparent. Goddard was first to show[17] that such a transition "state" ("structure") is allowed for concerted reactions if an additional lone-pair orbital takes part in the bonding system. This is illustrated further in Fig. 6, in which the transition structure is viewed as a resonance mixture of two linear, rather than cyclic four-electron systems.

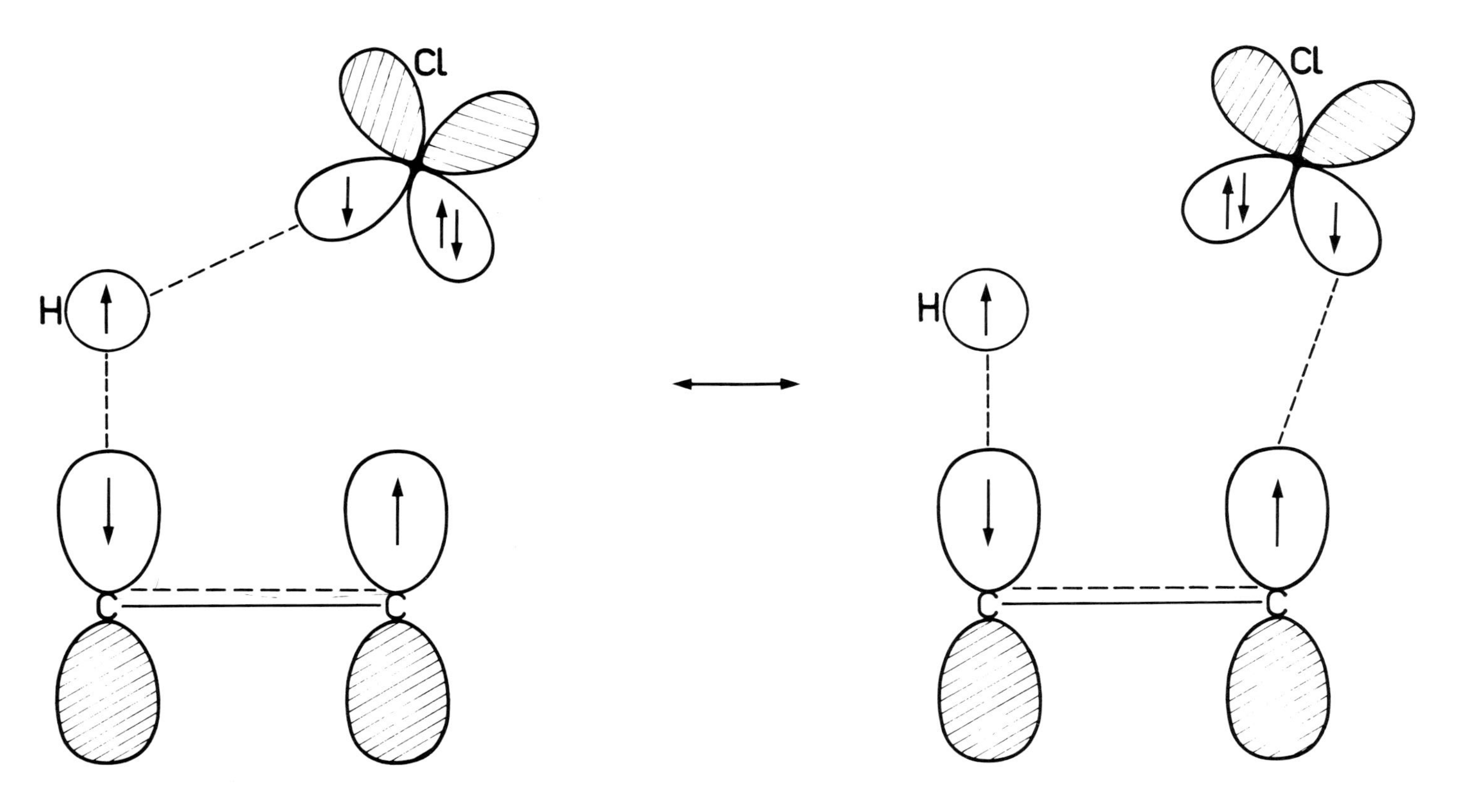

Figure 6 : Schematic transition "state" ("structure") for elimination of HCl from ethylene chloride.

Diels-Alder Reaction

Our study of the Diels-Alder reaction[18] has cost the same effort as that for the cyclopropane isomerization. Indeed the increased complexity of the reaction has been matched by an increasing sophistication in computer programs and computing techniques. We can describe our technique in a few words. Its main features are (a) appropriate choice of basis set ; (b) appropriate choice of Hamiltonian. For instance, in the Diels-Alder reaction, we are interested in comparing the synchronous concerted pathway (C_s symmetry) and the two-step pathway <u>via</u> the hex-2-ene-1,6 diyl diradical :

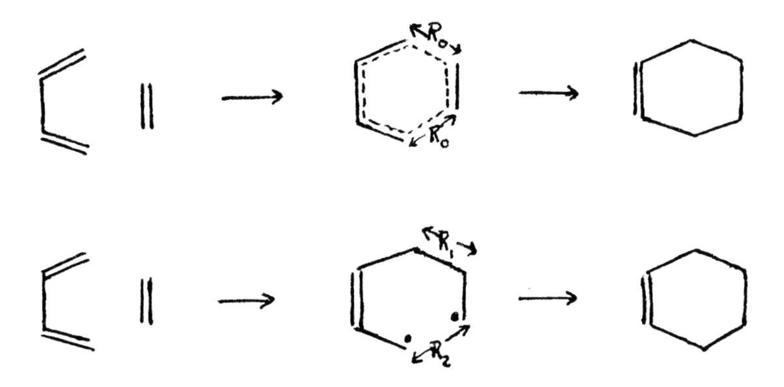

Such typical pathways are first explored with a minimal basis set of Gaussian orbitals ; next, specific points (cols, secondary minima, reactants, products) are optimized with the same basis ; finally pathways passing through the optimized geometries are recalculated with an extended basis set. Throughout two different Hamiltonians are used concurrently : a restricted closed-shell Hamiltonian more appropriate to "concerted" pathways, and a restricted open-shell Hamiltonian more appropriate for diradicals. In either case a three-by-three configuration interaction calculation is performed between the configurations built on the two non-bonding (or weakly bonding and weakly antibonding) molecular orbitals :

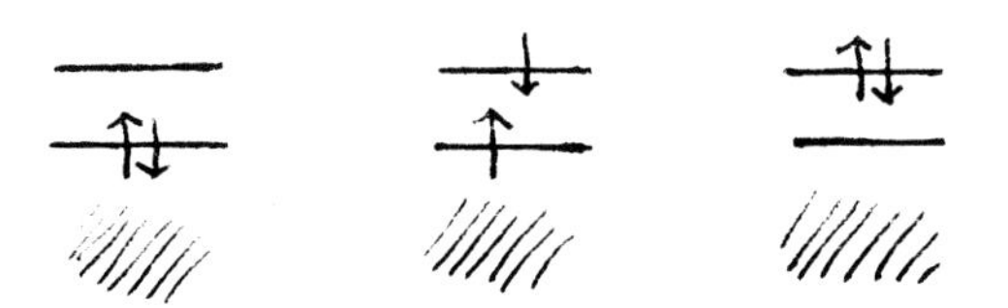

Thus the technique, while not perfect, correctly describes the covalent wave function of diradicals despite the ionic character in each of the molecular orbital configurations. The method is also capable of treating closed-shell systems (pericyclic transition states) and open-shell systems (diradicals) on a nearly equal footing. Finally the entire technique rests on an extremely powerful and versatile program[19]. It will never be emphasized enough how much such calculations owe to the scientists who have written these programs.

Our results for the Diels-Alder reaction are illustrated in Figure 7. The energy is plotted against the coordinates R_1 and R_2 defined above. Not surprisingly,the potential energy surface has the form of a complicated mountain range, with a low central col corresponding to the concerted pathway. The transition "state" ("structure") C corresponds to this perfectly synchronous pathway, with $R_1=R_2=R_0=2.21$ Å. At the edges of the surface are the two-step pathways which rise through higher cols (T_b, 3.4 Kcal/mole above C), and then descend to the hex-2-ene 1,6 diyl diradical which is simply a resting point on the side of the hill. The diradical D_b lies below the concerted col C ,in agreement with thermochemical estimates[20]. Yet to reach the diradical structure the molecule must climb higher than C, as shown by the two additional ridges which separate the concerted transition state C from the diradical D_b. Our results, which are confirmed by similar calculations by Leroy[21] (see however[22]), reconcile nicely the thermochemical calculations and the experimental stereospecificity.

We see that in essence Theoretical Chemistry is able to translate an experimental mechanism into the form of a potential energy surface, the heights and position of its cols, and the dynamical behavior of reactants in relation to this surface.

II - ELECTRONIC CONTROL BY ORBITAL SYMMETRY

We now ask the question : how is the potential energy surface determined ? Of course there are simple thermochemical factors (the energy cost for bond breaking, angle bending, twisting of double bonds,etc.) and steric factors. Such factors determine a good part of the potential "landscape", particularly for a reaction such as the geometrical isomerization of cyclopropane. However an additional factor, electronic control, is extremely important in many reactions. By electronic control proper we mean effects not due to the constitution of the reactant (bonds, angles,etc., which are related to the average potential energy of the electronic wave functions), nor to the presence of large groups (steric effects, corresponding to the

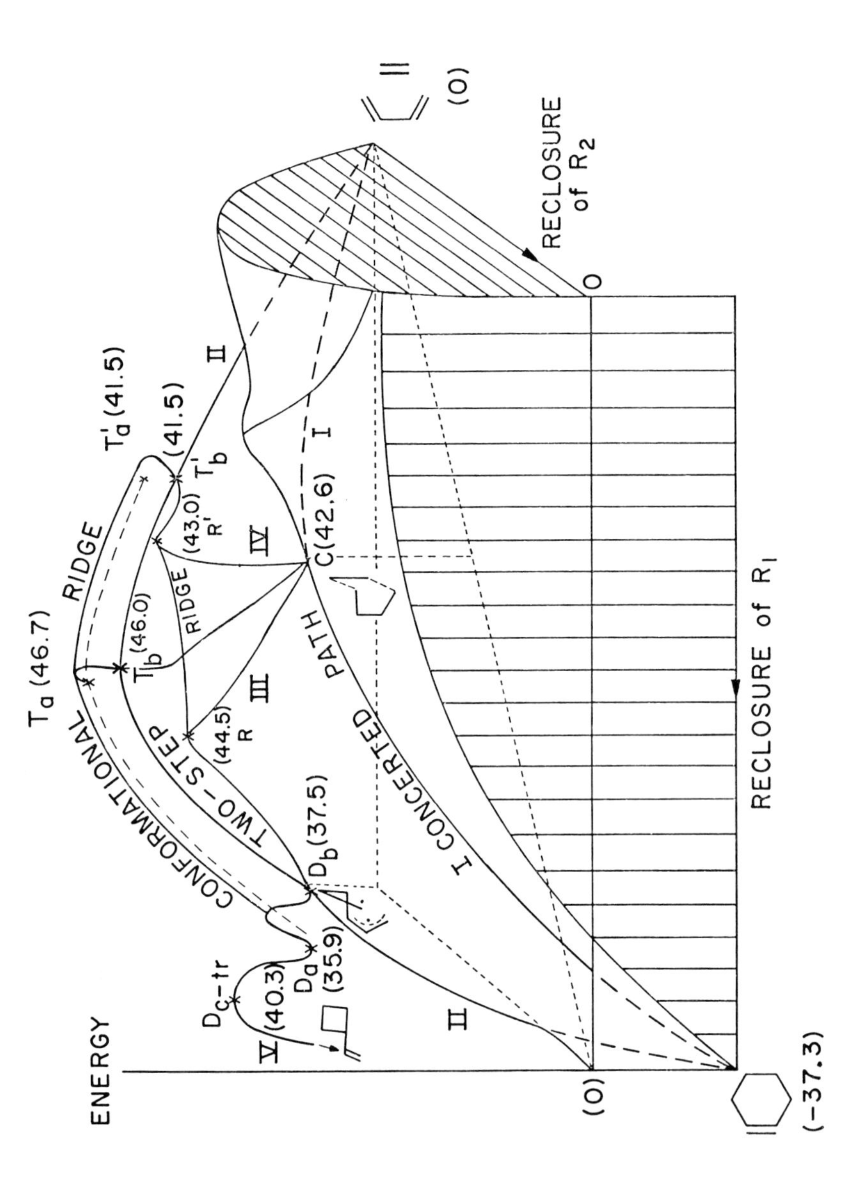

Figure 7 : Potential energy surface for the Diels-Alder reaction. From ref. 18. (Reprinted with permission from J.Amer.Chem.Soc. 98,0000 (1976). Copyright by the American Chemical Society).

Pauli exclusion principle), but which are inherent to its *topology* and *symmetry* (nodal properties of the wave functions)[23]. The control may be due to the symmetry and nodal properties of the one-electron *orbitals*, or to similar properties of the overall *states*.

The Woodward-Hoffmann rules[24] for the control of chemical reactions are both a historical and a text-book example of orbital symmetry control. The correlation diagram technique which they used so brilliantly has now been implemented, since 1965, for a wide number of chemical reactions.

Correlation Diagram for a Metal Catalyst

Let us give an additional modest example of orbital correlation diagram which may ultimately be helpful in understanding a reaction mechanism. In heterogeneous catalysis the approach of the substrate to the metal surface, and the subsequent chemical reactions, must depend strongly on the nature of the filled and empty d atomic orbitals encountered by the substrate. It is therefore important to know the ordering of atomic orbitals on each face of a metal catalyst. It is possible to obtain this pattern by using symmetry, together with a rough estimate of the atomic orbital levels (*via* the angular overlap model, for example). In this manner we have been able to correlate bulk orbitals and surface orbitals[25] of Nickel (Figure 8). This correlation diagram gives information on the nature of the "fuller" and "emptier" orbitals on the surface. But of course it is only one minute piece of information, since it is a correlation diagram within one of the two reaction partners.

III - ELECTRONIC CONTROL BY STATE SYMMETRY. THEORY OF PHOTOCHEMICAL REACTIONS

We now turn to another type of electronic control, which seems of paramount importance in determining the outcome of photochemical reactions. Electronic excitation of a reactant changes the nature of its electronic state. Decay to product also involves a net change of electronic state, with even the possibility of multiple state changes in between. Our purpose has been to investigate the manner in which electronic states control the outcome of photochemical reactions[26].

Now electronic excitation, in its very first act, generally separates two electrons which were initially paired (in a molecular orbital for instance). Hence the vertically excited molecule, with two odd electrons, has the likings of a diradical and can be thought of as a diradical precursor. It will generate

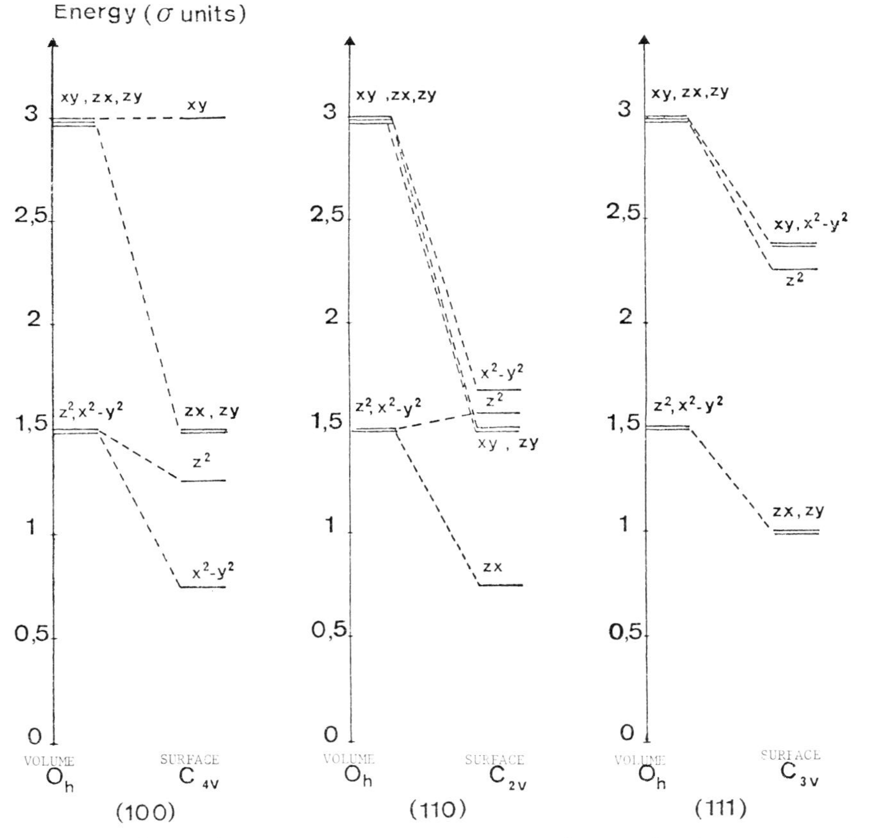

Figure 8 : Correlation diagram between bulk atomic orbitals and surface atomic orbitals for nickel metal. The correlations correspond to identical <u>local site</u> symmetries (two different site symmetries may be mixed in a given band). Note that the z axis of quantification does <u>not</u> have the same direction in the C_{2v} or C_{3v} groups as in the O_h group.

a primary product with all the electronic characteristics of a diradical. To understand photochemical reaction mechanisms we must therefore investigate the electronic properties of diradicals.

Electronic Properties of Diradicals

Diradicals have two centers, or "radical sites", each with a free residual valence, between which two electrons can be distributed. In practice there are four ways of doing this, and perforce diradicals have four low-lying electronic states[27]. Two of these are the diradical states (D) proper, in which one electron occupies separately each site. The electron spins can be parallel (^{3}D) or antiparallel (^{1}D). In the other two states the two electrons occupy simultaneously the same site, which becomes negatively charged, while there is a positive hole on the other site. An ion-pair is created and these singlet states are called zwitterionic (Z) states. This name has been chosen because zwitterions are very common intermediates in Organic Chemistry. In non-polar systems zwitterionic states have high energies. But if the system is intrinsically polar, and if the medium has appropriate polarity, a zwitterionic state becomes the ground state of the system - which precisely then is a zwitterion. Hence there is a duality diradical ↔ zwitterion : a diradical has excited zwitterionic states while a zwitterion has excited diradical states.

The simplest illustration of the four diradical states is given by 90°-twisted ethylene. The two diradical states lie well (79 Kcal/mole)[28] below the two zwitterionic states :

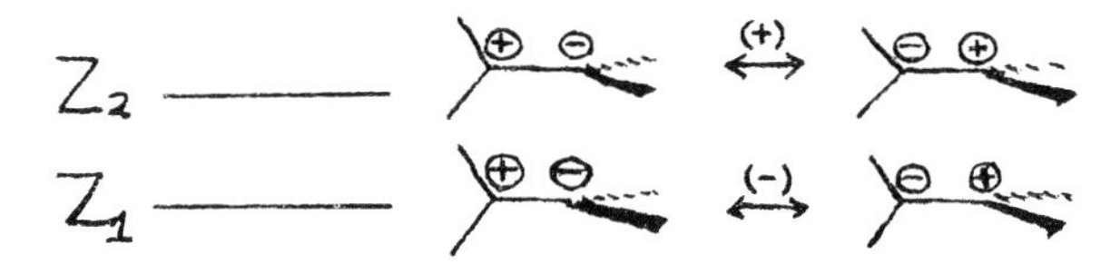

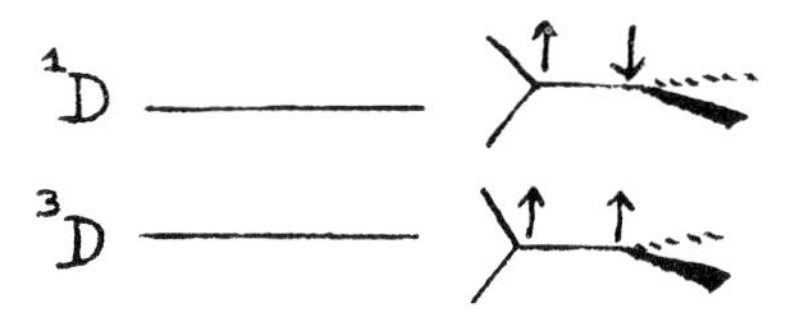

Although Hund's rule would be expected to place ^3D below ^1D, the ordering is uncertain since the recent calculations of Buenker and Peyerimhoff[28]. Similarly the (+) Z state may lie below the (-) Z state[28].

State Correlation Diagrams. Surface Crossings

Armed with knowledge of the available states of diradical intermediates, and knowing the low-lying excited states of reactants, we can <u>correlate</u> the states of reactants and primary products. Such a correlation is a natural thing to attempt. In 1972, in Baden-Baden, G. Quinkert had shown unambiguously[29] that photochemical α-cleavage of 2,4-cyclohexadienone leads to a ground state σ,π diradical :

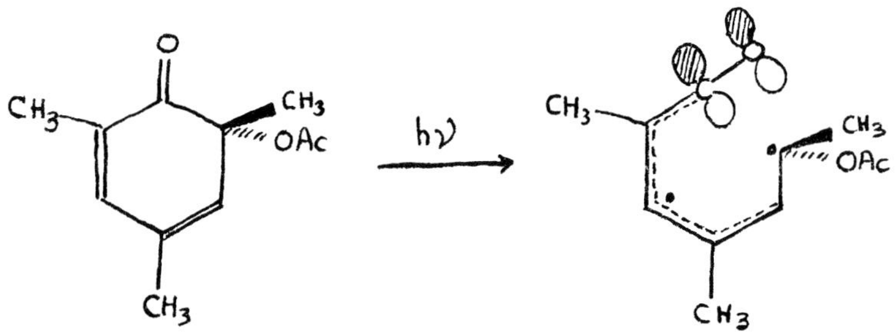

It was challenging to draw successive Pauling resonance structures for the excited state to try and discover at what point does the expected decay to ground state occur. I was not able to solve this problem but was fortunate enough to have learnt from Professor N.J. Turro another example of ketone photochemistry, the hydrogen abstraction reaction[30], for which it was possible to find a solution[31] (the first problem was thereafter easily solved). If a coplanar model reaction is chosen and if the common molecular and reaction plane is chosen as permanent symmetry element, it is possible to draw a state correlation diagram for the hydrogen abstraction (Norrish II) reaction of a ketone (Figure 9). The electron count shown in the Figure is just a simple manner of obtaining the symmetries of the states.

One striking result is the correlation of excited reactant with ground product. Furthermore, comparison with other reaction types shows a similarity between certain correlation diagrams, even though the reactions may be totally unrelated : surface crossing for hydrogen abstraction of ketones or for addition of electron-rich olefins to ketones, surface touching for α-cleavage of ketones or for ring opening of azirines. Clearly there must

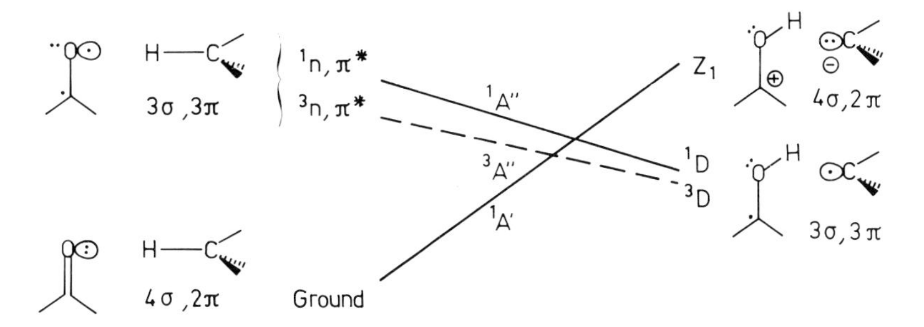

Figure 9 : State correlation diagram for hydrogen abstraction by a ketone.

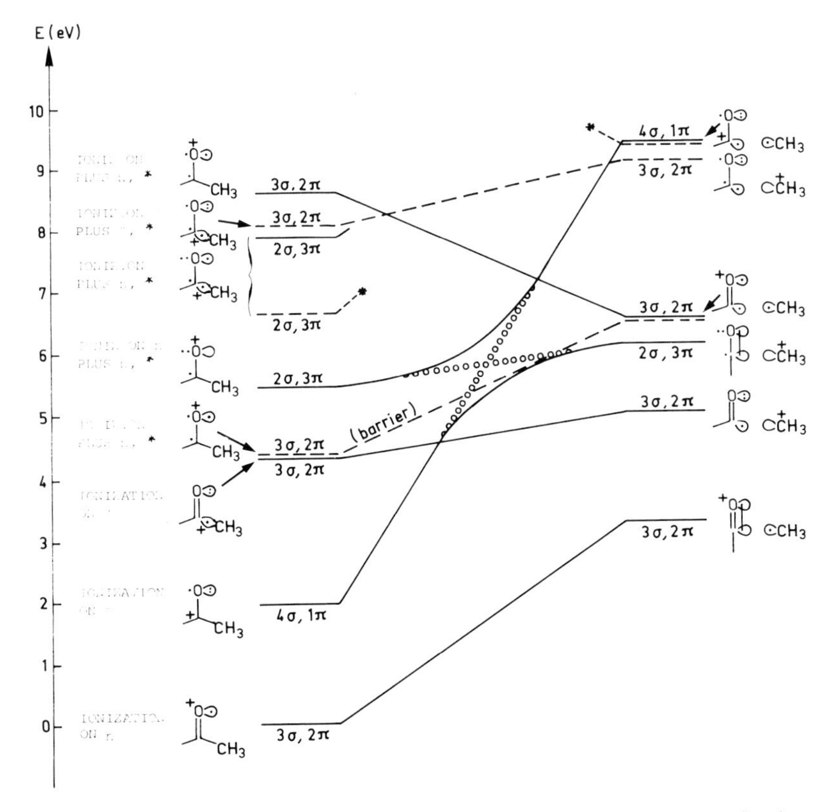

Figure 10 : Correlation Diagram for the α-cleavage of the acetaldehyde radical-cation. Full lines : doublet states. Dotted lines : quartet states. Cercles indicate an avoided crossing . From ref. 34. (Reprinted with permission from J.Amer.Chem.Soc., 98, 0000 (1976). Copyright by the American Chemical Society).

a unifying element which is the same for reactions with identical correlation diagrams.

Topicity

This unifying feature is the number and nature of radical sites created in the primary product - or topicity[31, 32]. The number of sites determines the number of low-lying states (diradical states ; zwitterionic states ; diradical ion-pair states which have mixed character). The label of the sites - generally a symmetry label like σ, or π - determines the spatial symmetry of the states. Topicity alone suffices to classify photochemical reactions[32]. If a photochemical reaction belongs to a given family, with a specific topicity, it should have a characteristic multidimensional mapping of its electronic states. In a first approximation the pathways of the reaction and its mechanism will obey the same requirements as those for the other photochemical reactions belonging to the same family. As an example, olefin isomerization and pericyclic reactions both are labeled $e^{\star}$, e° bitopic[32]. The potential energy maps are similar, and the manner in which excited reactant descends, from surface to surface, to ground product, should also be similar in both reactions. However we must not forget the ultimate discriminating role of dynamics.

Other State Correlation Diagrams. Mass Spectroscopic Fragmentation

We have seen that state correlation diagrams are of paramount importance in determining the electronic control of photochemical reactions. The date of the first state correlation diagram is hard to pin-point, but some pioneering diagrams have played an important role in Chemistry[33]. We have been able to extend the correlation diagram method to study mass spectroscopic fragmentation[34]. The correlation diagram for the α-cleavage of the acetaldehyde radical-cation is shown in Figure 10. Although this diagram has the disadvantage, over the preceeding ones, of involving a good number of states, it provides some new information. In particular the lowest state which can produce (CH_3^+) methyl cations corresponds to ionization of the first σ electron of acetaldehyde. This ionization is calculated to cost 3.8 eV more than ionization of the most labile n electron of acetaldehyde. The agreement with the 3.8 eV threshold[35] for appearance of CH_3^+ ions in the collision-induced dissociation of acetaldehyde ions by helium is remarkable.

Avoided Crossings

Other interesting features in Figure 10 are the avoided

crossings between the second doublet state and fourth double state, as well as the barrier in the lowest quartet, also due to an avoided crossing[36]. Avoided crossings are preferred regions for dynamical decay from a higher potential surface to a lower surface. This was clearly emphasized by Michl[37]. They are therefore of considerable interest to the study of chemical reaction mechanisms.

There are quite a few different types of avoided crossings[38]. The first type (A) corresponds to the region in the neighborhood of a real physical intersection, anywhere in the (3N-6)-dimensional hyperspace defined by the molecular degrees of freedom. Such an intersection generally corresponds to a situation with some symmetry - the intersecting surfaces have different symmetries - though _in principle_ any two surfaces intersect in a (3N-8)-dimensional region[39]. Figure 11 shows the intersection of a $^1A''$ state and of a $^1A'$ state (as in the Norrish type II reaction, Figure 9), and the corresponding avoided crossing when the plane of symmetry is destroyed. The avoided-crossing region is characterized by a rapid change in the pseudo-symmetry of the states. Actually, the gap between the adiabatic surfaces is an excellent measure of the deviation from C_s symmetry.

Avoided crossings of type B,C and D are "man made". An approximate model or Hamiltonian is chosen to describe the system : a crossing occurs. The model is improved : the crossing disappears ; only its "intention" and "avoidance", due to the model,subsist. The molecule really knows nothing about such types of avoidance. However the physics involved in the avoided-crossing regions may be important. Type B avoided crossings between an "ionic" and a "covalent" surface are characterized by an electron jump. Type C avoided crossings between two molecular orbital configurations are characterized by a switch in orbital symmetry for two electrons. Type D avoided crossings between a Rydberg orbital and a valence orbital are characterized by an orbital contraction or expansion. Recently another model-dependent avoided crossing has come to our attention[40]. Such type E avoided crossings occur between two surfaces corresponding to wave functions localized on separate fragments of a molecule. Then the entire wave function (for instance electronic excitation) jumps from one part of the molecule to another in the avoided-crossing region (see Figure 12).

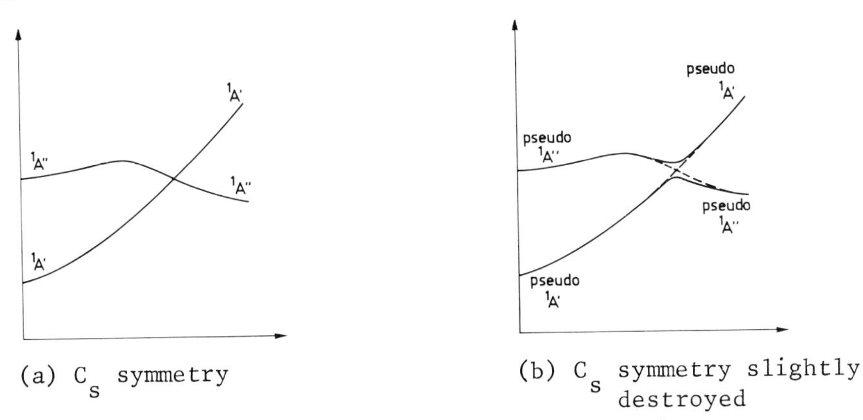

(a) C_s symmetry

(b) C_s symmetry slightly destroyed

Figure 11 : Type A avoided crossing

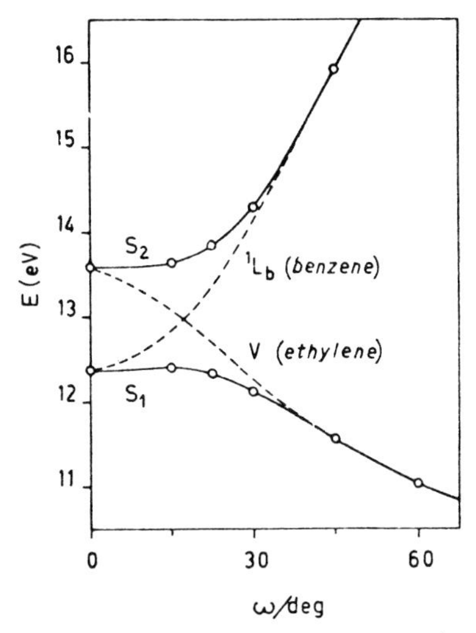

Figure 12 : Type E avoided crossing between the two lowest locally excited singlet configurations (one on benzene, one on ethylene) in styrene.From ref.40 (by permission of the North-Holland Publishing Co.).

Sudden Polarization Effect

In certain photochemical reactions the intended crossing is so strongly avoided that, due to the large energy gap in the region of primary product,the molecule remains excited for a certain length of time. Under such circumstances the reaction mechanism may be determined by the behavior of the excited state. For singlet-excited olefins, dienes, trienes, etc., such behavior should be zwitterionic. As seen previously the zwitterionic states occur in neighboring pairs, each a symmetrized combination of two ionic resonance structures :

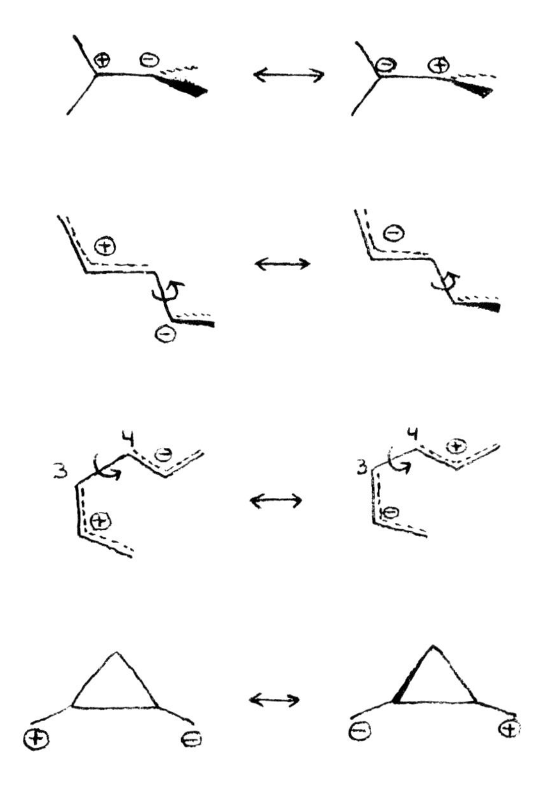

Full credit must be given to Dauben[41] for postulating a fully polarized ionic intermediate in the photocyclization of hexatriene. This stimulated us to study the charge distribution in excited di-allyl.

A calculation on the Z state of s-cis, s-trans diallyl showed all the negative charge to flow into the s-trans moiety when the 34 bond is twisted at 90°. This surprising result led us to the widespread phenomenon of sudden polarization[42]. Even though the overlap between the two radical sites a and b is zero , the charge can normally move back and forth between the two sites, thanks to the exchange integral K_{ab}. For twisted ethylene, $K_{ab} \approx 0.0021$ a.u. and it takes π/K_{ab} atomic units of time=4×10^{-14} seconds for the charge to go back and forth. Let us restrict ourselves to the two-electron part of the problem. The situation is one of two nearly-degenerate or degenerate systems - the localized ionic structures, with energies J_{aa} and J_{bb} - coupled by a matrix element (K_{ab}). The secular determinant for the problem is :

$$\begin{vmatrix} J_{aa} - E & K_{ab} \\ K_{ab} & J_{bb} - E \end{vmatrix} = 0 \qquad (3)$$

The coupling breaks down, and the solutions localize into the component polarized structures if :

$$K_{ab} << J_{aa} - J_{bb}$$

This requires very small (generally 0) overlap if the geometrical constitution and environment of sites a and b are similar. It also requires that these sites not be identical. For instance a nuclear distortion can carry a slightly dissymetric molecule through a zero-overlap situation, or make two weakly overlapping sites slightly non-equivalent. There is a sudden polarization of charge in the lowest zwitterionic excited state Z_1

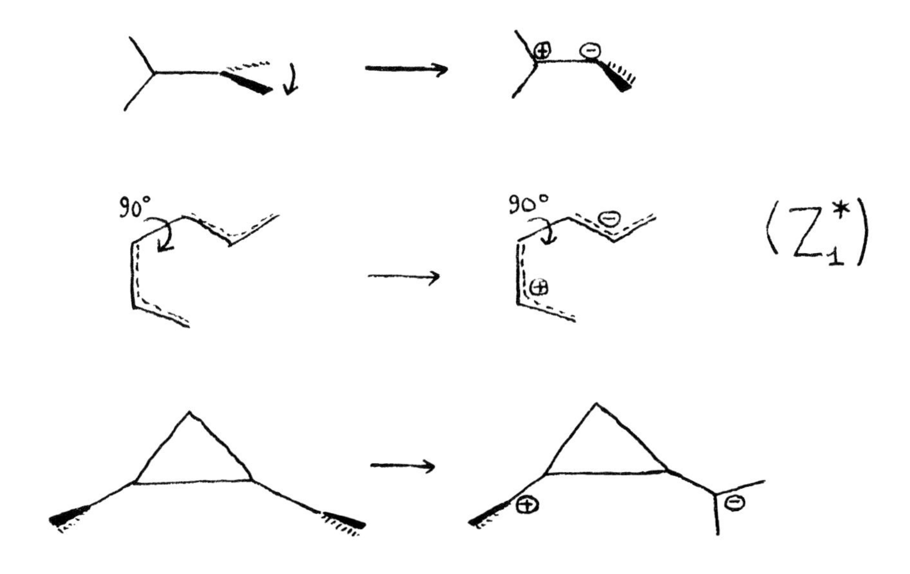

The second zwitterionic state Z_2 polarizes in the opposite direction. The charge separations calculated for pyramidalized 90°-twisted ethylene and for s-cis, s-trans diallyl are shown in Figure 13. Another example concerns retinal[43] and its protonated Schiff base. In the photoisomerization of the 11-cis, 12-s-cis form to the 11-trans, 12-s-cis form - the likely primary step in vision - passage through the 11-orthogonal, zero-overlap form creates a sudden polarization. The positive charge switches from one end of the polyene to the other. The absorbed photon is

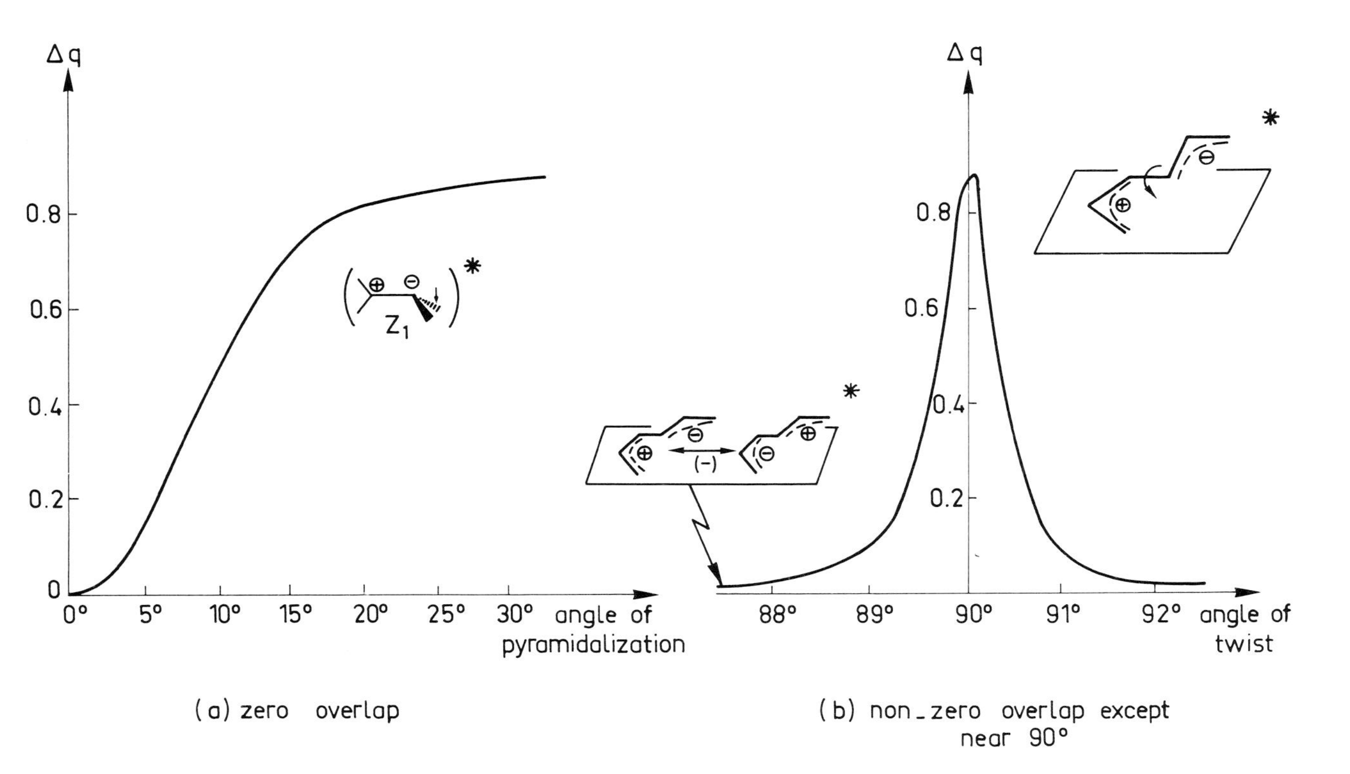

Figure 13 : L.h.s. : Charge separation Δq in Z_1 excited singlet state of 90°-twisted ethylene as a function of the pyramidalization angle Ø (Δq has been corrected for the intrinsic charge separation in the diradical ground state).
R.h.s. : Charge separation Δq in Z_1 excited singlet state of s-<u>cis</u>, s-<u>trans</u> diallyl as a function of twist angle θ. The curve is not symmetric about 90°. From ref.42.

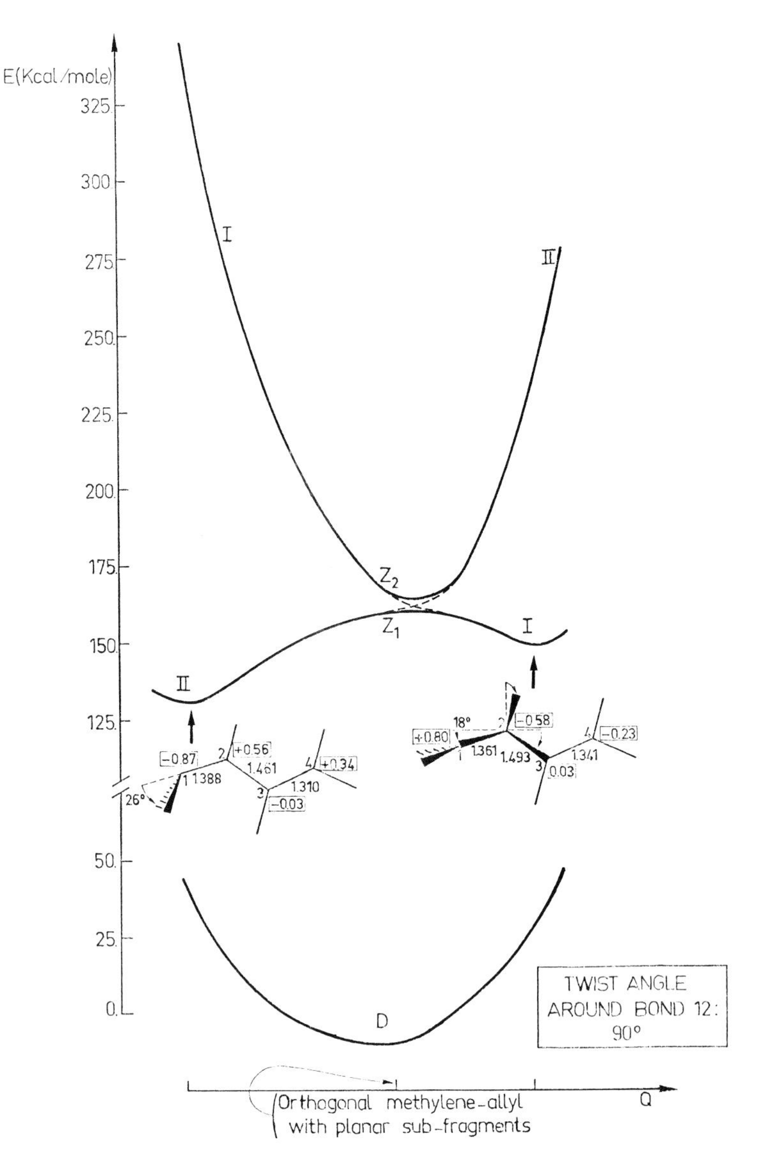

Figure 14 : Potential Surfaces for the lowest singlet states of 90° twisted methylene-allyl (minimal basis set). The coordinate Q is obtained by linear interpolation from the geometry of II to an intermediate geometry with planar subfragments, and from this intermediate geometry to the geometry of I. Bond lengths in this "half-way" skeleton are C_1C_2 = 1.375 Å, C_2C_3 = 1,475 Å, C_3C_4 = 1.325 Å (for this geometry II lies below I). Dotted lines show the avoided crossing between the configurations corresponding respectively to I and II. Note that the diradical D is a maximum, or a near-maximum, along the coordinate for twisting around bond 12 (i.e. vertical excitation does not occur from D, but from the untwisted butadiene).

thus transformed into an electric signal !

A prediction of the sudden polarization effect can be found in the remarkably farsighted work of Wulfman and Kumei[44]. It was a surprise for everyone that the effect occurs so easily. At the present time, however, we are still uncertain of the exact extent of the charge separation in the polarized state and of the energy separation between oppositely polarized Z_1 and Z_2 states. Our most reliable calculations use the restricted open-shell Hamiltonian which assumes initially one electron on site a and one electron on site b. The polarization of charge occurs only after the 3-by-3 configuration interaction step (see Section I), which corresponds to eq.(3) above. More extensive configuration interaction does not seem to decrease the charge separation ; nor do more flexible basis sets.

The calculated energy difference $E(Z_2) - E(Z_1)$ for fixed geometries varies over wide ranges, and we do not have definite values for this difference yet. However we expect the difference between $E(Z_2)$ at the optimized geometry of Z_2 and $E(Z_1)$ at the optimized geometry of Z_1 to be relatively small (1-10 Kcal/mole). Both these optimized states should lie on the same (S_1, lowest excited singlet) surface (see Figure 14). There may be interesting photochemical consequences of this double-minimum potential in the lowest excited singlet state.

IV - CONCLUSION

Our study has brought us from cyclopropane to retinal. The common link between these two systems is provided by the diradical nature of their reaction intermediates. This is true of many reaction intermediates involving the breaking of a single bond or the twisting of a double bond, and is one of the unifying threads in the study of photochemical reactions.

Another, less esoteric link has been the city of New Orleans. In 1970, after an eventful night session on diradicals at the 3rd IUPAC Photochemistry Congress in Saint-Moritz, I decided not to go forward with our work on diradicals. It was Roald Hoffmann - over a marvelous steak in a New Orleans restaurant - who, together with Luitzen Oosterhoff, convinced me to publish our work .

REFERENCES

1. Evans, M.G. and Polanyi, M., (1935), Trans.Far.Soc.31, 875
2. Professor R. Woodward and I have suggested that such a name be adopted. Our suggestion has been ardently rejected by numerous referees. An alternative is transition "point" (H. Bouas-Laurent, private communication)
3. Rabinovitch, B.S., Schlag, E.W. and Wiberg, K.B. (1958), J.Chem.Phys. 25, 504
4. Berson, J.A. and Pedersen, L.D. (1975), J.Am.Chem.Soc.97,238
5. Jean Y., Chapuisat, X. and Salem, L. (1975), J.Am.Chem.Soc. 97, 3646
6. Gelbart, W., Rice, S.A. and Freed, K., (1970), J.Chem.Phys. 52, 5718
7. Nordholm, K.S.J. and Rice, S.A. (1974), J.Chem.Phys. 61,203; (1974), J.Chem.Phys. 61, 768 ; (1975), J.Chem.Phys. 62, 157
8. Hoffmann, R. (1968), J.Am.Chem.Soc. 90, 1485
9. Jean, Y., Thesis,(1974), Faculté des Sciences d'Orsay, Université de Paris-Sud
10. Horsley, J.A., Jean, Y., Moser,C., Salem, L., Stevens, R.M. and Wright, J.S., (1972) J.Am.Chem.Soc. 94, 279
11. For a thorough discussion of this "thermochemical" postulate, which had been widely accepted, see Bergman, R.G., in Free Radicals (ed. by J. Kochi, Wiley, N.Y. (1973)), Vol.I,p.191
12. Chapuisat, X., and Jean, Y., (1975), J.Am.Chem.Soc. 97,6325
13. Wright, J.S., Tan, G., Laidler, K.J. and Hulse, J.E. (1975), Chem.Phys.Lett. 30, 200
14. Chapuisat, X., Jean, Y. and Salem, L. (1976), Chem.Phys.Lett. 37, 119
15. Barton, D.H.R. and Howlett, K.E., (1949), J.Chem.Soc. 155
16. Hiberty, P., (1975), J.Am.Chem.Soc. 97, 5975
17. Goddard, W.A., (1972), J.Am.Chem.Soc. 94, 793
18. Townshend, R.E., Ramunni, G., Segal, G., Hehre, W.J. and Salem, L. (1976), J.Am.Chem.Soc. 98,0000
19. Hehre, W.J., Lathan, W.A., Ditchfield, R., Newton, M.D., and Pople, J.A., Quantum Chemistry Program Exchange, n°236, Indiana University, Bloomington, Indiana
20. Benson, S.W., Thermochemical Kinetics, Wiley (N.Y., 1968), p. 93-95
    von E. Doering, W., (1971), Pure Appl.Chem.Suppl. (23rd. Congress) 1, 237
21. Burke, L.A., Leroy, G. and Sana, M., (1975), Theor.Chim.Acta, 40, 313
22. Dewar, M.J.S., Griffin, A.C. and Kirschner, S. (1974), J.Am. Chem.Soc. 96, 6225
23. Wilson, E.B.,(1975), J.Chem.Phys. 63, 4870
24. Woodward, R.B. and Hoffmann, R., (1969) Angew.Chem.Int.Ed. 8, 781

25. Kahn, O. and Salem, L., 6th International Congress on Catalysis, in press
26. For a review, see Salem, L., (1976), Science 191, 822
27. Salem, L. and Rowland, C., (1972), Angew. Chemie, Int.Ed., 11, 92
28. Buenker, R.J. and Peyerimhoff, S.D., (1975), Chem.Phys. 9, 75
29. Quinkert, G., (1973), Pure Appl.Chem. 33, 285
30. Turro, N.J., Dalton, J.C., Dawes, K., Farrington, G., Hautala, R., Morton, D., Niemczyk, M., and Schore, N., (1972), Acc.Chem.Res. 5, 92
31. Salem, L., (1974), J.Am.Chem.Soc. 96, 3486
32. Dauben, W.G., Salem, L. and Turro, N.J., (1975), Acc.Chem.Res. 8, 41
33. Longuet-Higgins, H.C. and Abrahamson, E.W., (1965), J.Am. Chem.Soc. 87, 2046
34. Minot, C., Nguyen Trong, A., and Salem, L., J.Am.Chem.Soc., submitted for publication
35. Giardini-Guidoni, A., Platania, R. and Zocchi, F., (1973), Int.J.Mass.Spectrom.Ion Physics, 13, 453
36. In both cases the avoided crossing is due to a correlation which, although allowed by state symmetry, is forbidden by orbital symmetry (interconversion of 2π electrons into 2σ electrons) and involves a crossing of two molecular orbital configurations. The avoided crossing is therefore of type C (see further).
37. Michl, J., (1972), J.Mol.Photochem. 4, 243
38. Salem, L., Leforestier, C., Segal, G., and Wetmore, R. (1975) J.Am.Chem.Soc. 97, 479
39. See Longuet-Higgins, H.C., (1975), Proc.Roy.Soc. A 344, 147, and references therein
40. Bruni, C., Momicchioli, F., Baraldi, I. and Langlet J. (1975), Chem.Phys.Lett. 36, 484
41. Dauben W.G., Kellogg, M.S., Seeman, J.I., Wietmeyer, N.D., and Wendschuh, P.H.,(1973), Pure Appl.Chem. 33, 197 ;
Dauben, W.G. and Ritscher, J.S., (1970), J.Am.Chem.Soc. 92, 2925 ;
Dauben, W.G., (Sept. 1972) private communication to the author
42. Bonacic-Koutecky, V., Bruckmann, P., Hiberty, P., Koutecky, J., Leforestier, C. and Salem, L., (1975), Angew.Chemie Int.Ed. 14, 575. A previous calculation by Dr. Hiberty on a 90°-twisted ethylene with the CH bonds of one CH_2 group slightly lengthened had shown a strongly polarized charge distribution in Z_1 but we did not understand its origin.
43. Salem, L. and Bruckmann, P., (1975), Nature, 258, 526
44. Wulfman, C.E. and Kumei, S., (1971), Science, 172, 1061.

SYMPOSIUM V. SURFACE QUANTUM CHEMISTRY AND CATALYSIS

Chairman : K. Fukui

Department of Hydrocarbon Chemistry
Kyoto University
Kyoto, Japan

INTRODUCTORY REMARKS ON THE QUANTUM THEORY OF CATALYSIS AND RELATED SURFACE PHENOMENA

Kenichi Fukui
Kyoto University, Kyoto, Japan

INTRODUCTION

The categories of interactions occurring in the transition metal catalysis spread over the range from coordination to chemisorption, and the range from inter-ligand reactions to inter-adsorbate reactions. The methods for theoretical approach to the elucidation of mechanisms of these interactions are tentatively classified into the following three groups based upon the "size" of the system concerned:

1. The molecular model approach,
2. The atomic cluster model approach and surface molecule model approach, and
3. The solid state theoretical approach.

The interrelation of these approaches can be represented by the following schemes:

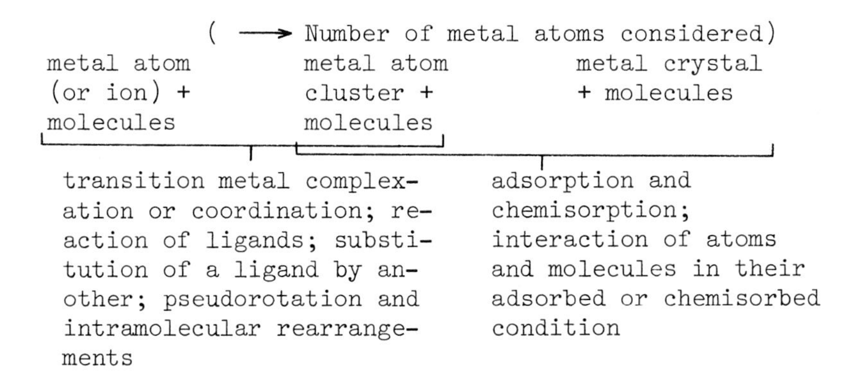

B. Pullman and R. Parr (eds.), The New World of Quantum Chemistry, 273–277.

The treatment of catalytic reactions in the molecular model approach is essentially not different from usual treatments of common chemical reactions. In this approach the role of a transition metal atom or compound, in the reaction of two molecules, can be studied by energy calculations or in an orbital interaction scheme. This model is applied most conveniently to the coordination catalysis. In the second group of methods the influence of interacting species is assumed to reach only to several atoms close to the site in the solid. Conventionally 5-10 or more atom clusters are considered to form a surface molecule together with the adsorbate atom or molecule. This model is important for the description of chemisorption. The third group belongs to the application of solid state theory to surface phenomena. This is for the most part based on the energy band calculation in regard to infinite crystals, in which the effect of chemisorbed species is incorporated into the band theory usually as a perturbation or by the use of resolvent technique.

We have three best persons to give most recent results covering the subject larger than molecules in size, so it is desirable to spare the remaining pages availble for a complementary description mainly on the molecular mechanism of catalysis.

ORBITAL INTERACTION IN CATALYSIS

In case of usual chemical interaction of one molecule with another, the electron delocalization between them plays an essential role. The most important is that occurring between the highest occupied molecular orbital (HOMO) of one molecule to the lowest unoccupied molecular orbital (LUMO) of the other molecule. The HOMO-to-LUMO delocalization scheme was used in the interpretation of orientation and stereoselection of common size molecule(1). Some quantum-chemical reactivity indices were derived in association with the electron delocalization theory, and used to explain the order of reactivity of various compounds(1). It was attempted to employ one of such indices, "delocalizability"(2), to interpret the effect of catalysts. A good correlation was found between the activation energy and the delocalizability at hydrogen atoms of hydrocarbons and alcohols in dehydrogenation and dehydration reactions by metal or metal oxide catalysts(2-4).

Imamura(5) interpreted the effect of catalysts on a molecule by the change of amplitudes of HOMO and LUMO of the molecule. Also the importance of the symmetry of HOMO and LUMO of metal compounds or metal catalysts were pointed out(6-8). Pearson(6) discussed the mechanism of oxidative additions by HOMO-LUMO interaction scheme. In the treatment of the metal-adsorbate system the importance of adsorbate HOMO was assumed and discussed(9). Baetzold(8) showed that the HOMO and LUMO of Pd and Ni clusters favor the orbital interaction in the hydrogen molecule adsorption. An interesting feature of HOMO and LUMO behavior was found in alloy clusters. Grimley(10) treated the weak chemisorption of CO on Ni by Mulliken's theory of donor-acceptor complexes, in which the HOMO and LUMO of adsorbate molecule played a significant role. This method was applied to the discussion of d-π bond in chemisorption and surface reactions of various hydrocarbons on Ni(11). Anderson(12) showed that the role of tungsten atom catalyzing the 1,3-hydrogen migration in allyl systems was to lower the activation energy by about 50% through admixture of metal d orbitals into the olefin HOMO and LUMO stabilizing the system at the transition state.

Mango used the HOMO-LUMO interaction scheme and the correlation diagram approach to explain the role of d orbitals, for instance, in the olefin dimerization and cycloaddition, and also in the disrotatory ring-opening of cyclobutane(13-15). Armstrong et al.(18) analyzed the electron distribution along the reaction coordinate corresponding to the Cossee mechanism(16,17) of ethylene polymerization to show that the important process was the interaction between the HOMO (highly localized at Ti-alkyl bond) and the LUMO (d_{zy}+ olefin π*).

Begley and Pennella(19) discussed the titanium trichloride-aluminum trialkyl catalyst system and calculated the catalyst HOMO-LUMO energy gap to be 0.66 eV which agreed with the experimental activation energy of 0.4-0.6 eV.

Fukui and Inagaki(20) determined the favorable phase relation of three-molecule HOMO-LUMO interactions and applied it to theoretical interpretation of the role of catalysts in various sorts of transition metal catalysis, such as dimerization and metathesis of olefins, regioselective insertion of coordinated olefins to metal-ligand bonds--- an elementary process involved in hydrogenation, hydroformylation, polymerization, etc.---1,3-hydrogen shift, simultaneous opening of two bonds in cyclobutane rings in polycyclic hydrocarbons, recombination of two ligands, and disrotatory ring opening of cyclobutanes.

ORIGIN OF CATALYTIC ACTIVITY

In usual chemical reactions it is dominantly the HOMO-LUMO electron delocalization that causes the bond exchange including the new bond formation and molecular deformation along the reaction path, which could be compared with the deformation of one-dimensional crystal due to Peierls transition(21).

It is the vibrational motion of nuclei that starts the electron delocalization between reactants by effectuating the spatial overlapping of orbitals and by making the HOMO-LUMO energy gap narrower. Consequently, a small gap of HOMO and LUMO levels having the order of magnitude comparable with vibrational energies in a compound would be most convenient for an effective occurrence of electron delocalization in a system in which this compound should be used as a catalyst. Obviously, the HOMO-LUMO gap becomes smaller as the number of metal atoms increases, and this might be correlated with the fact that a cluster composed of several metal atoms is usually needed to the revelation of catalytic activity(22).

As a consequence of various exemplifications mentioned above, it is evident that the appearance of high activity of transition metal catalysts can be closely connected with the following items:

1. Narrow HOMO-LUMO energy gap, or large state density at the Fermi level:

$$\rho(\varepsilon_F) \sim \frac{2}{\varepsilon_{LU}-\varepsilon_{HO}}$$

ε_F: energy of Fermi level
$\rho(\varepsilon)$: state density at energy ε
ε_{HO}: HOMO energy
ε_{LU}: LUMO energy

More generally speaking for the cases of clusters or crystals, high state densities near HOMO (high-lying occupied MO's) and LUMO (low-lying unoccupied MO's) levels.

2. Favorable symmetry situation of HOMO and LUMO composed largely of d orbitals which can produce various symmetries to meet those of reactant molecules.

The mechanisms of activation of catalysts and revivification of used catalysts, as well as the action of "promoter" were studied from many angles; one of essential origins could be correlated with the improvement of the nature of orbitals around the Fermi level. Conversely, one of the causes of deactivation of catalysts or the action of "inhibitor" would be the change of electronic states, that is, replacing d orbital nature by s or p nature in HOMO's or LUMO's, by adsorption of foreign matters on the catalyst surface or by substitution of surface atoms of catalysts by different atoms.

References:

1. K.Fukui, "Theory of Orientation and Stereoselection", Springer Verlag, Heidelberg, 1970 and 1975.
2. K.Fukui, H.Kato, and T.Yonezawa, Bull. Chem. Soc. Japan, 35, 1475(1962).
3. I.Mochida and Y.Yoneda, J. Catalysis, 9, 57(1967).
4. T.Hishida, T.Uchiyama, and Y.Yoneda, ibid., 11, 71(1968).
5. A. Imamura and T.Hirano, J. Am. Chem. Soc., 97, 4192(1975).

6. R.G.Pearson, Theoret. Chim. Acta, 16, 107(1970); Chem. Eng. News, 48, Sept. 28, p.66(1970).
7. R.Hoffmann, Acc. Chem. Res., 4, 1(1971).
8. R.C.Baetzold, J. Chem. Phys., 55, 4363(1971); J. Solid State Chem., 6, 352(1973); J. Catalysis, 29, 129(1973).
9. R.Gomer, Acc. Chem. Res., 8, 420(1975).
10. T.B.Grimley, "Molecular Processes on Solid Surfaces"(E.Drauglis, R.D.Gretz, and R.I.Jaffee, eds.), McGraw-Hill, New York (1969), p.299.
11. J.E.Demuth and D.E.Eastman, Phys. Rev. Lett., 32, 1123(1974).
12. A.B.Anderson, Chem. Phys. Lett., 35, 498(1975).
13. F.D.Mango, Adv. Catal., 20, 291(1967).
14. F.D.Mango and J.H.Schachtschneider, J. Am. Chem. Soc., 89, 2484(1967); 91, 1030(1969); 93, 1123(1971).
15. F.D.Mango, Tetrahedron Lett., 1969, 4813.
16. P.Cossee, "Advances in the Chemistry of the Coordination Compounds", Macmillan, New York(1961), p.241.
17. P.Cossee, J. Catalysis, 3, 80(1964); "The Stereochemistry of Macromolecules"(A.D.Ketley, ed.), Marcel Dekker, New York (1967), Vol.1, p.145.
18. D.R.Armstrong, P.G.Perkins, and J.J.P.Stewart, J.Chem. Soc. Dalton, 1972, 1972.
19. J.W.Begley and F.Pennella, J. Catalysis, 8, 203(1967).
20. K.Fukui and S.Inagaki, J. Am. Chem. Soc., 97, 4445(1975).
21. R.E.Peierls, "Quantum Theory of Solids", Oxford University Press, London(1955), p.108.
22. J.H.Sinfelt, J. Catalysis, 29, 308(1973).

METHODS FOR SURFACE QUANTUM CHEMISTRY

Jaroslav Koutecký

Institut für Physikalische Chemie und
Quantenchemie, Freie Universität Berlin,
1 Berlin 33

1. INTRODUCTION: SURVEY OF DIVERSE APPROACHES

If we look at the crystal as at the big molecule with principally identical interaction rules among constituent parts, it is natural to adopt and if necessary to adapt the quantum chemical concepts and methods in order to describe the chemical properties of a solid. The chemical point of view can be particulary useful for the theory of electronic structure of the surface because the environment of the surface atoms is inbetween the surroundings of the atoms in the molecule and that of the atoms in the bulk of the crystal. Therefore, the quantum chemistry concepts are useful for description of physical properties of clean surfaces. In the theory of chemisorption and catalysis the use of quantum chemical concepts is obviously necessary for the reason that one or more partners involved are normal chemical species. Therefore, the common method to handle both molecules and a crystal surface is needed.

In the solid state approach where only the crystal bulk is considered the high symmetry of the problem plays the crucial role. Even if only the clean crystal surface without structural changes is considered the translational symmetry in the direction which is not parallel to the surface is lost. The same kind of two dimensional translational symmetry can be mentained for the full surface coverage by chemisorbate. However, the qualitative structural changes can occur even for the

B. Pullman and R. Parr (eds.), The New World of Quantum Chemistry, 279–303.

clean surfaces as it is well known for Ge- and Si-crystal surfaces [1]. Of course, when chemisorption of one or more molecules of the chemisorbate takes place the translational symmetry is completely lost. In the theory of catalysis the emphasis of the surface symmetry can be even missleading, because the geometrical irregularities of the surface like edges and steps can play an important role as centers of high catalytic activity [2]. Nevertheless, the investigation of the ideal surfaces on one side and cluster of atoms on the other side is of importance because the comparison of their characteristic properties should help to understand the essential features of heterogeneous catalysis.

In quantum chemistry the wavefunction for the collective of electrons is usually constructed from the sum of products of finite number of more or less localized one-electron functions. Therefore, the incompleteness of the basis set is much more problematical than the overcompletness which can occur for the systems with infinite number of atoms. This is the case not only for Roothaan's approach to the SCF problem and its extension by CI methods but also for valence bond method as well as for the methods which during the determination of wavefunction allow changes of the one-electron basis set. The interpretation of the basis as atomic orbitals or states of quasi particles or simply basis kets is from our point of view of secondary importance. The salient point is that the localized functions or states built in the procedure make easier the chemical understanding of the molecular model and the chemical interpretation of the results. In very elaborated methods of quantum chemistry where the basis set is usually very large, the special procedures like for example mapping of one-electron densities or a posteriori construction of localized orbitals should be used in order to save the interpretative power. Very similar theoretical solid state methods are known under the name of tight binding methods. They are commonly used for the studies of electronic structures of intrinsic semiconductors and of d-electron properties of transition metals.

The typical methods of solid state theory calculate the electronic wavefunction or the one-electron density directly as a function of coordinates without the intermediate role of truncated basis set of localized functions. The examples of such methods are the free-electron approach for a jellium model, nearly free-electron methods for the periodic potential, the various versions of the electron scattering like orthogonal plane waves and Korringa or Kohn-Rostoker procedures [3]. The direct solution (if necessary numerical) of the Schrödinger equation for the elementary cell of a crystal with appropriate assumption upon the potential seen by an electron is the ultimate aim of these type of solid state methods.

The other class of approaches of the solid state theory works with the density-functional formalism of P.Hohenberg and W.Kohn [4] and studies a system of electrons moving in the electrostatic potential of atomic nuclei as well as in the Hartree and exchange electron-electron repulsion potential. The necessary estimate of the dependence of the non-electrostatic energy like exchange and correlation energy on the electron density is the decisive and most difficult point of the density-functional methods.

All mentioned methods following the line of solid state theory can be used for investigation of the surface effects if an appropriate model for the crystal surface within the framework of the given method is found. The proper description of crystal surface is usually reduced to the problem of fitting the wavefunction inside and outside the crystal. This is, of course, similar to the customary problem met in all methods which devide the whole system under study in the subspaces with corresponding different kind of potentials. The introduction of the appropriate model for chemisorbate layer or even more for the individual chemisorbed atoms is here difficult task because the importance of the local properties of the wavefunction in the neighbourhood of atoms is underestimated [5].

A typical and very elaborated example of the application of such kind of methods avoiding the mentioned difficulties is the work of Appelbaum and Hamann [6] on semiinfinite metal surface without and with chemisorbed atoms layer. The potential seen by an electron in the surface region is taken as a sum of Hartree potential, local exchange potential and electron-core pseudopotential. The wavefunctions are matched with the functions for the vacuum and the crystal bulk. The selfconsistency is required because the potential depends on electron density.

It is worth of mentioning that the X_{α} method [7] recently widely used in quantum chemistry is based on similar ideas as typical solid state approaches. The solutions of different regions in which the studied system is devided are joined by help of multiple-scattering-wave formalism. This method is recently applied for calculations of atomic clusters which should model the surface of catalist or its active centers.

As known from the literature the cluster calculations using various customary quantum chemistry methods became broadly explored field. The only but the serious limitation is the huge extent of the calculations if the methods applied should be so sophisticated that the desired results are reliable. The real difficulties are that our knowledge of the geometrical structure and the stability of the possible conglomerates of the atoms of interest is very limited and that the extent as well as the importance of interaction with the remaining crystal is not very well defined. Moreover the number of electrons which should be considered explicitely is unknown. The cluster calculations have been carried out with the simple Hückel [8], extended Hückel [9] without or with the iterative procedure, the CNDO and INDO [10] SCF methods for the clusters of lighter atoms and with the X_α approach. K.H.Johnson will report on Molecular Clusters and Catalytic Activity during this symposium. Transition between the class of calculations for the ideal crystal surface without or with chemisorption and the cluster type calculations is represented by the models with reduced number of atoms from order ∞^3 to ∞^2 or with the limited number of atoms explicitely considered but taking into account the appropriate boundary conditions. The example for the first type of such models is Bortolani's model [11] where the proper periodicity in the directions parallel to the semiconductor surface is respected but in the third direction of elementary crystal translations only finite number of layers (of the order of 20) is considered. The second type of mentioned models work with the simplified Born-Karmán's conditions. In general the Born-Karmán conditions indicate some kind of cyclic behaviour of the wavefunction in the j-th direction with very large number N_j of atomic cells in the cycle. Therefore, it is natural to decrease the number N_j so that the resulting cluster with the simplified Born-Karmán's conditions reflects the cyclic geometry. The use of the finite number of atoms for an infinite or semiinfinite elementary cell in Born-Karmán's conditions is assumed to be equivalent to the reasonable choise of the points in the Brillouin zone. Anyway some selection of the points in the Brillouin zone with appropriate weights for very large N_j must be made in order to performe concrete numerical calculations. The reasonable size of clusters makes the solution of the problem with customary semiempirical methods feasible. The extended Hückel calculations implimented on the described model have been carried out for transition metal surface by Van der Avoird and coworkers [12]. In a similar way Mesmer [13] is using the periodicity conditions in his CNDO calculations.

In the main part of this report I would like to compare different approaches using the basis set concept, mainly one-electron approximations (Hückel or Hartree-Fock type). The complementary approach is that of Schrieffer et al [14] which can be named in the language of quantum chemistry as valence bond like method. This approach will be discussed in the second contribution of this symposium. Therefore, the molecular orbital point of view in the quantum chemistry of crystal surfaces will be presented. The emphasis will be given to the common background of the corresponding methods. The analysis will be mainly made with the help of the Green-operator formalism.

2. ONE-ELECTRON APPROXIMATION

Let us shortly recall the one-electron approximation in a formalism which is suitable for discussion of the surface chemistry problems. Hamiltonian in the one-electron approximation or Hartree-Fock operator can be in general written as:

$$\hat{H} = \sum_{\mu} \alpha_{\mu} \hat{a}_{\mu}^{+} \hat{a}_{\mu} + 2\sum_{\mu<\lambda} \beta_{\mu\lambda} \hat{a}_{\mu}^{+} \hat{a}_{\lambda} \tag{2.1}$$

with creation and annihilation operators assigned to any orthogonal or nonorthogonal basis which is not necessarily built from the atomic orbitals. The quantities $\beta_{\mu\lambda}$ can be interpreted as probabilities of electron hopping from one to another more or less localized state. In the SCF procedure both α_{μ} and $\beta_{\mu\lambda}$ depend on one-electron densities.

If an ideal crystal is considered $\hat{H}$ commutes with all translation operators $t_{\vec{s}}$ from the translation symmetry group with

$$\vec{s} = \sum_{j=1}^{3} m_j \vec{a}_j \ . \tag{2.2}$$

The one-electron Bloch sum $|\vec{k},l>$ assigned to the irreducible representation of the translational group characterized by the wave vector $\vec{k}$ defined as [14]:

$$\vec{k} = \sum_{j=1}^{3} k_j \vec{a}_j^{*} \tag{2.3}$$

with

$$(\vec{a}_j \vec{a}_k^{*}) = 2\pi \delta_{jk} \tag{2.4}$$

is obtained by application of $\sqrt{N}$ multiple of the projector operator

$$\hat{P}_{\vec{k}} = N^{-1} \sum_{\vec{s}} e^{i\vec{k}\vec{s}} \hat{t}_{\vec{s}} \tag{2.5}$$

on appropriate ket because the following relation equivalent to Bloch theorem holds:

$$\hat{t}_{\vec{s}'} \hat{P}_{\vec{k}} = e^{-i\vec{k}\vec{s}'} \hat{P}_{\vec{k}} \quad . \tag{2.6}$$

In the LCAO approximation it is assumed that the Bloch sum which is the eigenfunction of $\hat{H}$ can be expressed in the terms of the same basis functions used for construction of $\hat{H}$ in Equation (2.1):

$$|k,l> = \sum_{\nu=1}^{M} C_{l\nu}(\vec{k}) \sqrt{N} \hat{P}_{\vec{k}} |\vec{0},\nu> \tag{2.7}$$

The label μ assigned to the basis functions in Equation (2.1) is now more specified: first symbol in the ket identifies the elementary cell and the second symbol distinguishes various states localized in the same elementary cell. The coefficients in the expansion of the Bloch sum satisfy the secular equations:

$$\sum_{\nu=1}^{M} C_{l\nu}(\vec{k})[H_{\nu'\nu}(\vec{k}) - E_l(\vec{k})S_{\nu'\nu}(\vec{k})] = 0 \quad , \quad \nu' = 1, \ldots M \tag{2.8}$$

where

$$H_{\nu'\nu}(\vec{k}) = N^{-1/2} \sum_{\vec{s}} e^{i\vec{k}\vec{s}} \langle \vec{s}, \nu' | \hat{H} | \vec{0}, \nu \rangle \qquad (2.9)$$

$$S_{\nu'\nu}(\vec{k}) = N^{-1/2} \sum_{\vec{s}} e^{i\vec{k}\vec{s}} \langle \vec{s}, \nu' | \vec{0}, \nu \rangle \qquad (2.10)$$

are Fourier transforms of the Hartree-Fock operator and overlap matrices, respectively.

The Wannier function localized around elementary cell s and assigned to the l-th band is

$$|w_{\vec{s}}, l\rangle = \sum_{\nu, \vec{s}} D_{l\nu}(\vec{s}' - \vec{s}) |\vec{s}', \nu\rangle \qquad (2.11)$$

where

$$D_{l\nu}(\vec{s}) = N^{-1} \sum_{\vec{k}} e^{i\vec{k}\vec{s}} C_{l\nu}(\vec{k}) \ . \qquad (2.12)$$

The Wannier functions are identical with the LCAO basis functions if only one basis ket is localized in every elementary cell. The coefficient in the Bloch sum depends on the point characterized by vector k in the Brillouin zone. In quantum chemistry terminology the kind of hybridization and the bonding situation are different in different parts of energy bands. If the periodicity is disturbed in one dimension as it is the case in all models for crystal surfaces only two components of vector $\vec{k}$ ($k_{\parallel} = (k_1, k_2)$) remain good quantum numbers. Then, the secular problem of the order of ∞^3 can be reduced only to the order of ∞ . This represents the basic difficulty of the LCAO approach.

3. SURFACE STATES

An interesting consequence of the broken symmetry in the direction which is not parallel to the surface is the possible existence of the localized one-electron states. These states can be labeled by the imaginary wave vector component k_3. The corresponding energies lay outside the two dimensional bands of volume states for an infinite crystal with the fixed component k_3. The energy bands of volume states of a semiinfinite solid characterized by $-\infty < m_1, m_2 < \infty$ and $0 \leq m_3 < \infty$ have the same position as the energy bands of the correspon-ding ideal infinite crystal. The presence of the surface as well as the presence of all local irregularities in the region of surface can be considered as a small perturbation because the amplitudes of the individual volume one-electron states are infinitesimally small.

The theoretical existence of surface states was discovered by Tamm [15] and later by Shockley [16] in the early days of quantum physics. The classification and interpretation of the surface states in late fifties and early sixties belong to the oldest chapter of the surface quantum chemistry [17,18]. At that time it was assumed that the surface states play the predominant role in the chemisorption because of their localization in the surface region. Later on it became clear that the local properties of the volume states in their total can be very important for the chemisorption, as well [17]. New interest in the surface states have been awaken recently by the new development in experimental spectroscopy [19].

The localized Coulombic perturbation of the local potential in the surface region gives rise to the Tamm states. The appearence of the Shockley states can be interpretated [20] as manifestation of dangling bonds when the interruption of the localized bonds occur due to the simple formation of the surface. The chemisorp-tion or Hoffmann states [21] arise if the bond between a surface atom and the chemisorbate molecule is strong enough. Characteristic properties of the surface states are the existence conditions which must be fulfilled for their appearance. The fulfillment of existence conditions for one kind of surface states often excludes the appearance of surface states of another kind. In nature, it is difficult to separate the effects which are responsible for fulfillment of different existence conditions and therefore, the possibility to observe the localized surface states is not very large especial-ly because they should be localized in relatively narrow energy gaps.

A surface state can in principle occur within the normal energy gap, in the volume state spectrum or within the two-dimensional energy gap between two bands with differently fixed k_3 of an ideal crystal. In the last case the surface state is degenerated with some volume states and the indication of its existence is the large probability of finding the electron near the surface in the given energy interval. A surface state can disappear from the "two-dimensional" energy gap and can be traced as a maximum of the local density of states inside the energy band.

If the one-dimensional chain of elementary cells is considered as the model for a crystal the secular problem can be relatively simply solved using the resolvent technique which will be discussed in the Section 4 or directly using algebraic means. Some of the basic notions important in the theory of electronic structure of the surfaces can be followed in a very transparent way already with a very simple one-dimensional model.

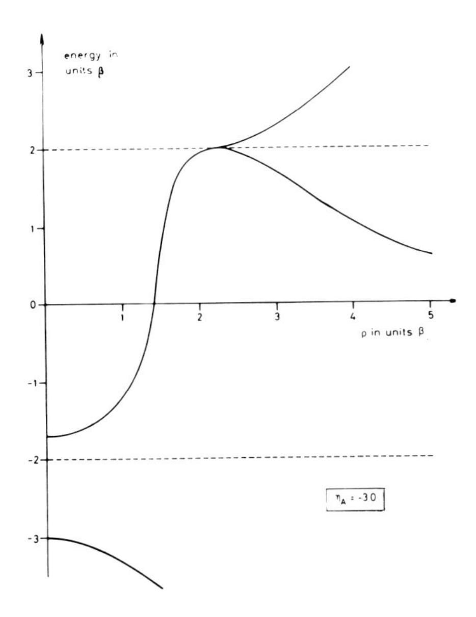

Fig.1. The change of energy of localized states and the position change of maxima of local density of states at the chemisorbed atom during approach of an adatom towards the linear atomic chain. Variable ρ is the hopping integral between the adatom and the chain. Coulomb integral for the adatom orbital has value $\alpha_A = \eta = -3$. Within the chain the hopping parameter $\beta = 1$ is fixed and all Coulomb integrals are $\alpha = 0$.

Characteristic development of the local density of states with the approach of an atomic orbital towards the end of the one-dimensional atomic chain will be shown using the Hückel type Hamiltonian. Such simple model of chemisorption neglects the influence of all other atoms in the surface region except of that atom with which the chemisorption bond is directly formed.

Figure 1 shows the position change of the surface state caused by approach of an adatom towards the atomic chain as the function of increasing hopping parameter $\rho = \beta_{AO}$. Coulomb integral of the adatom α_A is fixed and lays outside the energy band. The changes of the maximum of the local density of states at the chemisorbed atom is demonstrated as well. For very large ρ the bonding and antibonding Hoffmann or chemisorption surface states develop. In the Figure 2 analogous development of the surface state as in Figure 1 can be followed. The energy α_A now lays inside the upper half of the band of volume states. At first the maximum of the local density of states at the chemisorbed atom is

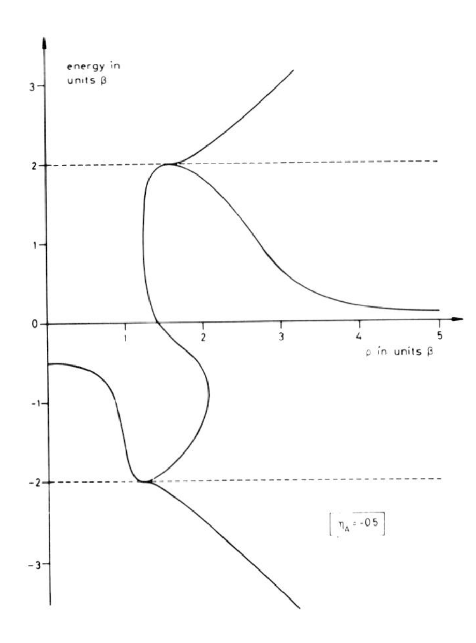

Fig.2. The change of energies and the position change of extreme of local density of states at the chemisorbed atom during approach of an adatom to the linear chain. Fixed parameters are: $\alpha_A = \eta_A = -0.5$ and for the chain $\beta = 1$ and $\alpha = 0$.

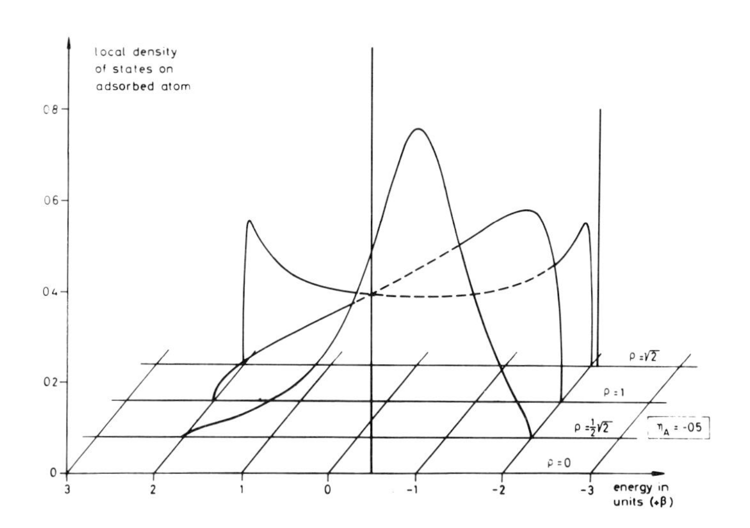

Fig.3. The local density of states for the model of Figure 2.

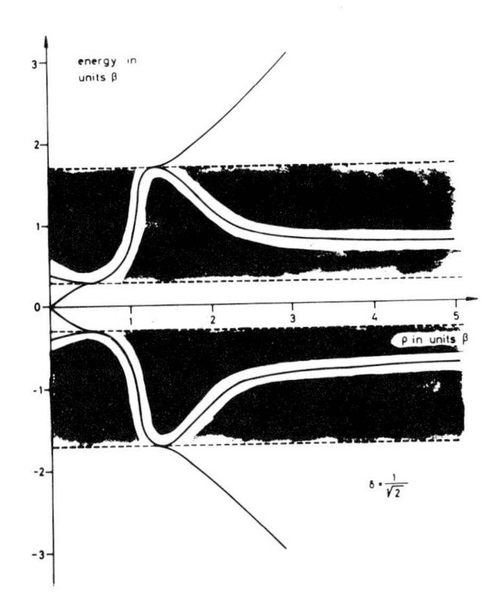

Fig.4. The change of energies and the position change of extrema of local density of states at the chemisorbed atom during the approach of an adatom to the linear chain. Within the chain the hopping parameter alternates between $\beta = 1$ and $\delta = 1/\sqrt{2}$ with the weaker bond at the end of the chain. $\alpha = \alpha_A = \eta_A = 0$

observed. With increasing bond strength modeled by the Hoffmann antibonding state is emerging from the top of the energy band and simultaneously two maxima of the local density of states are arising. For the closer approach of the adatom the bonding Hoffmann state emerges from the bottom of the band as well. The form of localized density of states at the chemisorbed atom for α_A laying inside the band is shown in the Figure 3.

The change in the position of surface states as an adatom approaches the semiinfinite chain with two atomic orbitals in one elementary cell is demonstrated in the Figure 4. This is the simplest model capable to describe localized bonds in a crystal. The Coulomb integrals for all orbitals of the chain atoms and of the adatom have the same value in this example. The localized bond is interrupted by formation of the chain end. Therefore, for the hopping parameter $\rho = \beta_{AO} = 0$ the Shockley state indicating a dangling bond appears in the energy gap between the bands of bonding and antibonding volume states. For this special case the Shockley state is degenerated with the energy of the electron localized at the adatom A. With increasing bond strength the Shockley state transforms gradually into bonding and antibonding localized states, then into the maxima of the localized density of states and finally for very large $\rho = \beta_{AO}$ develops into the Hoffmann chemisorption states.

This trivial analysis reveales in a very simple and transparent way many characteristic features of local density of states which were found in more realistic models of semiconductors [22] and transition metals [23]. It is also possible to understand why the localized surface states in energy gaps are so extremely sensitive to chemisorption and impurities forming the surface compounds.

4. GREEN OPERATORS. LONGUET-HIGGINS-COULSON INTEGRATION PROCEDURE AND KOSTER-SLATER METHOD

The main methodical difficulty met in the theory of surface phenomena is the necessity to handle simultaneously continuous or quasicontinuous and discrete energy spectra. In order to overcome this inconvenience the Green operator formalism is used. Therefore, the Green operator method will be shortly outlined in this part of the report [24]. The Green operator [25] assigned to the operator $\hat{H}$ with the eigenequation

$$\hat{H}|E_j,a\rangle = E_j|E_j,a\rangle \tag{4.1}$$

is defined as

$$\hat{G}(z) = (z\hat{I}-\hat{H})^{-1} = \hat{\Delta}^{-1}(z) \ . \tag{4.2}$$

The projector operator

$$\hat{P}_j = \sum_a |E_j,a\rangle\langle E_j,a| = \frac{1}{2\pi i}\oint_{\Gamma(E_j)} \hat{G}(z)dz \tag{4.3}$$

can be obtained by integration in the complex plain along contour $\Gamma(E_j)$ which includes only E_j point of the spectrum. The operator

$$\hat{P}_\Gamma = \frac{1}{2\pi i}\oint_\Gamma \hat{G}(z)dz \tag{4.4}$$

projects on the space of all eigenstates $|E_j,a\rangle$ with energies E_j laying inside the contour Γ. From the projection properties of $\hat{P}_\Gamma$ or from the spectral decomposition of $\hat{G}(z)$ follows that any operator function $f(\hat{H})$ can be expressed in the form:

$$f(\hat{H})\hat{P}_\Gamma = \frac{1}{2\pi i}\oint f(z)\ \hat{G}(z)dz \tag{4.5}$$

Two examples of special importance are:

$$f(\hat{H}) = \hat{H} \quad \text{and} \quad f(\hat{H}) = \hat{U}(t) = e^{-i\hat{H}t} \tag{4.6}$$

If the operator $\hat{H}$ is a sum of the operators $\hat{H}_o$ and the perturbation V the Green operator $\hat{G}$ assigned to $\hat{H}$ can be expressed by the help of Dyson equation through the Green operator $\hat{G}_o$ assigned to $\hat{H}_o$ as:

$$\hat{G}(z) = \hat{L}^{-1}(z)\ \hat{G}_o(z) \tag{4.7}$$

where

$$\hat{L}(z) = \hat{I}-\hat{G}_o\hat{V} \ . \tag{4.8}$$

From the rewritten form of Equation (4.1)

$$(E_j\hat{I}-\hat{H}_o)|E_j\rangle = \hat{V}|E_j\rangle$$

it is evident that the conditions for determination of eigenvalues E_j which are different from eigenvalues of $\hat{H}_o$

takes form:

$$|\mathbf{L}| = \left| \langle \mu | \hat{L}(E) | \nu \rangle \right| = 0 \tag{4.9}$$

Symbol $\mathbf{L}$ in the determinant $|\mathbf{L}|$ is the matrix representation of the operator L(E) in the basis $|\mu\rangle, |\nu\rangle$. The Equation (4.9) is direct consequence of Equation (4.7) because the poles of $\hat{G}(z)$ which are not the poles of $\hat{G}_o(z)$ must be zeros of the determinant $|\mathbf{L}|$. Until now no special assumption upon $\hat{H}$ has been made.

In the LCAO one-electron approach it is possible to reduce the order of the determinant $|\mathbf{L}|$ to the number of orbitals localized in the surface region Q where the perturbation $\hat{V}$ caused by existence of the surface or chemisorbed molecules is not negligible. Then, the Equation (4.9) using the projector operator $\hat{Q}$ on the space Q takes the following form [24,17]:

$$\left| \langle \mu | \hat{Q}\ \hat{L}(E)\ \hat{Q} | \nu \rangle \right| = 0 \tag{4.10}$$

which is of finite order if the symmetry parallel to the surface can be used. Equation (4.9) is Koster-Slater relation [26] which has been used in the adapted form of Equation (4.10) in the theory of surface states long time ago [17]. The one-electron Hamiltonian $\hat{H}$ can be written as:

$$\hat{H} = \sum_{j=1}^{N} \hat{h}(j) \tag{4.11}$$

The projector on the space $\underline{K}$ of doubly occupied eigenvectors $|\varepsilon_j\rangle$ of the one-electron operator h with eigenvalues ε_j is of course

$$\hat{P}_{\underline{K}} = 2 \sum_{j \in \underline{K}} |\varepsilon_j\rangle\langle\varepsilon_j| \tag{4.12}$$

Using expression (4.2) for the Green operator $\hat{G}(z)$ and the projection properties (4.4) having in mind the meaning of $\hat{P}_{\underline{K}}$ in one-electron approximation it is easy to show that the bond orders between the orthogonal orbitals $|\mu\rangle$ and $|\nu\rangle$ are:

$$P_{\mu\nu} = \frac{1}{\pi i}(-1)^{\mu+\nu} \int_{\Gamma(\underline{K})} \left\{ |\mathbf{\Delta}(z)|_{\mu\nu} \big/ |\mathbf{\Delta}(z)| \right\} dz \tag{4.13}$$

where $\boldsymbol{\Delta}(z)$ is the matrix representation of the operator $\hat{G}^{-1}$ (compare Equation (4.2)) and $|\boldsymbol{\Delta}(z)|_{\mu\nu}$ is the subdeterminant obtained from $|\boldsymbol{\Delta}(z)|$. Then, according Equation (4.5) the energy of the occupied orbitals is:

$$\varepsilon_{\underline{K}} = \frac{1}{\pi i} \int_{\Gamma(\underline{K})} z\mathrm{Tr}(G(z))dz = \frac{1}{\pi i} \int_{\Gamma(\underline{K})} z \left|\boldsymbol{\Delta}^{-1}\right| \frac{\partial|\boldsymbol{\Delta}|}{\partial z}\, dz \qquad (4.14)$$

Coulson and Longuet-Higgins [27] have derived Equations (4.13) and (4.14) for the Hückel model in different way using the integration path along the imaginary axis and a semicircle with infinite diameter for the integration contour Γ. These relations were generalized by Chirgwin and Coulson [28] for a set of nonorthogonal atomic orbitals. In addition to the basis of atomic orbitals the dual basis $|\mu_1>$ must be introduced:

$$|\mu_1> = \sum_{\nu} (S^{-1})_{\nu\mu} |\nu> \qquad (4.15)$$

and the representations $<\mu_1|\hat{G}|\nu_1>$ and $<\mu|\hat{G}|\nu_1>$ should be defined [24]. Grimley and coworkers have used the nonorthogonal basis [29] and they have considered also the case for which $(S^{-1})_{\mu\nu}$ does not exist because of the overcompleteness of the basis. The connection of the Coulson and Longuet-Higgins integral procedure with the theory of Green functions was shown by Linderberg and Öhrn [30].

5. APPLICATION OF GREEN OPERATORS ON THE THEORY OF SURFACE STATES AND CHEMISORPTION

The energies of the surface states can be determined by help of Equation (4.10). This relation arises either from the Koster-Slater method [17] or from the Green function theory [31] as conditions for determining the poles of Green function for the model of a crystal with the surface. For this purpose it is advantageous to consider $\hat{H}_o$ as Hamiltonian of the infinite ideal crystal and to respect the "cleavage" of the crystal by prohibiting the electron hopping from one semiinfinite half to another. Analogously by switching off and on the electron hopping we can distinguish the system S_o consisting from an ideal crystal without surface and a noninteracting molecule of a chemisorbate from the system S_1 of semiinfinite crystal with a surface being in interaction with the chemisorbate. In both cases the region with considerably large perturbation V is close to the surface and has the finite width. Therefore, it is possible to expect reasonable dimension of Equation (4.10).

The chemisorption energy in Hückel type model can be considered as a difference $\Delta\varepsilon$ between the energies of two systems S_1 and S_o. In S_o there is no interaction between the crystal surface and the chemisorbate. This interaction is switched on as a local perturbation $\hat{V}$ for S_1. Using Equation (4.14) the energy difference can be directly calculated from the following relation:

$$\Delta\varepsilon = \varepsilon(S_1) - \varepsilon(S_o) = \frac{1}{\pi i} \oint_{\Gamma(\underline{K})} z \frac{\partial |\mathbf{L}(z)|}{\partial z} \Big/ |\mathbf{L}(z)| \, dz$$

$$- \sum_{\varepsilon_j > \varepsilon_F} n_{Aj} \varepsilon_{Aj} + \sum_{\varepsilon_j < \varepsilon_F} (2 - n_{Aj}) \varepsilon_{Aj} \qquad (5.1)$$

where ε_F is the Fermi energy. The fact that the energy of the highest occupied orbital of an isolated chemisorbate generally differs from the Fermi energy gives rise to the last two terms on the right band side of Equation (5.1). As already mentioned determinant $|\mathbf{L}(z)|$ is of relatively low dimension.

If the SCF approach is attempted the difference of the one-electron density matrix elements must be calculated and used in the iterative procedure to achieve the selfconsistency. In that case instead of the effective one-electron Halmiltonian the Hartree-Fock operators should be used for constructing the Green operators. In order to maintain the advantage of the Green operator technique it is necessary to assume that the selfconsistency is of importance only in the region near to the surface. The Anderson model [32] limits the selfconsistency only at the adatom. Grimley's Hubbard type model [33] takes into account the selfconsistency at the adatom and in the crystal within very limited region neighbouring the chemisorbate [34].

6. COMPARISON OF PROCEDURES FOR CALCULATION OF ELECTRONIC SURFACE PROPERTIES

Large number of calculations for determination of electronic surface properties starts from a known solution of the related simple problem. For example, in one approach it is assumed that the solution of the ideal semiinfinite crystal is known and only the perturbation due to the chemisorption should be explicitely considered. In the other approach one starts with the solution for a single atom or atomic orbital and builds the whole system by allowing the electron hopping from one to another orbital successively. The important condition for practicable applicability of this schema is that the hopping parameter only between nearest neighbours are not neglected. The common feature of both approaches is the partition of the vector space spanned by the localized orbitals of the LCAO method in two subspaces P_m and $P_{m'}$ with corresponding projectors $\hat{P}_m$ and $\hat{P}_{m'}$:

$$\hat{I} = \hat{P}_m + \hat{P}_{m'} \quad , \quad \hat{P}_m \hat{P}_{m'} = \hat{0} \tag{6.1}$$

If the localized orbitals are orthonormal, the partition of the vector space of the one-electron function corresponds simply to the spatial devision of the system. An arbitrary one-electron operator can be partitioned:

$$\hat{A} = \sum_{j,k=m}^{m'} \hat{P}_j \hat{A} \hat{P}_k = \sum_{j,k=m}^{m'} \hat{A}_{jk} \tag{6.2}$$

For example, the Hamiltonian can be written as follows:

$$\hat{H} = (\hat{H}_{mm} + \hat{H}_{m'm'}) + (\hat{H}_{mm'} + \hat{H}_{m'm}) = (\hat{H}_m + \hat{V}_m) \tag{6.3}$$

The projection of the Green operator on the subspace P_m can be expressed in the terms of the Green operators $\hat{G}^m_{mm}$ and $G_{m'm'}$ assigned to $\hat{H}_{mm}$ and $\hat{H}_{m'm'}$ respectively describing the separated parts of the system:

$$\hat{G}_{mm} = \hat{P}_m \hat{G} \hat{P}_m = \left[\hat{\Delta}_m - \hat{H}_{mm'} \hat{G}_{m'm'} \hat{H}_{m'm} \right]^{-1} =$$
$$= \left[\hat{\Delta}_m - \hat{Q}_m \right]^{-1} \tag{6.4}$$

where in analogy to the definition $\hat{\Delta}(z)$ in Equation (4.2) $\hat{\Delta}_m$ is defined as:

$$\hat{\Delta}_m = z\hat{P}_m - \hat{H}_{mm} \tag{6.5}$$

The relation (6.4) can be derived by help of Dyson

Equation (4.7).

Such partition schema is mainly used in Grimley's work [29,33,34] where in the simplest case the space P_m is spanned by the atomic orbital of the chemisorbate and $P_{\bar{m}}$ is infinite crystal. The implementation of relation (6.4) is especially simple if P_m is formed only by a single orbital. The schema can be generalized by extending the space $\hat{P}_m$ inside the crystal assuming that the perturbation $\hat{V}$ defined as:

$$\hat{V} = \hat{H} - \hat{H}_o \tag{6.6}$$

has the component $\hat{P}_m \hat{V} \hat{P}_m \neq 0$ describing the local perturbation in the space P_m caused by chemisorption.

In the other mentioned approach the Green operator can be built in the following way: Let us start with a given atomic orbital $|0\rangle$ at the surface which spans P_o. Remaining space is $P_{o'}$. Then, we define P_1 as the space of all orbitals which are accessible to the electron by a single jump from $|0\rangle$. $P_{1'}$ is the remaining space within $P_{o'}$ to which the direct electron hopping from P_o is not allowed. In such a way it is possible to build the whole system for which the following relations among projectors hold:

$$\begin{aligned} &\hat{I} = \hat{P}_o + \hat{P}_{o'} \quad ; \quad \hat{P}_n \hat{P}_{n'} = \hat{0} \\ &\hat{P}_{n+1} + \hat{P}_{n'+1} = \hat{P}_{n'} \quad ; \quad \hat{P}_{n+1} + \hat{P}_{n'} = \hat{P}_{n+1} \end{aligned} \tag{6.7}$$

Using the Equation (6.4) and the above devision of the space the "local" Green function can be obtained:

$$\begin{aligned} \langle 0|\hat{G}|0\rangle &= \Big[z - \alpha_o - \sum_{n,n' \in P} \beta_{on}\beta_{n'o} \langle n|\hat{G}_{o'o'}|n\rangle\Big]^{-1} = \\ &\Big\{(z-\alpha_o) - \sum_{n,n' \in P} \beta_{on}\beta_{n'o} \Big[(\langle p|\hat{G}_{11}|p'\rangle \\ &- \sum_{r,r' \in P_2} \beta_{pr}\beta_{r'p'} \langle r|\hat{G}_{1'1'}|r'\rangle)^{-1}\Big]_{nn'}\Big\}^{-1} \end{aligned} \tag{6.8}$$

where β_{ij} are hopping parameters between the sites i and j and the symbol $[(c_{pp'})^{-1}]_{nn'}$ means the (n,n') matrix element of the inverse matrix to the matrix with the elements $c_{pp'}$. The process for construction of "local" Green function can be carried on and a continued fraction results if all spaces P_n are one-dimensional or if the hopping parameters are taken as small quantities. One way to make Equation (6.8) easy applicable is to transform the basis vectors into a linear chain. This has been demonstrated in an elegant way by Heine et al [35]. The connection of the described partition schema with the moment expansion method of Cyrot-Lackmann et al [23,36] can be also elaborated on the basis of relation (6.8).

An important source of information about electronic properties of the studied system are various kinds of densities of states. In order to determine the density of states the projection properties of the operator P_{Γ} defined as the integral of the operator $\hat{G}(z)$ over the contour Γ (compare Equation (4.4)) should be used. There are various possible choices of the integration contour Γ in the complex plane. One choice is the integration contour of Coulson and Longuet-Higgins [27] described in Section 4. The other integration way frequently used in solid state theory leads from $-\infty$ to the Fermi level ε_F parallel to the real axis at the distance -i0 and goes back from ε_F to $-\infty$ parallel to the real axis at the distance +i0. Using such integration contour the total density of states $\rho(E)$ in one-electron approximation is obtained (compare Equation (4.14)):

$$\frac{1}{2\pi}\oint_{\Gamma} \mathrm{Tr}\ \hat{G}(z)dz = -\frac{1}{\pi}\int_{-\infty}^{\varepsilon_F} \mathrm{Im}(\mathrm{Tr}\ \hat{G}(E))dE = +\int_{-\infty}^{\varepsilon_F} \rho(E)dE \quad (6.9)$$

The trace can be taken, of course, in the representation of atomic orbitals or of molecular orbitals of the subsystems obtained by partitioning. The local quantity connected with bond order in the quantum chemistry is defined as:

$$\rho_{\mu\nu}(E) = -\frac{1}{\pi}\,\mathrm{Im}\langle\mu|\hat{G}(E)|\nu\rangle \quad (6.10)$$

The local density of states is $\rho_{\mu}(E) = \rho_{\mu\mu}(E)$. The local densities of states in the region near to the crystal boundary can give especially important information about the physical and chemical properties specifically dependent on the local electronic properties of the crystal surface.

7. CONCLUSIONS

In this report it was attempted to survey the methodical background of various approaches used in the theoretical surface chemistry and surface physics. The main emphasis was given to the methods which are in principle capable to deliver an illustrative description of the chemical bonding in the surface region as well as between the chemisorbent surface and molecules or atoms of the chemisorbate. Knowledge about the local electronic character of crystal surfaces helps to understand the spectroscopic and geometric properties of surface layers. The estimate of the surface energy and chemisorption heat for purpose of comprehension of catalytic mechanism and catalytic activity is much more difficult task.

As it was shown in the previous Sections the chemisorption energy can be in principle directly calculated without carrying out the separate energy calculations when the interaction between the chemisorbate and chemisorbent is switched off and on. Nevertheless, the necessary transition from finite to infinite system gives rise to the difficulties in actual calculations. In contrast to the normalization conditions for MO's of a molecule which enforce the molecular charge to be an integer, the normalization conditions for semiinfinite crystal are not easy to formulate. Therefore, the charge in any finite surface layer can be fractional number of elementary charges at least if the appropriate charge screening is not taken into account in the model considered. This problem should be especially considered when the selfconsistent procedure is applied for the appropriate description of the charge distribution at the chemisorbate and its neighbourhood and in principle has nothing to do with spatial limitation of the region in which the selfconsistency is explicitly considered [37]. The work along this lines is in progress in our laboratory. As already mentioned, it is not simple to define the number of electrons within the appropriate models for surface chemistry. This circumstance underlines the necessity to consider the crystal surface as an open shell system with all precautions. Careful comparison with the valence bond like methods [14] is highly recommendable when the chemisorption properties or catalytic activity is considered.

The surprisingly good experience with the qualitative and semiqualitative discussions of the reaction mechanisms in organic chemistry based on extremely simple MO-approaches should encourage application of the simple quantum chemistry concepts in much more complicated field of catalytic activity. The attempts to apply the general notions which are successful in the theory of concerted reactions for understanding of heterogeneous catalysis, for example in the case of catalytic isomerisation [38] are of big interest. In such considerations it would be also desirable to respect the possible validity of physical models for chemisorption in which the electronic delocalization in layer parts of catalyst surface is accepted.

ACKNOWLEDGEMENTS

This work has been partly supported by a grant from Deutsche Forschungsgemeinschaft. The author would like to thank to Dr. V. Bonačić-Koutecký, Mr. W. Van Doorn and Mr. O. Fromm for cooperation in preparing this report. The illustrative examples in Section 3 are based on calculations by W. van Doorn.

REFERENCES

[1] Lander, J.J., Gobeli, G.W. and Morrison, J.:1963, J.Appl.Phys. 34, 2298.
Marsh, J.B. and Farnsworth, H.E.: 1964, Surface Science 1, 3.
Mac Roe, A.V.: 1966, Surface Science 4, 247.
Lander, J.J. and Morrison, J.: 1963, J.Appl.Phys. 34, 1403.
Farnworth, H.E., Schlier, R.E. and Pillow,Jr. J.A.: 1959, J.Phys.Chem.Solids 8, 116.

[2] Bond, G.C.:1968, "Homogeneous Catalysis, Industrial Applications and Implications", American Chemical Society Advances in Chemistry, Vol.70. American Chemical Society, Washington, D.C., p.25.
Kölbel, H. and Lenteritz, G., 1960, Ber.Bunsenges. Phys.Chem. 64, 437, 525.
Samorjai, G.A.:1972, Catalysis Rev., 7, 87.
Baron, K., Kely, D.W. and Samorjai, G.A.: 1974, Surface Science 41, 45.

[3] compare Taylor, P.L.: 1970, "A Quantum Approach to the Solid State", Prentice Hall, N.J..
Kittel, C.: 1967, "Quantum Theory of Solids", Wiley, New York.

[4] Hohenberg, P. and Kohn, W.: 1964, Phys.Rev. B134, 864.

Kohn, W. and Sham, L.J.: 1965, Phys.Rev. A140, 1133.
Lang, N.D. and Kohn, W.: 1971, Phys.Rev. B3, 1215, 6010.

[5] Heine, K.: 1972, Proc.Roy.Soc. London A331, 307.
Garcia-Moliner, G. and Rubio, J.: 1971, Proc. Roy.Soc. London A324, 257.
Bartoš, I. and Velický, B.: 1975, Surface Sci. 47, 495.

[6] Appelbaum, J.A. and Hamann, D.R.: 1974, Phys.Rev. B10, 559.

[7] Slater, J.C. and Johnson, K.H.:1972, Phys.Rev. B5, 844.
Johnson, K.H.: 1973, Advan.Quantum Chemistry 7, 143.
Batra, I.P. and Robaux, O.: 1975, J.Vac.Sci.Technol. 12, 242.
Rösch, N. and Rhodin, T.N.: 1974, Phys.Rev.Lett. 32, 1189.

[8] Polycyclic aromatic hydrocarbons like coronene or ovalene can be considered as simple Hückel models of a graphit cluster.

[9] Bennett, A.J., McCarroll, B. and Messmer, R.P.: 1971, Phys.Rev. B3, 1397; 1971, Surface Sci. 24, 191.
Fassaert, D.J.M., Verbeck, H. and van der Avoird, A.: 1972, Surface Sci. 29, 501.
Anderson, A.B. and Hoffmann, R.: 1974, J.Chem.Phys. 61, 4545.
Baetzold, R.C. and Mack, R.E.: 1975, J.Chem.Phys. 62, 1513.
Robertson, R.C. and Wilmsen, C.W.: 1971, J.Vac.Sci. Technol. 8, 53.

[10] Baetzold, R.C.: 1972, Surface Sci. 36, 123.
Bennett, A.J., McCarroll, B. and Messmer, R.P.: 1971, Phys.Rev. 35, 1397.
McCarroll, B. and Messmer, R.P.: 1971, Surface Sci. 27, 451.

[11] Bortolani, V., Calandra, C. and Kelly, M.J.: 1973, J.Phys. C6, L349.
compare also: Chadi, D.J. and Cohen, M.L.: 1975, Phys.Rev. B11, 732.
Blyholder, G.: 1973, J.Chem.Soc.Commun. 625.

[12] Fassaert, D.J.M. and van der Avoird, A.: 1976, Surf. Sci. 55, 291, 313.

[13] Bennett, A.J., McCarroll, B. and Messmer, R.P.: 1971, Phys.Rev. B3, 1397.
Dobrotvorskii, A.M. and Evarestov, R.A.: 1974, Phys.Stat.Sol. (b), 66, 83.

[14] Schrieffer, J.R. and Gomer, R.: 1971, Surface Sci. 25, 315.
Einstein, T.L. and Schrieffer, J.R.:1973, Phys. Rev. B7, 3629.

[15] Tamm, I.: 1932, Z.Physik, 76, 849.

[16] Shockley, W.: 1939, Phys.Rev. 56, 317.

[17] Koutecký, J.:1965, Advances Chem.Phys. 9, 85.
compare also: Koutecký, J.: 1957, Phys.Rev. 108, 13; 1958, Transactions Faraday Soc. 54, 1038.

[18] Davison, S.G. and Levine, J.D.: 1970, Solid State Physics 25, 1.
compare also: Henzler, M.:1971, Surface Sci. 25, 650.
Levine, J.D. and Freeman, S.:1970, Phys.Rev. B2, 3255.
Freeman, S.: 1970, Phys.Rev. B2, 3272.

[19] Brundle, C.R.: 1975, J.Vac.Sci.Technol. 11, 212; 1975, Surface Sci. 48, 98.
compare also: Eastman, D.E. and Nathan, M.I.: April 1975, Physics Today, p.44 and references herein.

[20] Koutecký, J.: 1960, Phys.Rev. 120, 1212; 1961, Czeck.J.Phys. B11, 566; 1964, Angew.Chem.internat. Ed. 76, 365.

[21] Hoffmann, T.A. and Kónya, A.: 1948, J.Chem.Phys. 16, 1172.
Hoffmann, T.A.: 1950, J.Chem.Phys. 18, 984.

[22] Appelbaum, J.A. and Hamann, P.R.: 1972, Phys.Rev. B6, 2166.
Appelbaum, J.A., Baraff, G.A. and Hamann, D.R.: 1975, Phys.Rev. B11, 3822.

[23] Desjongueres, M.C. and Cyrot-Lackmann, F.: 1975, J.Phys.(Paris), L45.

[24] Koutecký, J.: Prog.in Surf.and Membrane Sci., in press.

[25] Messiah, A.: 1964, "Quantum Mechanics II," North-Holland, Amsterdam, p.712.

[26] Koster, G.F. and Slater, J.C.: 1954, Phys.Rev. 95, 1167.

[27] Coulson, C.A.: 1940, Proc.Cambridge Phil.Soc. 36, 201.
Coulson, C.A. and Longuet-Higgins, H.C.: 1947, Proc.Roy.Soc. London A191, 39.

[28] Chirgwin, R.H. and Coulson, C.A.: 1950, Proc.Roy. Soc. London A201, 196.

[29] Grimley, T.B.: 1970, J.Phys. C3, 1934.

[30] Linderberg, J. and Öhrn, Y.:1973, "Propagators in Quantum Chemistry", Academic Press, London.

[31] Kalkstein, D. and Soven, P.: 1971, Surface Sci. 26, 85.
Foo, E- Ni and Wong, H.: 1974, Phys.Rev. B9, 1859.
Grimley, T.B.: 1975, Prog.Surface Mem.Sci. 9, 71 and references herein.

[32] Anderson, P.W.: 1961, Phys.Rev. 124, 41.

[33] Grimley, T.B.: 1967, Proc.Phys.Soc. 90, 751; 1967, ibid. 92, 776.
Newns, P.M.: 1968, Phys.Rev. 178, 1123.

[34] Thorpe, B.J.: 1972, Surface Sci. 33, 306.
Hayman, E.A.: 1975, Phys.Rev. B11, 3739.
Grimley, T.B.: 1971, Ber.Bunsenges. 75, 1003.
Kelly, M.J.: 1974, Surface Sci. 43, 587.

[35] Haydock, R., Heine, V. and Kelly, M.J.: 1972, J.Phys. C5, 2845.
compare also: Haydock, R. and Kelly, M.J.: 1973, Surface Sci. 38, 139.
Gadzuk, J.W.: 1974, Surface Sci. 43, 44.

[36] Cyrot-Lackmann, F.: 1967, Adv.Phys. 16, 393; 1968, J.Phys.Chem.Solids 29, 1235; 1970, J.Phys. Paris, Suppl. 61, 67.
Cyrot-Lackmann, F., Ducastelle, F. and Friedel, J.: 1970, Solid State Commun. 8, 685.
Cyrot-Lackmann, F. and Ducastelle, F.: 1970, J.Phys. Chem.Solids 31, 1295; 1971, ibid. 32, 285; 1971, Phys.Rev. B4, 2406.
Gaspard, J.P. and Cyrot-Lackmann, F.: 1973, J.Phys. C6, 3077.

[37] Grimley, T.B. and Pisani, C.: 1974, J.Phys. C7, 2831.

[38] Compare for example: McKervey, M.A., Rooney, J.J. and Samman, N.G.: 1973, J.Catal. 30, 331.

THE CHEMISORPTION BOND*[+]

J. R. Schrieffer

Department of Physics, University of Pennsylvania, Philadelphia, Pennsylvania

1. Introduction

The understanding of the chemisorption bond between an adsorbate and the surface of an extended solid requires the marriage concepts familiar in quantum chemistry and in solid state physics[1]. The quasi localized nature of the bond is complemented by the continuum nature of the one-electron spectrum in such systems. As in other areas of quantum chemistry and solid state physics, a number of techniques of varying reliability are available, with the computational effort required to obtain a given level of accuracy increasing rapidly as this level is increased. General questions, such as "how localized is the chemisorption bond?" are easily answered in a qualitative sense, yet the reliable prediction of bond energies as a function of the nuclear coordinates is difficult at present. The situation concerning one-electron spectra is intermediate between these two extremes.

Quantities of experimental interest concerning the chemisorption bond can be divided into two classes. The first class concerns properties which are in essence determined by the ground electronic state of the system. They include: 1) equilibrium bonding geometry, 2) binding energy, 3) charge transfer and change of work function, 4) surface diffusion energy, 5) vibrational spectra, 6) Born-Oppenheimer potential energy for one (many) adsorbates, determining overlayer phases, surface chemical reactions and heterogeneous catalysis.

The second class of experimental quantities has to do with electronic excited states. There exists a host of experimental techniques for observing the electronic excitations of chemisorption systems. Each technique measures its own peculiar response function. In part it is the role of theory to interrelate these

B. Pullman and R. Parr (eds.), The New World of Quantum Chemistry, 305–315. *Reserved.*

various excitation spectra as well as relating the spectra to the ground electronic state properties listed in the above paragraph. Several surface sensitive spectroscopies are: 1) field emission; 2) photo emission; 3) Auger emission; 4) ion neutralization emission; 5) inelastic (electron, ion) scattering; 6) field desosorption spectra. As is well known from other systems, there is no simple or direct route to relate observed one-electron spectra and the binding energy of the system. Furthermore, relaxation effects (correlation energy and self-consistent field effect shifts between the ground and excited states) are often sizable compared to shifts of the spectrum due to bond formation in the ground state. Hence, theory must play a strong role in gaining a microscopic understanding of the chemisorption bond in order to unravel these many effects.

II. One-Electron Schemes

Most calculations of chemisorption bonding have utilized an effective one-electron potential V, the details of which depend upon the particular method in question. For a given potential V, there are two general approaches for determining the one-electron energy eigenstates and eigenvalues. These are the LCAO-MO schemes and the multiple scattering (or Green's function) schemes. In the former method, the one-electron state $\psi_k(r)$ is represented as a linear combination of localized functions $\varphi_i(r)$ thus,

$$\psi_k(r) = \sum_i c_{ik} \varphi_i(r) \tag{1}$$

The coefficient c_{ik} are determined by minimizing the expectation value of the effective Hamiltonian, leading to the well-known conditions

$$\sum_j (H_{ij} - E_k S_{ij}) c_{jk} = 0, \tag{2}$$

$$\det (H_{ij} - E_k S_{ij}) = 0, \tag{3}$$

where

$$S_{ij} = \int \varphi_i^*(r) \varphi_j(r) d^3r. \tag{4}$$

The indices i and j label both the atomic site and the orbital quantum numbers.

In the multiple scattering method, one casts the Schrodinger in integral form

$$\psi_k(r) = \int G(r,r',E) V(r') d^3r' \tag{5}$$

where G is the free electron Green's function. The energy eigenvalues are determined by

$$\det \left[t_{ij}^{-1}(E) - G_{ij}(E) \right] = 0, \tag{6}$$

where t(E) is the single-site t-matrix, which describes the scattering of a free electron by the potential associated with the site in question. Thus eqns. (1) and (3) correspond to eqns. (5) and (6) in the two methods. Typically, the size of the determinants are smaller in the multiple scattering method than in the LCAO-MO scheme.

There are a number of methods for determining the effective potential V. Within the LCAO-MO schemes, H_{ij} is determined by the Hartree-Fock self-consistent method, or by semi-empirical methods, i.e., fitting the parameters either to experiment or to the results of more accurate calculations. In the multiple scattering method, the effective potential is frequently determined by a suitably chosen function of the local electron density, a topic we will discuss in more detail later.

III Bond Locality

An important question is how localized is the chemisorption bond? To address this question, we need some measure of the perturbation in the local electronic structure produced by the chemisorption bond. A convenient measure of this perturbation is provided by the projected density of states $\rho_{ij}(E)$, which within the LCAO-MO scheme is defined by

$$\rho_{ij}(E) = \sum_k c_{ik}^* c_{jk} \delta(E-E_k) \tag{7}$$

for orthonormal states φ_i. These quantities satisfy the sum rule

$$\int_{-\infty}^{\infty} E^n \rho_{ij} (E) dE = \langle i|H^n|_j\rangle . \tag{8}$$

From eqn. 7, we see that ρ_{ij} is simply the sum over all molecular orbital states k of the Mulliken population multiplied by a dirac delta function. This projected density estate is the natural generalization of the Mulliken populations to a continuous spectrum. Discreet, or so-called split-off states, lead to a delta function contribution to ρ_{ij}.

The change of ρ_{ii} due to chemisorption measures the spatial fall-off of the bond as i moves away from the adsorption site. Calculations of Soven and Kalkstein[2] for ρ_{ii} on a clean (100) surface of a simple-cubic s-band solid were extended by Einstein to calculate the change $\Delta\rho_{ii}$ due to absorbing an adatom (a) atop of a surface atom (1). For a weak coupling V between the adsorbate and atom (1), a narrow Lorentzian peak occurs for ρ_{aa} and a small change of ρ_{ii} occurs in the atoms surrounding the absorption site. As V is increased the adsorbate resonance splits into two resonances, corresponding to the bonding and anti-bonding states of the diatomic surface complex formed from atoms on a and 1, with the states being weakly admixed into the continuum of the indented solid, as first discussed by Grimley and later by Newns. Einstein's[3] calculations of $\Delta\rho_{ii}$ show that for neighbors and next neighbors of 1 in the surface plane, the change of ρ_{ii} is small compared to that for ρ_{aa} and ρ_{11}. Similarly, the $\Delta\rho_{ii}$'s decrease rapidly as one moves into the crystal from the surface plane. Thus, the chemisorption bond is relatively well localized in space.

Nevertheless, significant errors are introduced if one terminates the cluster at the first coordination shell surrounding the absorption site.

IV. Effective Potential V

Aside from empirical schemes, there are two principle methods for determining V, namely the Hartree-Fock (HF) and the local density functional (LDF) schemes. For HF, one has for spin orientations.

$$\langle i|H|j\rangle = \langle i|p^2/2m + V_1|j\rangle + \sum_{ks'} \langle ik|e^2/r_{12}|jk\rangle n_{ks'} - \sum_{k} \langle ik|e^2/r_{12}|kj\rangle n_{ks}, \qquad (9)$$

where V_1 is the nuclear potential and $n_{ks'}$ is the occupation number for the molecular orbital state ks'. While the HF scheme has the advantage of treating direct and exchange interactions self-consistently, it has several serious disadvantages for chemisorption problems. Firstly, HF calculations for large systems are very time consuming computationally, particularly if well converged results are desired. Secondly, electron-electron correlations are neglected in this scheme, leading to incorrect dissociation limits for open shell systems, poor bond energies when measured on the scale of chemical accuracy (~0.1 eV error). Furthermore, relaxation effects are neglected for delocalized excitations, a situation which can lead to large unphysical shifts of one-electron levels in chemisorption problems, particularly for narrow resonance levels which are partly filled.

The density functional scheme, originally developed by Kohn and Hohenberg represents the total energy E as a functional of the electron density n(r) and determines the actual density by the variational principle

$$\frac{\delta E}{\delta n(r)} = 0. \qquad (10)$$

Kohn and Sham[5] suggested approximating $E(\{n(r)\})$ by introducing effective one-electron wave functions $\psi_k(r)$ such that n (r) is expressed as

$$n(r) = 2\sum_{k(occ)} |\psi_k(r)|^2, \qquad (11)$$

where the factor of 2 accounts for the spin sum and k is restricted to occupied states. Within the LDF scheme E is approximated by

$$E = 2\sum_{k(occ)} \langle k|p^2/2m + V_1|k\rangle + \tfrac{1}{2}\int n(r_1)n(r_2)(e^2/r_{12})d^3r_1 d^3r_2 + E_{xc}(\{n(r)\}), \qquad (12)$$

where

$$E_{xc} = \int n(r)\mathcal{E}_{xc}\left[n(r)\right]d^3r, \qquad (13)$$

with $\mathcal{E}_{xc}(n)$ being the exchange plus correlation energy of a uniform electron gas of density n. By combining eqns. (10)-(13), one has an effective Schrodinger equation

$$\left[\frac{-\hbar^2\nabla^2}{2m} + V_1(r) + V_d(r) + V_{xc}(r)\right]\psi_k(r) = \epsilon_k\varphi_k(r), \tag{14}$$

where the direct potential is

$$V_d(r) = \int n(r')\frac{e^2}{|r-r'|}d^3r' \tag{15}$$

and the exchange-correlation potential is

$$V_{xc}(r) = \frac{d}{dn}n\mathcal{E}_{xc}(n)\Big|_{n=n(r)} = \mu_{xc}\left[n(r)\right],$$

$\mu_{xc}(n)$ is the exchange plus correlation part of the chemical potential for the uniform electron gas of density n.

The $X\alpha$ scheme of Slater is formally analogous to the LDF scheme, except that V_{xc} is replaced by

$$V_{Xc} = \frac{3}{2}\alpha\frac{d}{dn}n\mathcal{E}_x(n)\Big|_{n=n(r)} = \frac{3}{2}\alpha\mu_x\left[n(r)\right], \tag{16}$$

where $\mu_x(n)$ is the exchange part of the chemical potential of a uniform electron gas of density n. There are a number of methods for choosing α, with values in the range of 0.70-0.8 being typical, depending on the problem in question. Since α=2/3 would correspond to pure exchange, the somewhat larger values of α in effect include part of the correlation energy, although not in a first principles manner.

The LDF scheme has been generalized to spin and particle density (corresponding to unrestricted Hf) by introducing effective spin-orbitals ψ_{ks}, where

$$n_s(r) = \sum_{k(occ)}|\psi_{ks}(r)|^2 \tag{17}$$

and

$$n(r) = n_\uparrow(r) + n_\downarrow(r) \tag{18}$$

$$\xi(r) = \left[n_\uparrow(r) - n_\downarrow(r)\right] \tag{19}$$

In terms of these variables, the exchange-correlation energy is given by

$$E_{xc} = \int n(r)\mathcal{E}_{xc}\left[n(r),\xi(r)\right]d^3r \tag{20}$$

with $n\mathcal{E}_{xc}(n,\xi)$ being the exchange plus correlation energy for a uniform electron gas of density n and fractional spin polarization ξ.

In table I the results of calculations by Gunnarsson and Johansson are shown for the HF, LDF and LSDF (local spin density functional) schemes for H, H_2 and H_2^+. For H, the ground state energy is 1.33 eV too high for the LDF scheme, while this error is reduced to 0.20 eV for the LSDF scheme. These incorrect val-

TABLE I

		LDF	LSDS	HF	Exact
H-atom	E_o =	-12.25eV	-13.38 eV	-13.60 eV	-13.60 eV
H_2	E_B =	4.79 eV	4.79 eV	3.64 eV	4.75 eV
	R_o =	1.44 a_o	1.44 a_o		1.40 a_o
H_2^+	E_b =		2.96 eV	2.81 eV	2.81 eV
	R_o =		2.16 a_o	2.00 a_o	2.00 a_o

ues reflect a basic shortcoming of local density schemes, namely, that they do not preserve the exact cancellation of the self-direct interaction by the self-exchange interaction, as is inherent in the Hartree-Fock scheme[9]. Thus, within the LDF and LSDF schemes, each electron experiences a self-force which leads to an unphysical (upward) energy shift. This fictitious self-force vanishes for truly delocalized states in a large system and generally leads to no difficulties if the electron density varies slowly in space. However, for localized or "split-off" states, such unphysical effects appear in both the total energy as well as in the eigenvalues, if one attempts to interpret the ϵ_k as one-electron energies to be used in calculating excitation spectra.

In table I, the values of the binding energy and equilibrium bond lengths for H_2 and H_2^+ are seen to be of high accuracy, considering the relative simplicity of the calculations in comparison with alternative methods for achieving results of this quality (configuration interaction or variational methods). The Xα result for the binding energy of H_2 at equilibrium is close to the HF value for the usual value of α for H, i.e., 0.777.

In attempting to apply the LDF approach to larger systems, one encounters the usual difficulty of accurately solving the self-consistent field equations. If one uses the multiple scattering method, it is traditional to restrict the self-consistent potential to the "muffin-tin" form, namely spherically symmetric about each atom and about a large sphere enclosing the entire complex, and a constant in the intersphere region.[10] While the muffin tin approximation leads to qualitatively correct eigenvalues ϵ_k, the binding energy is often qualitatively incorrect, as shown by Danese and Connolly.[11] However, by using the orbitals ψ_k of the muffin-tin problem as zero-order states, these authors treated the non-muffin-tin effects in first order and found dramatic improvement for the binding curve for C_2. Subsequent calculations by Danese[12] have shown that quadratic terms in the complete first order perturbation result lower the equilibrium binding energy so that the total binding energy is roughly one-half the observed binding for C_2 and CO, and 95% of the observed value for N_2, using the traditional values of α in the $X\alpha$ method. The LDF did not give a significantly improved binding when applied to CO.

Thus, while the LDF and $X\alpha$ methods are highly efficient and give qualitatively reasonable results for the binding curves, at least for small molecules, there appear to be quantitative difficulties arising from the rapid spacial variation of the electron density in such systems so that the exchange-correlation effects are not accurately reproduced by a local density approximation. Further numerical study of this question is warranted.

V. Excitations

Turning to the one-electron excitation spectrum, the Hartree-Fock scheme has a number of shortcomings, some of which were mentioned above. To include correlation (and "relaxation") effects in the excitation spectrum, one must consider the appropriate response function which enters in determining the observed spectrum for each type of experiment. For the field emission experiment, the one-body Green's function enters, while for photoemission a three-body Green's function enters in determining the observed spectrum, a considerably more complicated situation. We restrict the discussion to the one-body Green's function $G(r,r',E)$, defined by

$$G(r,r',E) = \int_{-\infty}^{\infty} i\langle T\psi(r,t)\psi^{+}(r,o)\rangle e^{iEt}dt, \qquad (21)$$

where ψ is the electron field operator and the average value is taken in the exact ground state of the system. T is the time ordering operator. G satisfies Dyson's equation

$$(E-H_o)G(r,r',E) - \int \Sigma_{xc}(r,r'',E)G(r'',r'E)dr'' = \delta(r-r'), \qquad (22)$$

where Σ_{xc} is the exchange-correlation self-energy, and the direct potential is included in H_o.

One can write a spectral representation for G:

$$G(r,r',E) = \sum_k \frac{\psi_k(r,E)\psi_k^*(r',E)}{E-E_k(E)}, \qquad (23)$$

where the eigenfunctions ψ_k satisfy

$$(E_k(E)-H_0)\psi_k(r,E) - \int \Sigma_{xc}(r,r'',E)\psi_k(r'',E)dr'' = 0 \qquad (24)$$

If $E_k(E)$ is slowly varying with E, the pole of G at E_k is identified with a quasi-particle excitation. More generally, the imaginary part of G determines the field emission current energy distribution.

As for the ground state energy, Kohn and Sham[13] proposed using a local density approximation for Σ_{xc}, namely

$$\Sigma_{xc}(r,r',E) = \delta(r-r')\mu_{xc}\left[n(r)\right], \qquad (25)$$

where μ_{xc} is the exchange-correlation part of the chemical potential for the uniform electron gas. If the electron density varies slowly in space over the range of nonlocality of the self-energy Σ_{xc}^{o} for the uniform gas, then this should be a good approximation, so long as E is in the vicinity of the Fermi energy. Uniform electron gas calculations show that Σ_{xc}^{o} is much better localized in space than the exchange self-energy Σ_{x}^{o} alone, in essence because of screening of the exchange interaction. Further, these calculations show that Σ_{xc}^{o} is weakly varying with E within the hole band. In this case eqn. (24) reduces to eqn. (14) and the ψ_k and ϵ_k discussed in section III have the interpretation of quasi-particle wave functions and energies respectively, with the many-body effects being absorbed into the effective potential V_{xc}.

Similarly, the $X\alpha$ scheme interprets the ψ_k and ϵ_k in terms of one-electron excitations for "transition" states (see below).

VI. Limitations of the Local Density Approximation

There are several situations in which the LDF approximation breaks down. Firstly, if n(r) varies significantly over the range of non-locality of $\Sigma_{xc}(r,r',E)$, i.e., the size of the exchange-correlation hole, then a non-local Σ_{xc} must be used. A simple example is the self-energy of a core hole state on an atom a distance d above a metal surface. In this case, in addition to the conventional atomic relaxation a screening of the hole's electric field by the metal occurs, and this screening is characteristic of the metallic electron density, not the density of charge on the ion. Thus, the non-locality of Σ_{xc} is on the scale of d not the inter-electronic spacing on the ion, $n_{ion}^{-1/3}(r)$, as in the LDF scheme.

A second difficulty occurs for discrete localized states or resonant states imbedded in a continuum. As discussed above, within the LDF or $X\alpha$ schemes, a self-force acts on an electron in a localized state φ_α.[9] Crudely speaking, the direct plus exchange-correlation energy for this state varies with the occupation number n_α for φ_α as

$$E_\alpha = \tfrac{1}{2}\left\{U_\alpha n_\alpha^{\,2} - J_\alpha n_\alpha^{\,4/3}\right\}, \tag{26}$$

where the effective exchange integral J_α is of order U_α so that E_α is small for $n_\alpha = 1$ or 0. However, the effective potential entering the eigenvalue equation for φ_α is

$$V_\alpha = \frac{\partial E_\alpha}{\partial n_\alpha} = U_\alpha n_\alpha - \frac{2}{3} J_\alpha n_\alpha^{\,1/3}. \tag{27}$$

For $n_\alpha = 1$ V_α is of order $U_\alpha/3$ and the eigenvalue ϵ_α is shifted upward by this large unphysical quantity. To eliminate this (and other) effects. Slater[14] introduced the transition state, in which the occupation numbers are averaged between their initial and final state values. The potential V_α entering the transition state $n_\alpha = \frac{1}{2}$ is according to eqn. (27), small, as desired.

Unfortunately, the transition state scheme does not remove the above difficulty for resonant states, in which the exact states ψ_k are truly delocalized. In this case, the transition state and ground state potentials are identical (the analog of Koopman's theorem). Furthermore, it is not only the self-interaction which causes the difficulty, but also the incorrect treatment by the LDF scheme of the correlation effects in a resonance level. Clearly, a uniform electron gas is not a good model system for describing a resonance imbedded in a continuum. This difficulty is apparent in the one-electron spectra calculated for chemisorption on a jellium surface when the adsorbate is removed from the surface.[15-17] There, one observes large unphysical upward shifts of the adsorbate resonance levels. Nevertheless, if the adsorbate levels are strongly broadened in the equilibrium bonding position, the LDF spectrum may be reasonably accurate. Again, the criterion for the validity of the LDF approximation is whether the range of the exchange-correlation hole for the state in question is less than the distance over which the electron density varies significantly, i.e., whether the system behaves locally like a uniform electron gas.

Returning to the localized state φ_α, the transition state has the added advantage of including so-called relaxation effects, in which the self-consistent effective potential changes in going from the initial to the final state. This relaxation corresponds, however, to the time average distribution of hole charge and therefore vanishes in the limit of a truly delocalized hole. Physically, the relaxation or screening cloud follows the hole as it wanders throughout its spatial probability distribution $|\psi_k(r)|^2$, just as the phonon cloud follows the electron in the polaron problem. Thus, the relaxation effects vary smoothly as one goes from a localized to a delocalized excitation. For example, for a model in which the hole is coupled to electronic screening variables whose resonance energies are E_o (the plasmon energy $\hbar\omega_p$ in an electron gas model), one finds the hole relaxation energy E_{relax} varies as

$$E_{relax}(x) = E_{relax}(0)\,\frac{\log(1+x)}{x}\,, \qquad (28)$$

for weak coupling, where x is the ratio of the hole band width W to E_o . In eqn. (28), the hole band is taken to have a constant density of states, although similar results obtain for different band shapes. For a localized excitation x=0 and the relaxation energy properly reduces to the value $E_{relax}(0)$ for a localized hole. For small x, this result is reduced by the factor (1-x/2), showing that E_{relax} decreases smoothly as the excitation becomes delocalized. There appears to be a good deal of confusion about this point in the literature, although a detailed discussion has been given by Doniach[18]. The total relaxation of a delocalized hole is not zero even though the mean field relaxation energy goes to zero in this case. Dynamical relaxation joins smoothly onto the mean field relaxation as the hole delocalizes.

A word of caution is that Σ_{xc} automatically includes all relaxation effects, by definition, so that one should not use a transition state if the true Σ_{xc} is included. The LDF approximation for Σ_{xc} includes the proper dynamical relaxation effects so long as the distance over which the hole is locally screened is small compared to the distance over which n(r) varies significantly. Thus, if one uses the traditional LDF or $X\alpha$ schemes for energy band calculations, relaxation effects are already included in the eigenvalues ϵ_k so long as n(r) does not vary rapidly in the region where $|\psi_k(r)|^2$ is large. For narrow band solids with orbitals well localized spatially within each unit cell, the LDF is not expected to give a good description of the one-electron spectra associated with these narrow bands. Similar statements hold for molecules.

How then is one to proceed beyond the LDF schemes to include non-local self-energy effects (i.e., the proper exchange-correlation hole) for rapidly varying n(r)? One approach is to carry out diagrammatic perturbation theory based on orbitals ψ_k generated from a zero-order scheme, such as LDF. Another approach is to treat narrow resonances in zero-order as isolated sharp states, whose energy depends on the ionization state of that level, such as in the induced covalent bond scheme[19]. Hopping interactions then broaden these ionization and affinity states as well as quench the unpaired spin in the case of a free radical adsorbate, as in the Kondo effect. Green's function equations of motion[20] and variational methods may also be of value[21]. While these problems are difficult they are of great importantce in developing a quantitative theory of bonding in solids, large molecules and in chemisorption systems.

Acknowledgements

The author gratefully acknowledges stimulating conversations with Drs. H. Capellman, J. B. Danese, J. W. Davenport, H. Ehrenreich, E. W. Plummer and P. Soven.

REFERENCES

+This work supported in part by the National Science Foundation Grant No. DMR 73-07682 A03 and DMR-75-09491, as well as the Laboratory for Research on the Structure of Matter, University of Pennsylvania, the Advanced Research Project Agency of the Department of Defense as monitored by the Air Force Office of Scientific Research under Contract No. F 44620-75-C-0059, the National Science Foundation under Grant DMR 72-03025, and the American Gas Association.

1. See "The Physical Basis for Heterogeneous Catalysis," E. Drauglis and R. I. Jaffee, eds., Plenum Press, N.Y. (1975); J. R. Schrieffer and P. Soven, Physics Today, Apr. 1975, P. 24; and J. C. Slater and K. H. Johnson, Physics Today, Apr. 1974, p.34.

2. D. Kalkstein and P. Soven, Surf. Sci. 26, 85 (1971).

3. T. L. Einstein, Surf. Sci. 26, 713 (1974).

4. P. C. Hohenberg and W. Kohn, Phys. Rev. 136, B864 (1964)

5. W. Kohn and L. J. Sham, Phys. Rev. 140, A 1133 (1965)

6. L. Hedin and B. I. Lundqvist, J. Phys. C4, 2064 (1971)

7. J. C. Slater, Advances in Quantum Chemistry, Vol. 6, edited by P. O. Lowdin (Academic, New York, 1971).

8. O. Gunnarsson and P. Johansson, Int. J. Quantum Chem. (in Press)

9. J. R. Schrieffer, J. Vac. Sci. and Technol. Vol. 13, 335 (1975)

10. J. C. Slater and K. H. Johnson, Phys. Rev. B5, 844 (1972); K. H. Johnson and F. C. Smith, Jr., Phys. Rev. B5, 831 (1972)

11. J. B. Danese and J. W. D. Connolly, Int. J. Quantum S7, 279 (1973); J. Chem Phys. 61, 3063 (1974); J. B. Danese, J. Chem. Phys. 61, 3071 (1974)

12. J. B. Danese, to be published; J. B. Danese and J. R. Schrieffer, Int. J. of Quantum Chem. (in press)

13. W. Kohn and L. J. Sham, Phys. Rev. 145, 561 (1966).

14. J. C. Slater, Advances in Quantum Chemistry, Vol. 6 edited by P. A. Lowdin, (Academic, New York, 1971).

15. N. D. Lang and A. R. Williams, Phys. Rev. Lett. 34, 531(1975)

16. O. Gunnarsson and H. Hjelmberg, Physica Scripta 11, 97 (1975); with B. I. Lundqvist (unpublished).

17. J. A. Appelbaum and D. R. Hamam, Phys. Rev. Lett. 34, 806 (1975); with G. A. Baraff, Phys. Rev. B11, 3822 (1975)

18. S. Doniach, Computational Methods in Band Structure, edited by P. M. Marcus, J. F. Janak and A. R. Williams, Plenum, New York (1971), p. 500.

19. R. H. Paulson and J. R. Schrieffer, Surf. Sci. 48, 329 (1975)

20. W. Brenig and K. Schonhammer, Z. Phys. 267, 201 (1974).

21. K. Schonhammer, Z. Phys. B21, 389 (1975).

MOLECULAR CLUSTERS AND CATALYTIC ACTIVITY

Keith H. Johnson

Department of Materials Science and Engineering
Massachusetts Institute of Technology
Cambridge, Massachusetts 02139

INTRODUCTION

In the broadest perspective, the science of catalysis is as diverse as chemistry itself. For example, it includes reactions catalyzed heterogeneously by metallic and nonmetallic surfaces, reactions catalyzed homogeneously by transition-metal complexes in solution, acid or base catalyzed polymerization of organic molecules, certain reactions catalyzed in the upper atmosphere, photocatalytic reactions, and biochemical reactions catalyzed by enzymes.

For the theorist who wishes to investigate catalysis from a fundamental electronic point of view, the problems are enormous. For example, if one concentrates on heterogeneous catalysis and, in particular, on catalysis by metallic or bimetallic particles dispersed on a support, should the metal particles be considered from the band-theory point of view of the solid-state physicist or from the molecular point of view of the chemist? Since chemisorption is a precursor in heterogeneous reactions on supported catalysts, what are the most important adsorbate-particle configurations? What are the effects of the supporting material on chemisorption and catalytic activity? Since the theorist investigating electronic structure is accustomed to starting with the arrangement and positions of atoms in a solid or molecule, structural information is crucial. Unfortunately, such information is often lacking for heterogeneous catalysts. This is in contrast to the isolated transition-metal complexes involved in homogeneous catalysis and enzymes or proteins involved in biocatalysis, where the molecular structure and coordination chemistry are often known to reasonable accuracy.

B. Pullman and R. Parr (eds.), The New World of Quantum Chemistry, 317–356.

Assuming that one has at least some idea of the molecular structure of the catalyst, a new set of questions can be asked. What is the mobility or lability of the adsorbates on the catalyst surface? What reaction intermediates or decomposition products of the adsorbates, if any, are important after the initial step of chemisorption? How do these intermediates convert or interact with other adsorbed species to yield the reaction products? These questions are related to the reaction kinetics and, from a theoretical point of view, are much more complex than those regarding simple chemisorption. In view of the rather poor status of first-principles theories for the kinetics of even the simplest chemical reactions, it is unlikely that accurate quantitative calculations for the "potential surfaces" of catalytic reactions will be practical for many years to come. However, with the proper guidance from experiment, calculations on a few configurations and not the entire potential surface may be sufficient to elucidate key features of the reaction kinetics.

With these limitations in mind, one may take a more optimistic outlook and consider what quantum chemistry can presently accomplish to understand catalysis at the electronic level. Since it is clear that chemisorption is a precursor in heterogeneous catalysis, it is important to try to understand the possible structures and bonding of adsorbates on clusters of atoms which simulate the active centers of supported metallic and bimetallic catalysts or the active centers of oxide and other nonmetallic catalysts. It is also important, from the theoretical point of view, to try to understand the coordination chemical bonding of ligands in isolated transition-metal complexes of the type important in homogeneous catalysis and the electronic structure of the active centers of enzymes and proteins involved in biocatalysis. As we will attempt to show in this paper, there are striking analogies between the chemisorption of molecules on the surfaces of transition-metal aggregates and the chemical bonding of the same molecules as ligands in isolated transition-metal coordination complexes, as well as equally striking analogies between the active centers of certain metalloproteins and the active centers of supported metal catalysts. Such analogies are probably _not_ fortuitous. They should be investigated and their common basis elucidated.

TRANSITION- AND NOBLE-METAL CLUSTERS

The electronic structures of small transition- and noble-metal aggregates are of considerable current interest and importance. For example, the active centers of commercial heterogeneous catalysts often consist of small metallic, bimetallic, or multimetallic clusters, typically based on Group-VIII and IB elements, supported in a porous refractory material such as silica or alumina.[1-3] The electronic structures of small metal clusters are also intrinsically interesting in the ways they are related to the bulk electronic band structures and surface states of the corresponding crystalline metals.

In comparison with the large number of band-structure calculations for crystalline metals and alloys,[4,5] relatively little fundamental work has been directed to the electronic structures of small metallic clusters until recently. Band theory, in its conventional form, is based on the assumptions of long-range crystalline order, Bloch's theorem, and reciprocal- or $\vec{k}$-space representation, which do not apply to small clusters, where there is at most only short-range order.

Molecular-orbital (MO) theory, on the other hand, is well suited, in principle, for describing metal-cluster electronic structure but is limited by the accuracy and practicality of available MO computational techniques. Thus far, conventional Hartree-Fock MO theory, based on representing the orbital wavefunctions as linear combinations of atomic orbitals (HF-MO-LCAO theory), has been applied exclusively in semiempirical form to transition- and noble-metal clusters. Baetzold[6] has reported semiempirical MO-LCAO calculations for Ag, Pd, Cd, Cu, and Ni clusters, utilizing both the extended-Hückel (EH)[7] and complete-neglect-of-differential-overlap (CNDO)[8] methods. Blyholder[9] has also used the CNDO method to investigate small nickel clusters.

Following the original EH and CNDO cluster models of Bennett, McCarroll, and Messmer[10] for the chemisorption of first-row atoms and molecules on a graphite surface, various workers have used these methods to construct cluster models of chemisorption on transition-metal surfaces. Examples include EH calculations for hydrogen chemisorption on nickel[11] and tungsten,[12] nitrogen chemisorption on tungsten,[13] carbon monoxide chemisorption on nickel,[14] and the dissociative chemisorption of first-period diatomic molecules and ethylene on tungsten and nickel.[15] CNDO calculations for hydrogen chemisorbed on nickel clusters have also recently been reported.[16]

It is important to assess the relative efficacies of various MO methods in the study of elemental transition- and noble-metal clusters *per se* before attempting to investigate the interactions of such clusters with adsorbates. Therefore, systematic MO studies of transition- and noble-metal clusters have been undertaken using the recently developed self-consistent-field-Xα-scattered-wave (SCF-Xα-SW) method,[17] in which the results have been compared with the bulk band structures of the corresponding crystalline metals and with the results obtained for metal clusters by the EH and CNDO methods. The SCF-Xα-SW approach to MO theory is based on the combined use of Slater's Xα density-functional approximation to exchange and correlation[18] and the multiple-scattered-wave method of solving the one-electron Schrödinger equation.[19] This technique has been successfully applied to a wide range of polyatomic molecules and to clusters simulating local molecular environments in solids, such applications being summarized in recent review articles.[20,21]

Small lithium clusters were the first metallic aggregates investigated by the SCF-Xα-SW method.[22] Results of these studies that are of possible general significance to the properties of small metal aggregates are: (1) there is a gradual increase of cohesive energy with increasing cluster size, approaching the value for the bulk solid; (2) there is a gradual decrease of first ionization potential (calculated by the SCF-Xα "transition-state" method[17,18]) with increasing cluster size, approaching the bulk work function; (3) the cluster with icosahedral geometry is the most energetically stable of all the clusters considered.

Because the required computation time does not increase inordinately with the number of electrons per atom, the SCF-Xα-SW method is ideally suited for the study of small transition- and noble-metal clusters. During the past year, such calculations have been carried out, with initial emphasis on elemental Cu, Ni, Pd, Pt, and Fe clusters having cubic, cubo-octahedral, and icosahedral geometries. At the time of writing of this paper, calculations have been completed and analyzed for Cu, Ni, Pd, and Pt clusters and are in progress for Fe clusters.

As an illustrative example, one may consider the results obtained for 13-atom copper and nickel clusters having the cubo-octahedral geometry shown in Fig. 1, the structure characteristic of the local arrangement of atoms in a face-centered-cubic (fcc) crystal. First of all, the MO representation of the electronic wavefunctions of the cluster and the use of orbital symmetries immediately allow one to distinguish between those orbitals which have no components on the "interior" central atom of the cubo-octahedron and which therefore are localized on the 12 equivalent "surface" atoms (hereafter called "surface orbitals"), and those orbitals which have components on all 13 atoms or are predominantly localized on the central atom (hereafter called "bulk orbitals"). Obviously, the distinction between "surface" and "bulk" orbitals of such a small cluster is academic since most of the atoms are actually surface atoms. Nevertheless, we shall adopt this terminology because it allows one to discuss the cluster results in relation to the bulk and surface electronic states of the corresponding crystals. Indeed, the rotational equivalence of the surface atoms of the cubo-octahedron is analogous to the translational equivalence of the surface atoms on an fcc crystal.

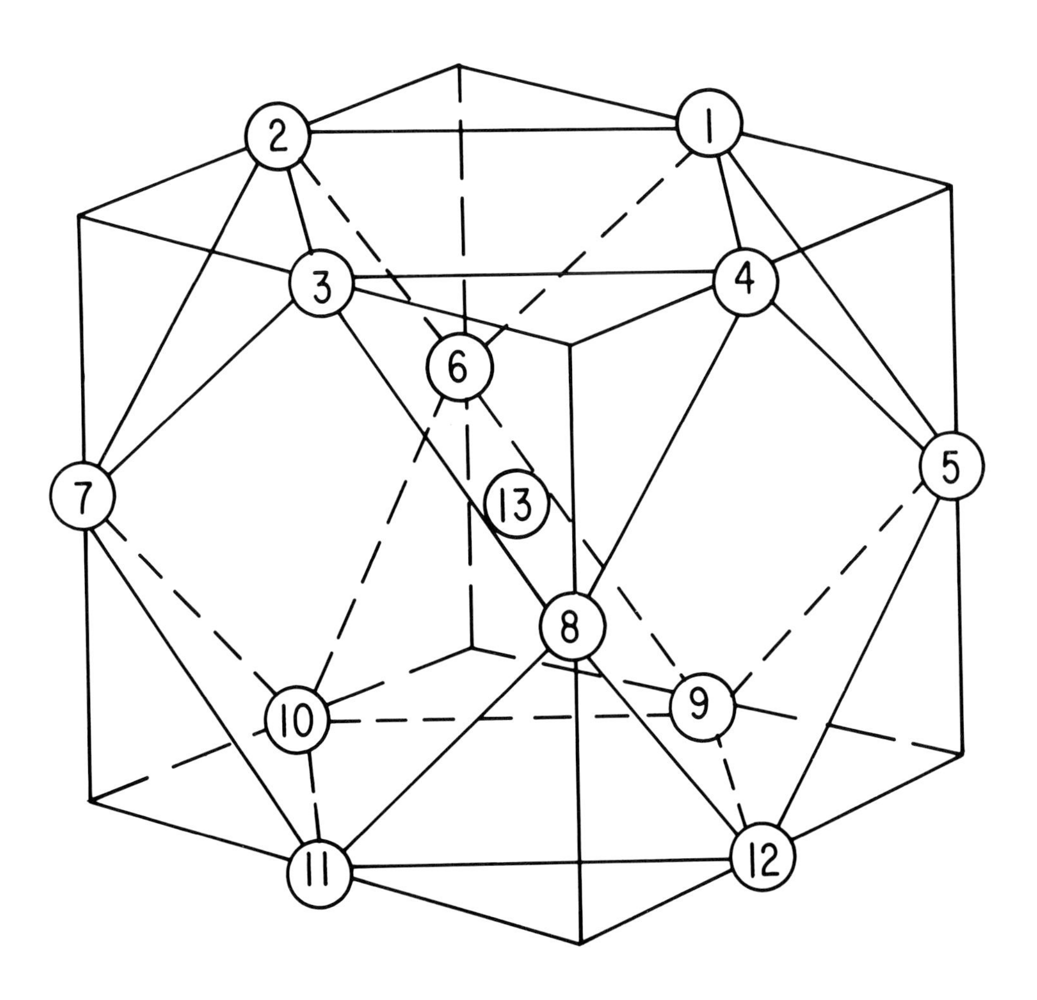

Figure 1. Cubo-octahedral cluster containing 13 atoms.

However, the cubo-octahedral cluster includes both "(100)" and "(111)" type surface planes (see Fig. 1), so that it is actually analogous to a faceted or polycrystalline surface, rather than a single-crystal one.

Using the orbital symmetries associated with the electronic energy levels of Cu_{13} and Ni_{13} clusters, we can separate the "bulk" and "surface" orbitals, resulting in the energy-level diagrams shown in Fig. 2. The levels are labeled according to the irreducible representations of the cubic (O_h) symmetry group. The first thing to note is the high density of surface orbitals in the vicinity of the Fermi level in the case of Ni_{13}. These are antibonding and nonbonding d-orbitals which are spatially oriented away from the surface of the cluster, as exemplified by the orbital contour maps shown in Figs. 3 and 4. In Fig. 3, the occupied a_{2g} orbital of Ni_{13} just below the Fermi level is mapped in an equatorial plane containing four surface atoms and the central atom of the cubo-octahedron (atoms numbered 5 through 8 and 13, respectively, in Fig. 1). The equatorial map clearly shows the absence of any contribution of the central atom to the wavefunction, consistent with the surface localization of the a_{2g} orbital.

The unoccupied Ni_{13} a_{1u} orbital lying above the Fermi level in Fig. 2 is mapped in Fig. 4 in the plane of a square face of the cubo-octahedron (atoms numbered 1, 2, 3, and 4 in Fig. 1). The splitting off of this d-orbital from the manifold of closely spaced d-levels around the Fermi energy may be viewed, in the simplest approximation, as a "surface ligand-field" effect arising from the repulsion of neighboring surface atomic d-orbitals. Because it is an empty "surface state" which is spatially directed away from the surface, it could possibly overlap and effectively accept electrons from interacting adsorbate atoms and molecules. On the other hand, in Cu_{13} the a_{1u} orbital is fully occupied and lies below the Fermi energy (see Fig. 2). In fact, the entire manifold of surface and bulk d-levels in Cu_{13} lies well below the Fermi energy, which itself coincides with an energy level (t_{2g}) that corresponds, not to a localized d-orbital, but to a highly delocalized s,p-like orbital. This is very similar to the well known band structure of crystalline copper, where the d-band is completely occupied and lies well below the Fermi level. Thus, it is not expected that the d-orbitals will play a dominant role in the interaction of adsorbates with a small copper aggregate or crystallite, as they do in the case of the surface reactivity of transition metals such as nickel, where the density of surface d-orbitals around the Fermi level is high. How this situation changes as a function of "alloying" copper with nickel, to form a bimetallic cluster, is currently under investigation (see following section).

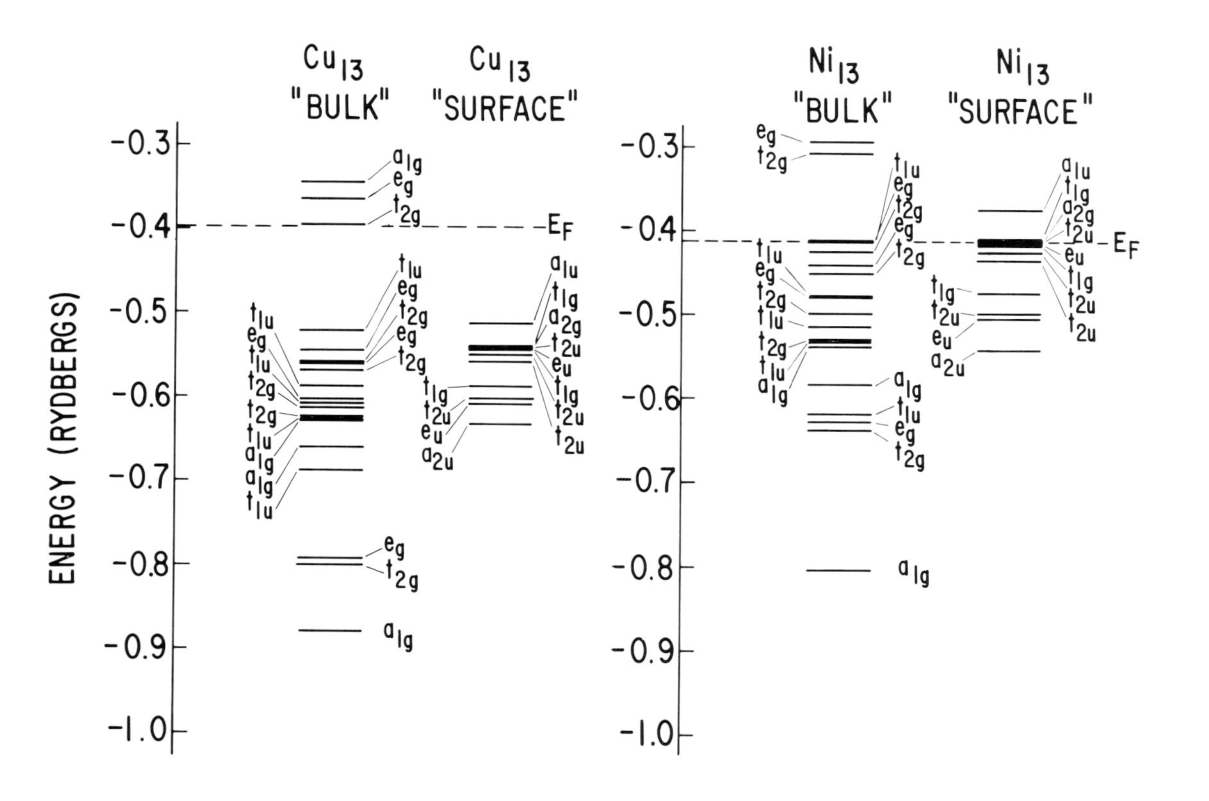

Figure 2. Separation of the SCF-Xα-SW electronic energy levels for Cu_{13} and Ni_{13} clusters into "bulk" and "surface" components.

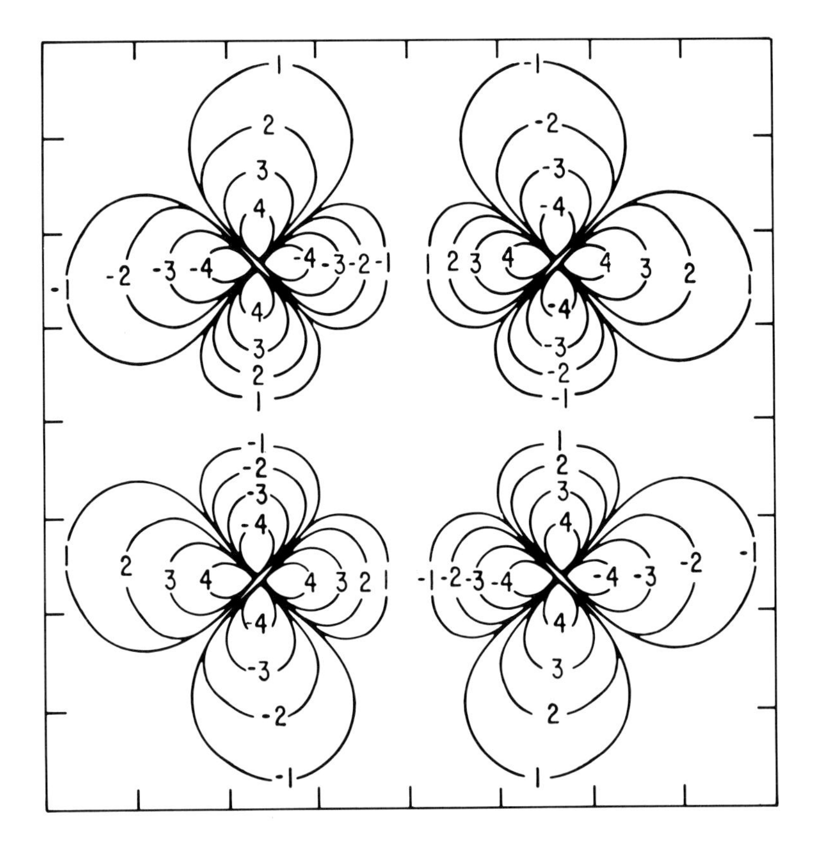

Figure 3. Contour map for the occupied a_{2g} antibonding "surface" orbital of a Ni_{13} cluster, corresponding to the energy level -0.413 Ry shown in Fig. 2, plotted in the equatorial plane containing atoms 5, 6, 7, 8, and 13 of the cubo-octahedron illustrated in Fig. 1.

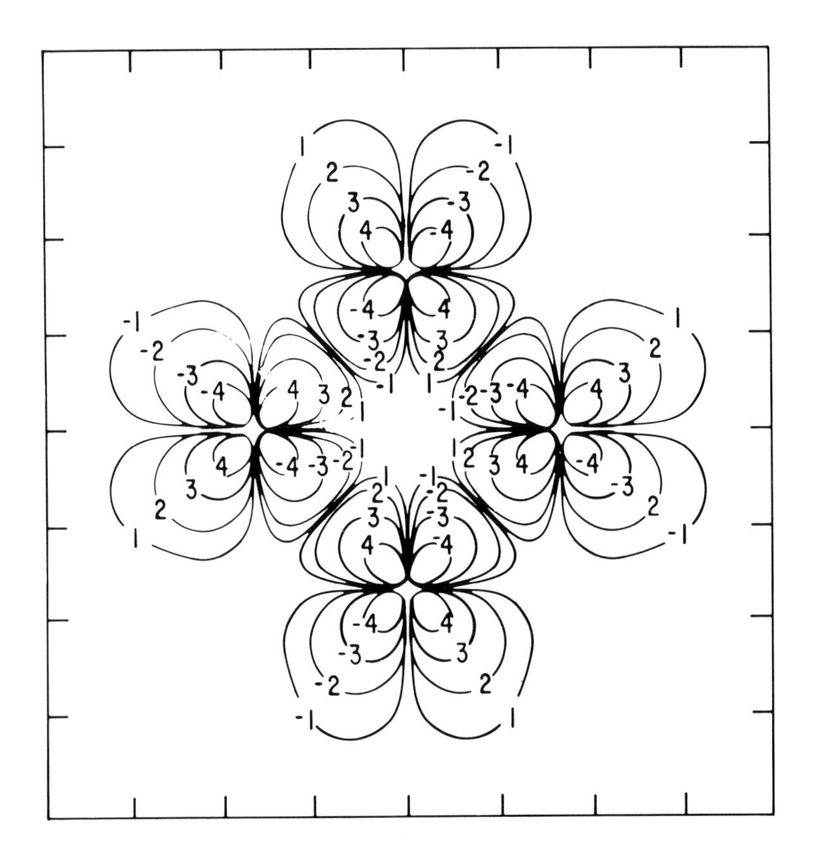

Figure 4. Contour map for the unoccupied a_{1u} antibonding "surface" orbital of a Ni_{13} cluster, corresponding to the energy level -0.376 Ry shown in Fig. 2, plotted in the plane of the square face containing atoms 1, 2, 3, and 4 of the cubo-octahedron illustrated in Fig. 1.

The calculations also suggest a slight depletion of electronic charge and therefore a slight positive charge on the surface atoms of the cluster, compensated for by a buildup of excess electronic charge on the central atom of the cubo-octahedron to preserve total electrical neutrality of the cluster. The effective surface charge (+0.06 e/atom) on Ni_{13} is somewhat larger than the value (+0.04 e/atom) for Cu_{13} and would be expected to be especially significant on transition-metal aggregates where there is a large percentage of surface atoms with relatively low coordination number, such as those at "corners" and "edges" of facets. The number of surface sites of low coordination number will tend to decrease rapidly with increasing particle size, for ideally shaped crystallites.

Further details of these results, the effects of magnetic spin polarization in nickel clusters, results for palladium and platinum clusters, and a critical comparison of SCF-Xα-SW results with extended-Hückel and CNDO MO studies of metal clusters are described in recent publications.[23,24] Spin-polarized SCF-Xα-SW calculations for small iron clusters are in progress, and preliminary results have recently been reported.[25] Relativistic SCF-Xα-SW calculations for small palladium and platinum clusters, carried out within the framework of Dirac theory, have also been carried out and preliminary results reported.[26]

One frequently hears the statement that small metallic clusters are intrinsically "different" in their catalytic behavior from larger particles or crystalline surfaces, especially in regard to so-called "demanding" or structure-sensitive reactions.[1] The higher fraction of corner and edge atoms on the surface of a small cluster, as compared with an ideal "flat" crystal surface, and the presence of spatially directed surface d-orbitals at these corners and edges (e.g., see Figs. 3 and 4) could possibly contribute to the structure sensitivity of catalytic reactions through the preferential chemisorption of reacting molecules on these sites. Furthermore, the effective positive charges of these sites and their associated electric fields could enhance the interaction of the reacting molecules through electrostatic polarization or attraction of the adsorbate charge distributions. Other things being equal, the relatively high ionization potentials (∿ 7 eV) and electron affinities (∿ 4 eV) calculated for small transition-metal clusters by the SCF-Xα-SW "transition-state" procedure[17,18] will also tend to promote electron transfer from a nucleophilic adsorbate to cluster. The cluster ionization potential and electron affinity gradually approach the bulk work function (∿ 5 eV) with increasing cluster size.[23]

On the basis of the SCF-Xα-SW MO calculations of the type described above, one is struck by the systematic similarities between the electronic structures of small transition- and noble-metal clusters and those of the corresponding crystalline metals, including the effects of "surface" electronic structure. On the other hand, this observation cannot be made regarding recent EH and CNDO MO calculations for transition- and noble-metal clusters,[6,9-16] and in fact the contrast between the SCF-Xα-SW results and these semiempirical LCAO results is rather profound in most cases. As has been discussed in detail in Refs. 23 and 24, these differences are probably due to the somewhat arbitrary parametization of the EH and CNDO methods for transition metals and to the lack of self-consistency, which leads to the improper representation of electronic charge transfer. No *ab initio* Hartree-Fock MO studies of transition- or noble-metal clusters, beyond isolated diatomic molecules, have yet been reported.

The importance of having a reliable description of the electronic structures of metallic clusters *before* using these clusters to model chemisorption and catalysis thereon cannot be overemphasized. The systematic relationship of the electronic structures of small metallic clusters to the band structures of the corresponding crystalline metals, as revealed by SCF-Xα-SW studies,[23,24] suggests that this technique is a reasonable starting point for the investigation of the active centers of supported metal catalysts. Semiempirical MO methods, such as the EH and CNDO techniques, may also be useful in such studies if the physical parameters are chosen properly for transition and noble metals. A possible systematic approach to this problem would be to determine parameters by matching the occupied bands as obtained from an EH or CNDO band-structure calculation for the bulk metal with those obtained from more rigorous band-structure calculations or from experiment. These parameters would then provide a reasonable starting point for MO calculations on metallic clusters. Alternatively, EH and CNDO parameters might be chosen by matching to the results of SCF-Xα-SW calculations for clusters.

The value of *ab initio* Hartree-Fock studies of metallic clusters is questionable, except as the starting point for configuration-interaction (CI) calculations, since Hartree-Fock theory is well known to fail to describe the band structures and cohesive energies of bulk crystalline metals. Valence-bond theory is probably a more promising *ab initio* approach to metallic clusters than Hartree-Fock theory since it explicitly accounts for effects of electron-electron correlation. Unfortunately, valence-bond theory, even in its generalized (GVB) form, is very costly in computer time to apply to complex transition- and noble-metal systems without further simplification.

BIMETALLIC CLUSTERS AND THE EFFECTS OF ALLOYING

Bimetallic and multimetallic clusters have become increasingly important in the design of active centers of supported heterogeneous catalysts, particularly those that are selective for specific reactions.[2] For example, the hydrogenolysis of ethane and dehydrogenation of cyclohexane are reactions which are both activated by elemental nickel catalysts. However, as copper is alloyed with nickel, the rate of hydrogenolysis decreases precipitously with increasing copper concentration, while the dehydrogenation reaction remains relatively constant over a wide range of composition.[27] Two factors of possible relevance to this behavior are that the hydrogenolysis reaction is structure sensitive, whereas the dehydrogenation reaction is normally structure insensitive, and there may be surface segregation of copper in the Cu-Ni aggregates.

There have been very few applications of quantum chemistry reported for bimetallic clusters of catalytic importance and none for multimetallic clusters. For example, Baetzold[6] has carried out semiempirical MO-LCAO calculations for very small Cu-Ni and Pd-Ni aggregates and has attempted to relate the calculated electronic structures to catalytic behavior. However, the lack of self-consistency of these calculations and the profound differences between the results of similar semiempirical MO-LCAO studies of elemental Cu, Ni, and Pd clusters and those obtained by the SCF-Xα-SW technique (see Refs. 23 and 24) suggest that the latter approach is likely to give a more realistic description of the electronic structures of bimetallic clusters.

To illustrate how the SCF-Xα-SW method can be applied to a bimetallic cluster, consider the host aggregate to be a Cu_{13} cluster having the cubo-octahedral geometry shown in Fig. 1. One can then form Cu-Ni clusters by selectively substituting Ni atoms for Cu atoms in the cubo-octahedron, allowing for possible relaxations of interatomic distance, and then calculating self-consistently the electronic structure of the composite cluster. The effects of "surface segregation" can be modeled by embedding small "islands" of Ni atoms, e.g., Ni_2 diatoms, in a "sea" of surface Cu atoms. The effects of alloying on chemisorption can be modeled, in turn, by allowing adsorbates to interact with the bimetallic cluster and comparing the results with those obtained for chemisorption on pure Cu_{13} and Ni_{13} clusters. This procedure can, in principle, be carried out over a range of cluster compositions, sizes, and geometries, including the effects of typical supporting environments (see later section). SCF-Xα-SW calculations are currently being carried out for 13-atom and 19-atom Cu-Ni and Cu-Fe clusters, preliminary results of which have recently been reported.[28]

For example, the SCF-Xα-SW results for copper-rich Cu-Ni clusters, corresponding to dilute nickel impurities in a copper environment, indicate that the manifold of copper cluster d-levels is largely unperturbed by the impurities, although there is significant hybridization of Ni d-orbitals with delocalized Cu s,p orbitals at the Fermi level (see Fig. 2). This is the cluster analogue of the virtual bound d-resonances predicted by Friedel[29] and detected in photoemission spectra by Spicer and co-workers[30] for dilute nickel impurities in crystalline copper.

An important implication of this result is that the electronic effects of surrounding copper atoms must be taken into consideration when describing the chemisorptive and catalytic activities of isolated nickel sites embedded in a copper cluster or on a copper-segregated surface of a Cu-Ni alloy. Such effects may also be relevant to the catalytic *selectivity* observed for Cu-Ni and other IB-VIII alloys.[27]

MAGNETIC EFFECTS IN METALLIC AND BIMETALLIC CLUSTERS

It is possible to study the effects of magnetic spin polarization on the electronic structures of transition-metal clusters and the changes in magnetic behavior with alloying, using *spin-unrestricted* MO theory. For example, the spin-polarized SCF-Xα-SW orbital energies for a cubo-octahedral Ni_{13} cluster are shown in Fig. 5. The levels are displayed on two energy scales, one encompassing the entire d-band and overlapping s,p-levels, and the other resolving the upper part of the d-band on a much finer energy mesh. The "Fermi level," E_F, separates the unoccupied spin orbitals from the occupied ones and passes through a high density region of minority-spin levels. Note that the spin splitting of the energy levels corresponding to orbitals which are predominantly s,p-like is smaller than that of the d-levels. The spin magneton number per atom in Ni_{13} is 0.46, assuming the net spin density, arising from the six unpaired spins in the topmost occupied $t_{1u}\uparrow$(-0.425 Ry) and $t_{1g}\uparrow$(-0.427 Ry) spin-orbitals, to be delocalized uniformly over all 13 atoms. This value may be compared with the magneton number (0.25) per atom calculated for the smaller cluster, Ni_8,[23] and with the value (0.54) characteristic of ferromagnetic crystalline nickel. Although spin-polarized SCF-Xα-SW calculations have thus far been carried out for nickel clusters of rather limited size and geometry, the results are consistent with the experimental observation of Carter and Sinfelt[31] that the paramagnetic magneton number for small nickel crystallites decreases continuously from the bulk value as the crystallite size is reduced from 100 Å to approximately 10 Å.

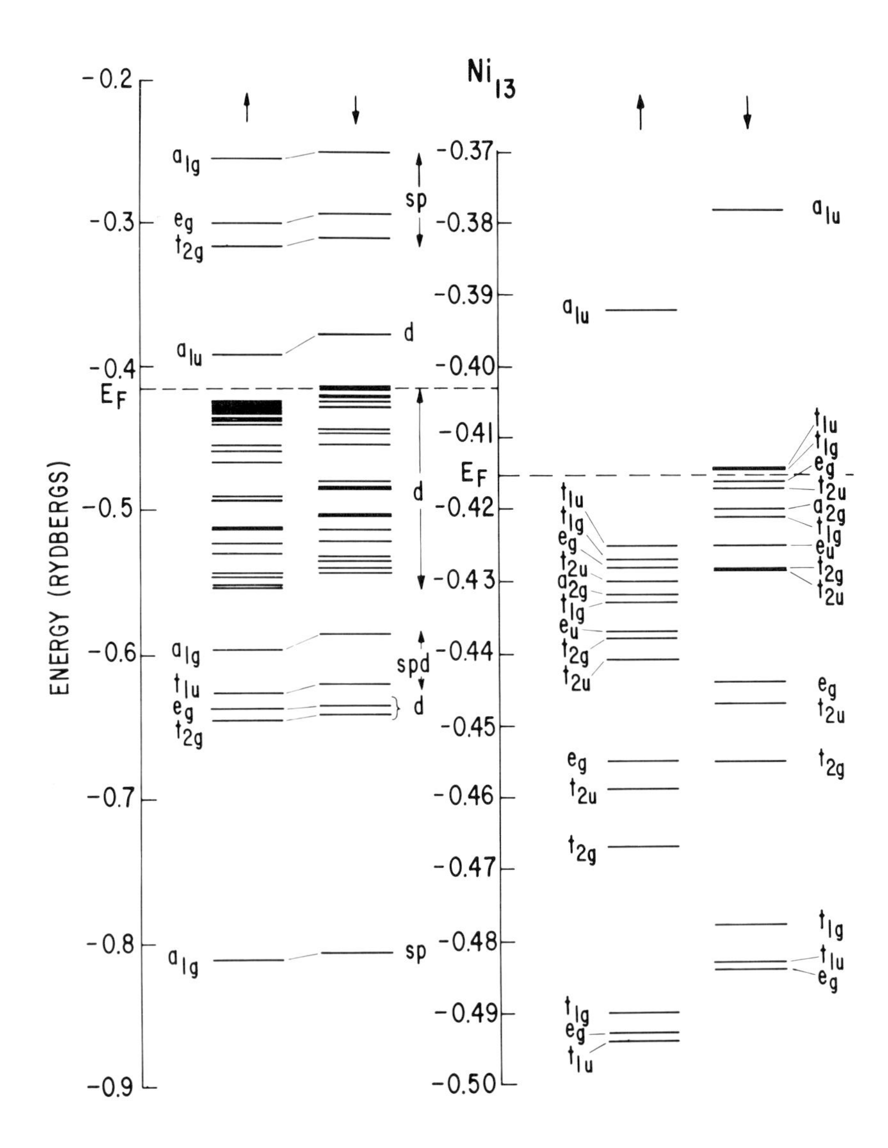

Figure 5. Spin-polarized SCF-Xα-SW electronic energy levels for a cubo-octahedral Ni_{13} cluster with nearest-neighbor internuclear distance equal to that for bulk crystalline nickel. The levels are labeled according to the irreducible representations of the O_h symmetry group. The "Fermi level" E_F separates the occupied levels from the unoccupied ones.

The so-called "superparamagnetism" of such metal aggregates is often utilized as a measure of the particle size and dispersion of supported metal catalysts.[2]

One of the most interesting results obtained from such theoretical studies is the effect of alloying on the spin polarization and magnetic properties of the clusters. For example, spin-unrestricted SCF-Xα-SW calculations for a Ni impurity at the center of a cubo-octahedral Cu_{13} cluster indicate that the magnetic moment of Ni is quenched out by the copper environment, whereas for an Fe impurity at the center of the Cu_{13} cluster, a strong magnetic moment remains whose magnitude depends on the ligand-field splittings of the Fe d-orbitals.[28] The latter result is particularly important since dilute Fe impurities in a crystalline copper environment are known to constitute a Kondo system.[32]

EFFECTS OF THE SUPPORTING ENVIRONMENT

When utilized as the active centers of a heterogeneous catalyst, small metallic or bimetallic clusters are usually supported in a porous refractory material such as silica or alumina.[3] If the sensitivity of a catalytic reaction cannot be explained solely in terms of the surface morphology or intrinsic electronic structure of the clusters themselves, then it is possible that the supporting material itself may be involved.[1] Most supporting materials are oxides and may effectively withdraw electrons from the small metal clusters in contact with them, much in the same way as the ligands of a simple transition-metal complex effectively withdraw electrons from the transition-metal atom. The electron depletion could then allow the small metallic cluster to interact with adsorbates in a different manner from the isolated clusters themselves. On the other hand, for larger metal aggregates or particles in a support, the electron withdrawing effect of the surrounding material may be largely screened out by the intervening metal atoms, from the surface of the particle. Thus for these larger particles the main effect on catalytic behavior might be determined by the effective surface area.

For example, clusters of fewer than six platinum atoms supported in the supercages of a Y-zeolite containing multivalent cations exhibit a catalytic activity per surface Pt atom which is enhanced by a factor of about five over that of other platinum catalysts for the hydrogenation of ethylene, a reaction that is normally structure insensitive.[33] They also chemisorb less oxygen than larger platinum crystallites. This fact and the enhanced catalytic activity have been attributed to electron transfer from the small platinum clusters to the zeolite support, as if platinum were behaving more like its neighbor, iridium, to the left in the periodic table.[33]

Theoretical studies of the effects of supporting environments on the electronic structures of small metallic and bimetallic clusters are currently in progress. For example, in modeling a small platinum cluster supported in Y-zeolite, the cluster is embedded in a local molecular environment simulating that of the zeolite, and SCF-Xα-SW calculations are carried out for the composite system. The effects of substituting multivalent cations such as Mg^{2+} and Ca^{2+} in the zeolite environment[33] are also under theoretical investigation. In constructing models for Y-zeolite, which has a silica-alumina network structure, one can make use of earlier SCF-Xα-SW results for silica and alumina,[34] which indicate that only relatively small representative clusters are needed to reproduce the electronic structure and related properties of these materials.

An important effect of electron transfer from a small metallic cluster to its supporting environment is to increase the effective work function and electron affinity of the cluster, thereby promoting its interaction with nucleophilic adsorbates but diminishing its interaction with electrophilic adsorbates. This is the simplest explanation of the observed effect of Y-zeolite in reducing the capacity of platinum clusters to chemisorb oxygen.[33]

IRON-SULFUR PROTEINS

Somewhat related to the fundamental issue of metal-support interactions is the nature of the electronic structure of the 4-Fe active sites in ferredoxin and other "high-potential" iron-sulfur proteins which catalyze electron transfer in the metabolism of bacteria, plants, and animals.[35] The active site, an $Fe_4S_4(S\text{-}Cys)_4$ cluster (see Fig. 6), consists of an approximately cubic array of Fe and S atoms on alternate vertices, with four outer sulfur atoms attaching the cluster to the surrounding protein via cysteinyl groups. The sulfur atoms may be viewed as providing a "support" for the Fe_4 cluster, and therefore possible analogies with the effects described in the preceding section are obvious.

The results of recent SCF-Xα-SW calculations for the electronic structure of $[Fe_4S_4(SCH_3)_4]^{2-}$, a prototype model of oxidized ferredoxin and reduced "high-potential" proteins, in which methyl groups have been substituted for the cysteinyl moieties, suggest significant amounts of charge transfer and orbital delocalization of the Fe_4 electrons onto the surrounding sulfur ligands, with each iron site in the effective $Fe^{+2.5}$ valence state.[36] The distorted cubane geometry of the active centers of oxidized ferredoxin, the reduced high-potential proteins, and their analogues are explained

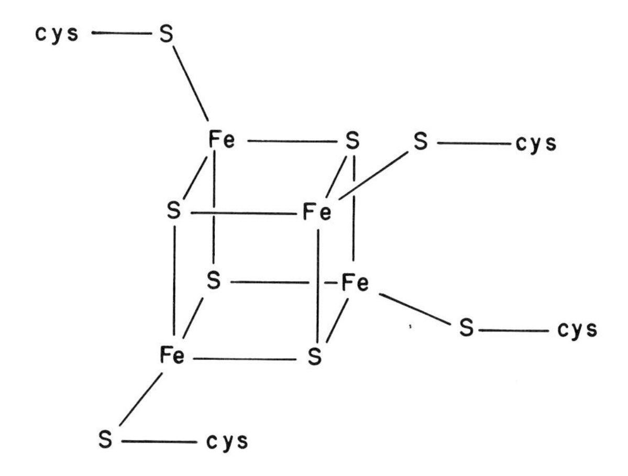

Figure 6. The approximately cubic cluster of iron and sulfur atoms that is the active center of bacterial ferredoxin. The four cystein molecules attach the cluster to the polypeptide chain of the protein.

in terms of the Jahn-Teller instability of the cubic (T_d) symmetry, and spin-orbital occupancy in the lower (D_{2d}) symmetry is consistent with measured temperature-dependent magnetic properties.[36]

To subject the SCF-Xα-SW results for ferredoxin to further experimental test, properties such as quadrupole splittings, electric-field gradients, and magnetic susceptibility, many of which have been measured,[35] are currently being calculated using the computer program recently developed by Case and Karplus[37] for the determination of electronic properties directly from SCF-Xα-SW orbital wavefunctions. Case and Karplus[37] have already applied this procedure with good results to molecules as simple as NO_2 and as complex as copper and iron porphyrins (e.g., hemoglobin). Studies of 4-Fe clusters are being closely coordinated with SCF-Xα-SW calculations for 1-Fe and 2-Fe iron-sulfur proteins (e.g., rubredoxin) currently being carried out by Norman and co-workers.[38]

In connection with recent SCF-Xα-SW studies of the interaction of molecular nitrogen with iron clusters and the activation of ammonia synthesis by iron-based catalysts,[25] it is important to note that iron-sulfur proteins like ferredoxin have been suspected to play a key role in the reduction of dinitrogen to ammonia by the enzyme, nitrogenase, the fundamental process of nitrogen fixation by soil bacteria.[39] It has been suggested that both a molybdenum-iron protein and iron protein are necessary for this reduction, the former binding the reducible N_2 molecule and the latter supplying low-potential electrons to effect the reduction. Although the role, if any, of 4-Fe complexes in this process is uncertain, it may be speculated that the iron sites in the pertinent proteins are the active sites of interaction and electron transfer with dinitrogen.

In view of the SCF-Xα-SW results described in Ref. 36 for $[Fe_4S_4(SCH_3)_4]^{2-}$, the analogue of oxidized ferredoxin and reduced high-potential iron-sulfur proteins, and further work on the reduced analogue $[Fe_4S_4(SCH_3)_4]^{3-}$ still in progress,[40] one may begin to investigate the interaction of N_2 with these 4-Fe complexes and then to compare the results with those obtained for the interaction of N_2 with isolated Fe_4 clusters. This may allow one to establish whether transfer and delocalization of the tetrametal electrons onto the surrounding sulfur ligands, as described in Ref. 36, can significantly influence the interaction of the 4-Fe sites with N_2, in analogy to the effects on chemisorption and catalytic activity of electron transfer between a small metal cluster and local environment in a supported metal catalyst.

There are probably many useful analogies to be made between supported metal cluster catalysts and the active centers of metallo-proteins and metallo-enzymes like ferredoxin. Such analogies are probably not fortuitous. They should be sought after and the common basis of understanding elucidated.

CHEMISORPTION AND REACTIVITY ON METALLIC CLUSTERS

In regard to the nature of chemisorption on transition-metal clusters of catalytic importance, it is important to note the occurrence in inorganic chemistry of metal cluster complexes[41] such as the iron carbonyl carbide[42] illustrated in Fig. 7. This is a case where linear and bridged carbonyl groups occur, both of which have been speculated to form in the chemisorption of CO on transition-metal surfaces and to be important as surface species in catalytic reactions involving CO. The existence of such cluster complexes suggests how small metal clusters of the type that may be found as the active centers of heterogeneous catalysts can form more or less normal chemical bonds on the cluster periphery and particularly at corners and edges of a faceted cluster surface. The cluster complex shown in Fig. 7, which is only one of many examples,[41] is a true molecular analogue of CO chemisorbed on iron carbide, since there is a central carbon atom coordinated octahedrally by Fe atoms at almost the precise Fe-C distance characteristic of bulk iron carbide.[42] Therefore, it may be no coincidence that iron carbide is an excellent catalyst for the Fischer-Tropsch synthesis of CO to hydrocarbons.[43]

It is also important to observe that the study of metal clusters "saturated" by adsorbates, as shown in Fig. 7, is likely to be relevant to the high surface coverage of adsorbates characteristic of catalytic reactions which normally occur on heterogeneous catalysts at relatively high pressures, whereas theoretical studies which focus on a single adsorbate bonded to an extended substrate are likely to be relevant to low-pressure chemisorption on a single-crystal surface. A molecular-orbital treatment of a metal cluster more or less symmetrically covered with adsorbate atoms automatically includes the effects of adsorbate-adsorbate interactions along with metal-adsorbate bonding.

To illustrate the type of information an MO study of chemisorption provides, when implemented properly, we consider recent SCF-Xα-SW studies of prototype transition-metal-olefin systems of catalytic importance, carried out in collaboration with R. P. Messmer and N. Rösch.[44-46] Olefins are involved in many chemical reactions which are catalyzed heterogeneously and/or homogeneously. For example, in Fig. 8 we schematically represent possible cata-

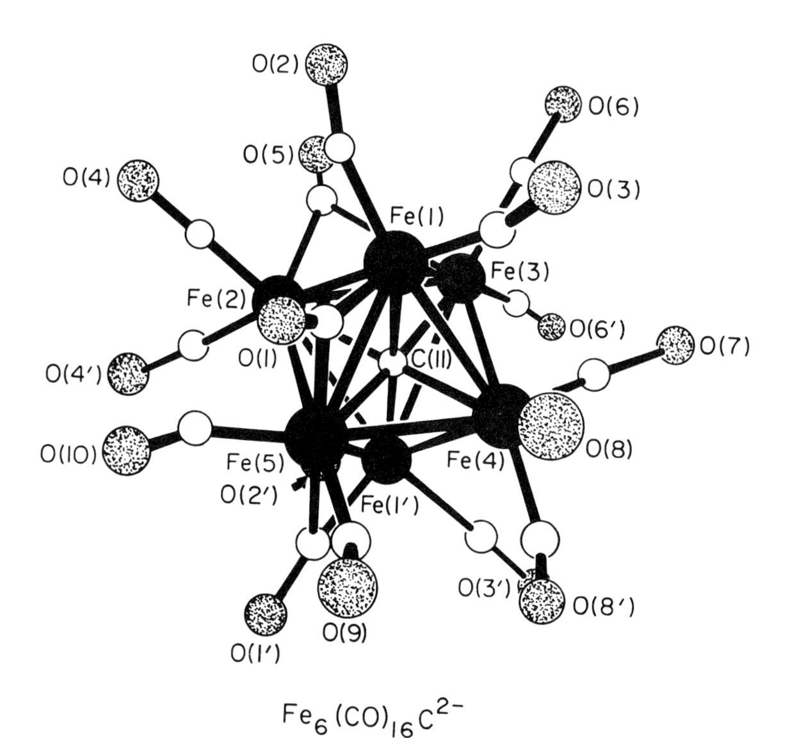

Figure 7. Structure of the iron carbonyl carbide molecular ion, $Fe_6(CO)_{16}C^{2-}$, which is a true molecular analogue of carbon monoxide chemisorbed on iron carbide.

lytic precursors and reaction intermediates in the heterogeneous hydrogenation of ethylene (C_2H_4) on a supported transition-metal cluster and homogeneous hydrogenation of ethylene by a "square-planar" transition-metal coordination complex. The major steps in the heterogeneous reaction shown in Fig. 8(a) are: the molecular chemisorption of ethylene and dissociative chemisorption of hydrogen; transfer of chemisorbed hydrogen atoms to chemisorbed ethylene; desorption of ethane. The major steps in the homogeneous reaction shown in Fig. 8(b) are: the molecular addition of ethylene and dissociative addition of hydrogen as ligands of the transition metal, with ligand exchange; the transfer of hydrogen ligands to the ethylene ligand; dissociation of the hydrogenated ethylene ligand as ethane. It is clear that there is a one-to-one correspondence between the chemisorption of ethylene and hydrogen on the metal cluster and the addition of ethylene and hydrogen as ligands of the isolated metal complex. It is therefore important to determine the degree to which chemisorption and heterogeneous reactivity on the active sites of supported metal or bimetallic catalysts are governed by the same principles of coordination chemistry which underlie the chemical bonding and homogeneous reactivity of isolated transition-metal complexes. Since transition-metal complexes tend to be more selective or specific with respect to the reactions they catalyze and are generally more resistant to catalyst poisoning than are the extended surfaces of much larger metal aggregates,[47] a fundamental understanding of how the local site symmetry and coordination of a transition-metal atom affect its chemical bonding could ultimately aid in the design of very small metal or bimetallic cluster complexes which exhibit optimum catalytic activity, selectivity, and stability.

As a first step toward this long-range objective, one may study the bonding of ethylene to very small aggregates and complexes of Ni, Pd, and Pt. The intent of such a study is to establish how the metal-olefin chemical bonding varies with the composition and size of the metal aggregate, supporting environment, and presence of coordinated nonmetallic ligands.

The simplest case to start with and the one that is also most directly comparable with isolated transition-metal complexes that homogeneously catalyze the hydrogenation of ethylene[47] is that of a single ethylene molecule bonded to a single atom of Ni, Pd, or Pt. In Fig. 9 we show the resulting SCF-Xα-SW orbital energies for these systems, calculated for a metal-ethylene bond distance equal to that used in earlier theoretical studies of the platinum-olefin complex, Zeise's anion $[Pt(C_2H_4)Cl_3]^-$, by the SCF-Xα-SW method.[44] This bond distance was chosen for comparison of the bonding of C_2H_4 to Ni, Pd, and Pt because many ethylene complexes of these Group-VIII metals have bond distances which

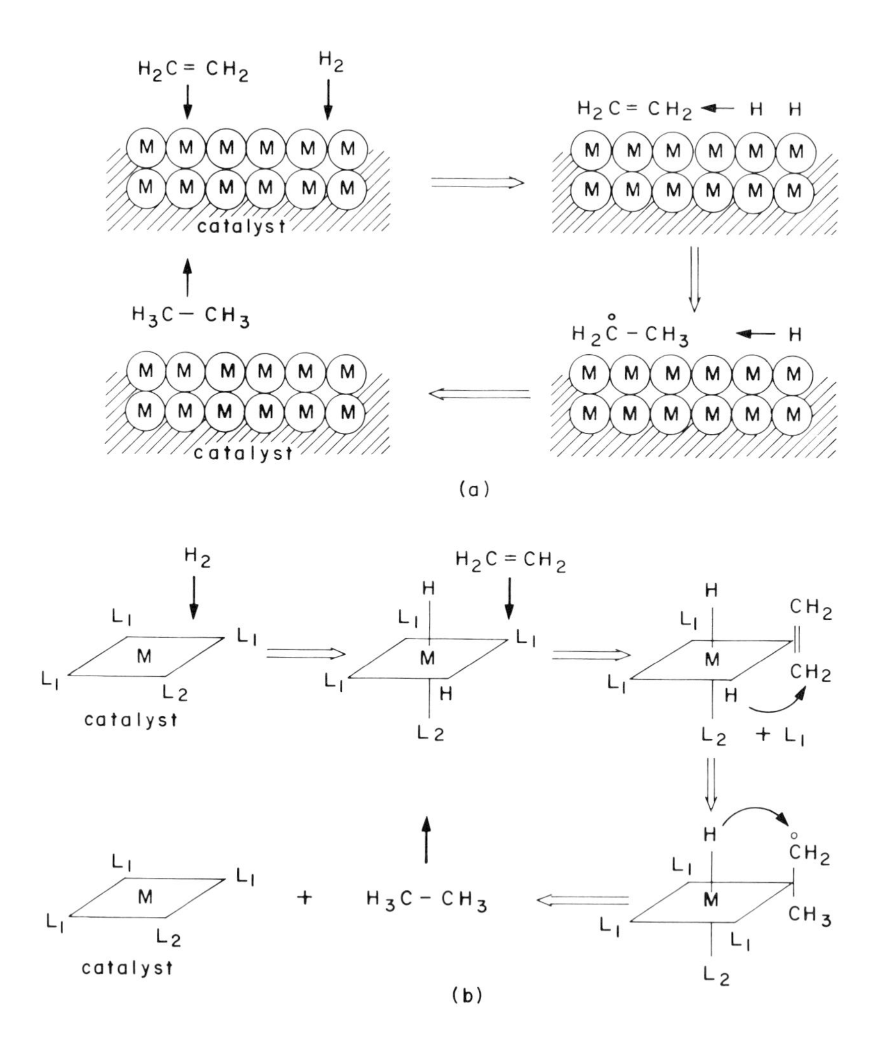

Figure 8. Heterogeneous vs. homogeneous catalysis. (a) Schematic representation of heterogeneous catalysis of ethylene hydrogenation on the surface of a supported transition metal aggregate; (b) Schematic representation of the homogeneous catalysis of ethylene hydrogenation by an isolated transition-metal complex.

are within 0.1 Å of this value. The metal-ethylene orbital energies in Fig. 9 are labeled according to the C_{2v} symmetry group, and the highest occupied level in each case is the one labeled $5a_1$. Also shown for comparison are the SCF-Xα-SW orbital energies for the isolated ethylene molecule, labeled according to the symmetry group.

In Fig. 9, the related orbitals are connected by broken lines. The ethylene orbitals from $2a_g$ up to $1b_{3g}$ undergo an almost uniform shift when ethylene interacts with platinum, with only a slight upward shift on going from Pt to Pd to Ni. The π and π* levels of ethylene on the other hand experience a rather profound change on interacting with an atom of Pt, Pd, or Ni. Thus the $3a_1$ and $2b_2$ orbitals are in each case primarily responsible for the bonding, through the overlap and interaction of metal d orbitals (d_{z^2} and d_{yz}, respectively, if the metal-ethylene and carbon-carbon bond directions are along the z and y axes, respectively) with the ethylene π and π* orbitals. The metal s and p orbitals also participate in the bonding, but to a lesser extent. The $2a_2$, $4a_1$, and $2b_1$ levels correspond principally to nonbonding d_{xy}, $d_{x^2-y^2}$, and d_{xz} atomic orbitals, whereas the $5a_1$ level corresponds to a metal d_{z^2} orbital that is slightly antibonding with respect to the ethylene π orbital. These characteristics of the orbitals are revealed in Figs. 10(a)-10(d) and 11(a)-11(d), where contour maps of the $3a_1$, $4a_1$, $5a_1$, and $2b_2$ wavefunctions are shown in the y-z plane containing the metal atom and carbon atoms of ethylene. The overlap of the axially symmetric metal d_{z^2} orbital with the ethylene π orbital in the $3a_1$ wavefunction is evident in Figs. 10(a) and 11(a), corresponding to the classic "σ-donation" or ethylene π electrons into the unoccupied metal d_{z^2} atomic orbital as originally postulated for transition-metal-olefin bonding by Dewar, Chatt, and Duncanson.[48] In Figs. 10(d) and 11(d) for the $2b_2$ wavefunction, we see the overlap of the four-lobe d_{yz} orbital of the metal with the ethylene π* orbital, corresponding to the "π-backdonation" of metal d_{yz} electrons into the unoccupied π* orbital originally postulated by Dewar, Chatt, and Duncanson.[48]

The increasing stabilization of the $2b_2$ orbital and the increasing overlap of the metal d_{yz} orbital with the ethylene π* orbital in the $2b_2$ wavefunction as one goes from Ni to Pd to Pt [see Figs. 9, 10(d), and 11(d)] suggest increasing metal-to-ethylene backdonation from Ni to Pt. In this regard, it is interesting to note that Pd, which exhibits a degree of back-bonding to ethylene intermediate to that of Ni and Pt, is an order of magnitude more catalytically active for the hydrogenation of ethylene than are Ni and Pt.[49] It has been suggested that metal-olefin chemical bonding of "moderate" strength, rather than bonding that is "too weak" or "too strong," is the optimum

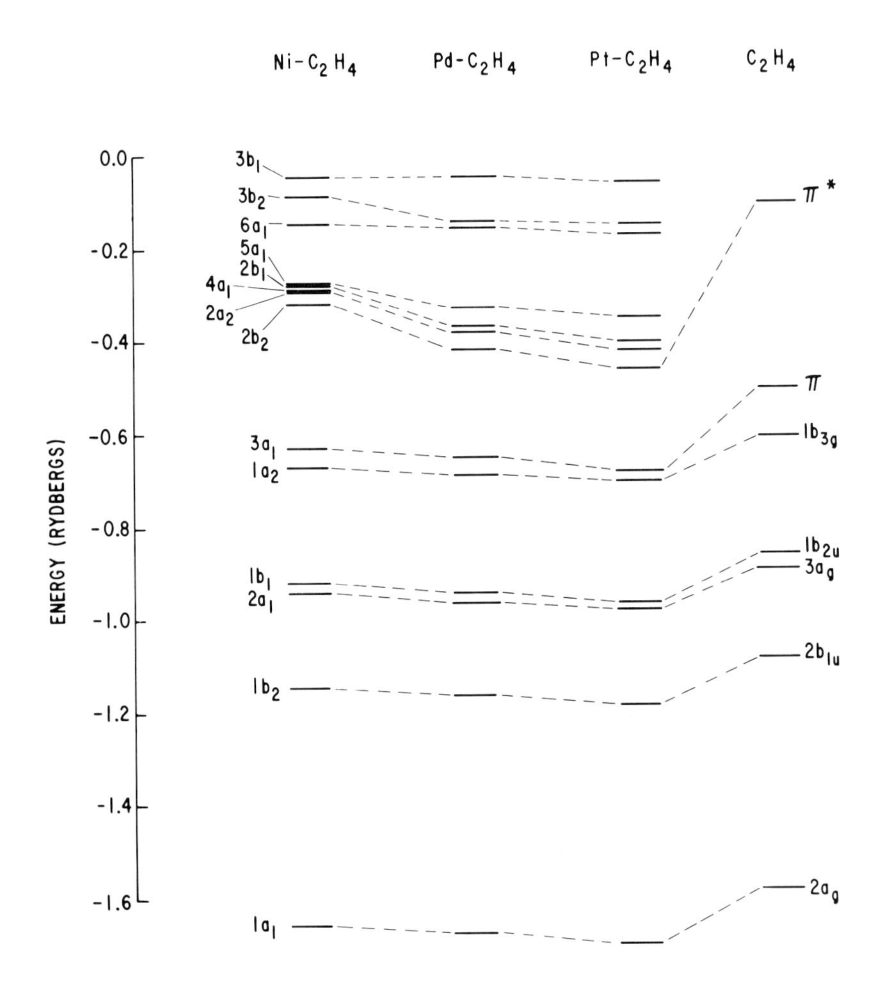

Figure 9. Comparison of calculated electronic energy levels for $Ni-C_2H_4$, $Pd-C_2H_4$, $Pt-C_2H_4$, and C_2H_4. The same geometry was employed for all calculations.

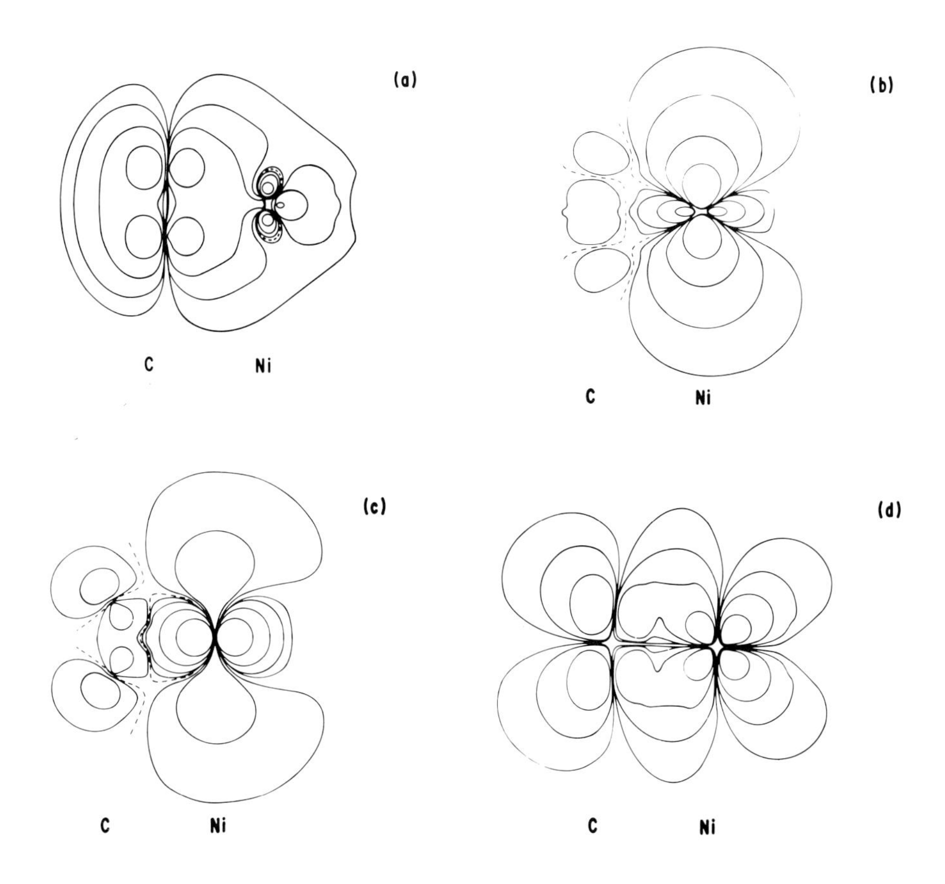

Figure 10. Contour maps of the (a) $3a_1$; (b) $4a_1$; (c) $5a_1$; and (d) $2b_2$ orbitals of Ni-C_2H_4. Nodes are indicated by dashed curves.

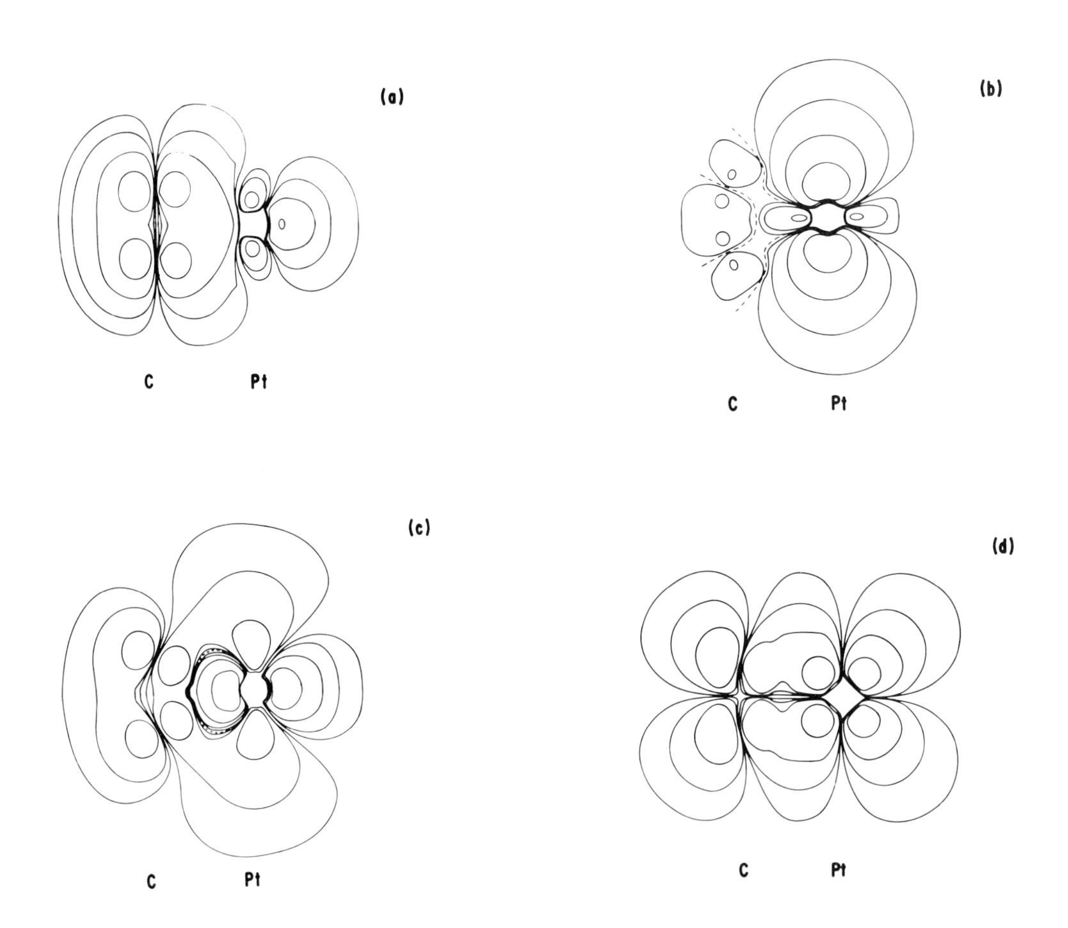

Figure 11. Contour maps of the (a) $3a_1$; (b) $4a_1$; (c) $5a_1$; and (d) $2b_2$ orbitals of $Pt-C_2H_4$. Nodes are indicated by dashed curves.

precursor in transition-metal catalyzed olefin hydrogenation.[50]

Of course, the above described study of the molecular orbitals for ethylene bonded to a single transition-metal atom at fixed internuclear distance, while instructive as to the principal components of metal-olefin bonding and their systematic variation from Ni to Pd to Pt, is rather far removed from the interaction of olefins with metal clusters and complexes of the type illustrated in Fig. 8 which are important in heterogeneous and homogeneous olefin catalysis. As a preliminary study of the effects of gradually increasing the size of the metal aggregate [see Fig. 8(a)] and adding electronegative ligands [see Fig. 8(b)] on metal-olefin chemisorption, the following prototype examples may be investigated: (1) adding another metal atom to the M-C_2H_4 complex and (2) adding three planar chlorine ligands to the M-C_2H_4 complex. In the former case, the M_2-C_2H_4 complex may be considered in two alternative configurations: (1) with the C_2H_4 molecule attached to the metal diatomic "end-on" (the "linear" configuration) and (2) with the carbon-carbon bond direction parallel to the metal-metal bond (the "bridged" configuration). Both configurations are speculated to be important in the chemisorption and reactivity of olefins on larger transition-metal aggregates and extended surfaces. In the case of Ni_2-C_2H_4, such calculations have already been carried out, using the SCF-Xα-SW method.[51]

A direct comparison of the SCF-Xα-SW orbital energies calculated for Ni-C_2H_4 and Ni_2-C_2H_4 (in the "linear" bonded configuration) is shown in Fig. 12. The broken lines in Fig. 12 connect orbitals which are related in the two cases and are of particular importance in understanding the effects of adding the extra metal atom on the bonding of ethylene to the first metal atom. The $1a_2$ level, which corresponds to an ethylene σ orbital, remains relatively unchanged. The $3a_1$ level, which corresponds to the ethylene π orbital that effectively "donates" electrons into the metal d_{z^2} orbital, is slightly lowered in energy in the case of Ni_2-C_2H_4, indicating that the second Ni atom, although not directly bonded to ethylene in the "linear" configuration, contributes to the stabilization of this orbital. The $2b_2$ level, which corresponds to the metal d_{yz} orbital that effectively "backdonates" electrons into the ethylene π^* orbital, is also somewhat lowered by the presence of the second metal atom. However, the most dramatic change occurs in the $5a_1$ level which, while slightly antibonding in the case of Ni-C_2H_4, is significantly stabilized by the addition of the second Ni atom. This orbital is lowered in energy below the nonbonding d orbitals and contributes to the metal-ethylene bonding in Ni_2-C_2H_4. Similar results, differing however in important details, have also been obtained for Ni_2-C_2H_4 in the "bridged" configuration. Contour maps for the most important molecular orbitals of Ni_2-C_2H_4, in both the "linear" and "bridged" configurations, may be seen in Ref. 51 and compared directly with the

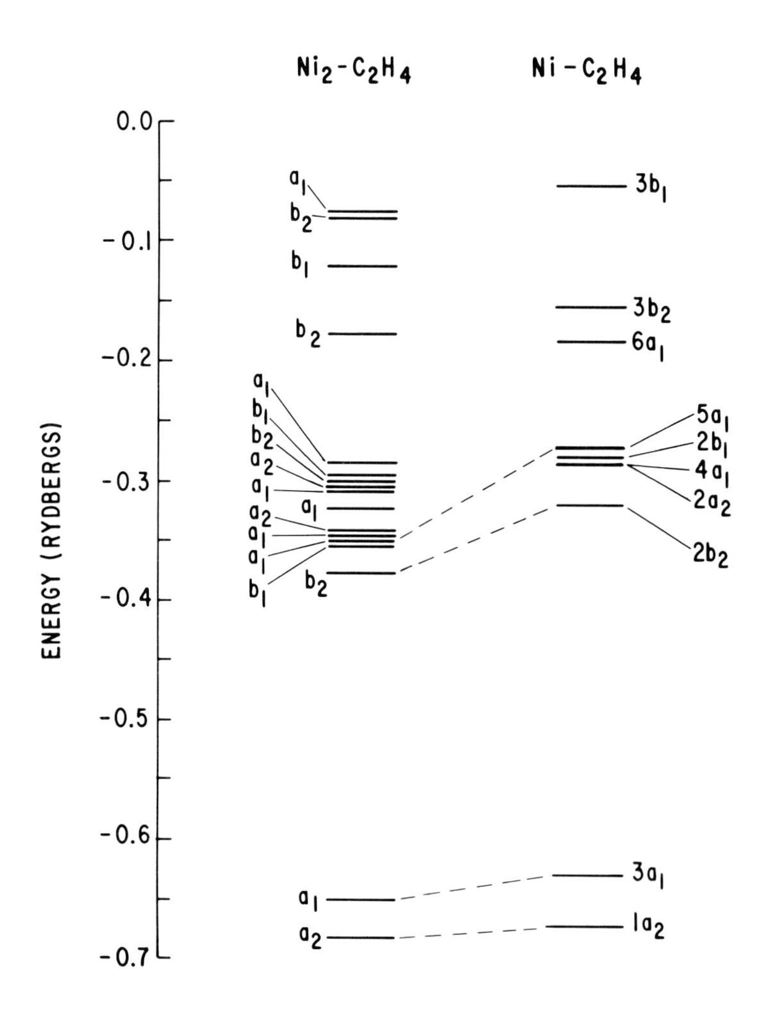

Figure 12. Comparison of calculated electronic energy levels for $Ni-C_2H_4$ and $Ni_2-C_2H_4$; the latter one from Rösch and Rhodin.[51]

orbital contour maps for Ni-C_2H_4 shown in Figs. 10(a)-(d).

It is, of course, highly desirable to establish, from first principles, the relative energetic stabilities of the linear, bridged, and other possible geometric configurations in metal-olefin bonding, especially in regard to the precursors, reaction paths, and kinetics of olefin catalysis. The same issues also arise in the chemisorption and catalytic reactions of other molecules, such as CO, on transition metals. This requires the calculation of total energies and their differences as a function of molecular geometry, often to an accuracy of a few kilocalories per mole. One can significantly improve the accuracy of total energies and molecular geometries calculated with the SCF-Xα-SW technique by starting with the recently developed "overlapping-sphere" extension of this method,[52-54] which allows one to compensate for errors resulting from the "muffin-tin" approximation without sacrificing computational efficiency. Indeed, overlapping atomic spheres were used in the metal-ethylene molecular-orbital calculations which have been described above. Whether this approach is ultimately accurate enough to serve as the basis for calculating catalytic reaction paths remains to be seen, particularly for larger metallic and bimetallic clusters based on Ni, Pd, and Pt, where the total energies are very large numbers and the pertinent energy differences are very small numbers.

Therefore, one should also attempt to develop alternative concepts of catalytic reactivity, based purely on orbital quantities that can be computed accurately and efficiently with the SCF-Xα-SW method for both small and large systems. An example of the latter is Slater's "transition-state" concept,[18] which has already been successfully applied to a wide range of molecules and clusters[20,21] to determine electronic excitation energies, including orbital relaxation effects. Electronic excitations of optical energies are, of course, crucial to photocatalytic and other photochemical behavior.[55] Such excitations redistribute electronic charge in the molecule, often making it more susceptible to chemical reaction than the molecule in its ground state. Electronic excitations of lower energy (e.g., thermal energy) and those effectively induced through charge transfer between adsorbate and catalyst probably underlie other types of catalytic activity. For example, the combined effects of donating ethylene π electrons into the d orbitals of a transition-metal catalyst and backdonation of d electrons into the ethylene π* orbital can be viewed as an effective partial excitation of electrons between the π and π* levels in ethylene. In other words, the d orbitals of the transition metal provide a pathway for electrons to flow between the π and π* orbitals, redistributing charge in ethylene so as to make it more susceptible for chemical reaction with certain other atoms and molecules.

Orbital symmetry rules, such as those introduced by Woodward and Hoffmann[56] and Fukui[57] and further developed by Mango and Schachtschneider[58] and Pearson[59] can also be used to elucidate certain features of chemical reaction paths. The basic idea is that, for a chemical reaction to occur with reasonable activation energy, there must be a low-lying unoccupied molecular orbital (LUMO) for the reacting system having the same symmetry as the highest occupied molecular orbital (HOMO), and the energy gap between the HOMO and LUMO must be reasonably small (a few eV or less). The requirement that the HOMO and LUMO have the same symmetry is equivalent to the requirement that these two orbitals have a net positive overlap. Implicit in the application of these rules is the notion of electron flow between the HOMO and LUMO.

A catalyst can often be viewed as providing a pathway for electrons to flow between the reactants, where such flow is normally prevented to occur by orbital symmetry restrictions. In other words, a primary role of the catalyst is to _break_ the orbital symmetry restrictions through the coordination chemical bonding of reactants and to permit the reaction to occur at low activation energy. A good example is olefin hydrogenation, e.g.,

$$C_2H_4 + H_2 \rightarrow C_2H_6$$

where the transfer of electrons directly from the filled π orbital of the olefin to the empty σ^* orbital of H_2 is forbidden by the orbital symmetries. However, hydrogen atoms that have first been dissociated by a transition-metal catalyst can be readily transferred to an olefin molecule that is d_{yz}-π^* backbonded to the metal, resulting in the formation and evolution of the hydrogenated species (ethane). As shown in Fig. 8, this process can take place heterogeneously on the surface of a supported metallic or bimetallic aggregate or homogeneously through exchange of ligands by an isolated transition-metal coordination complex. As emphasized, however, in the recent excellent review of heterogeneous catalysis by Madey _et al._,[60] the application of orbital symmetry rules to catalysis is based on the assumption of _concerted_ reactions with high symmetry transition states, whereas alternative mechanistic schemes are sometimes more likely.[61]

In regard to the chemisorption and reactivity of hydrogen on transition-metal aggregates, it has recently been suggested that delocalized metal s orbitals, and _not_ localized d-orbitals, are responsible for the chemical bonding and catalytic reactivity of hydrogen.[62] While metal s orbitals may indeed contribute significantly to the chemisorption and lability of atomic hydrogen (SCF-Xα calculations indicate that metal s orbitals are indeed important), the d orbitals are essential for the catalytic dissociation of molecular hydrogen on transition metals and for the coordination chemical bonding and reactivity of olefins with hydrogen on transition metals. Recent theoretical studies of the interaction of CO with transition metals also show that metal d orbitals are primarily responsible for the bonding, whereas metal s and p orbitals participate in the bonding to a lesser extent.[63]

The SCF-Xα-SW method is particularly well suited for the quantitative implementation of orbital symmetry rules in catalysis, since the proper ordering of both the occupied and unoccupied molecular orbitals (HOMO and LUMO) and the corresponding orbital energies as a function of occupation number are automatic consequences of the self-consistent-field Xα-density-functional representation of molecular-orbital theory.[17,18] In this regard, it should be emphasized that the orbital energies in the SCF-Xα theory are rigorously equal to derivatives of the total energy with respect to occupation number, i.e.,

$$\varepsilon_{iX\alpha} = \partial\langle E_{X\alpha}\rangle/\partial n_i$$

and therefore should not be identified with Hartree-Fock orbital energies, which are equal to the differences

$$\varepsilon_{iHF} = \langle E_{HF}(n_i = 1)\rangle - \langle E_{HF}(n_i = 0)\rangle$$

between total energies calculated when the ith orbital is occupied and when it is empty (fixing the remaining occupied orbitals).[18] In the conventional Hartree-Fock approach, moreover, the unoccupied ("virtual") orbitals are represented less accurately than the occupied ones, whereas in the Xα method the unoccupied and occupied orbitals are treated to the same degree of accuracy. The Xα orbital energies, in fact, correspond closely to the "orbital electronegativities" defined by Hinze et al.[64] as a generalization of Mulliken's definition of electronegativity. Therefore, the relative positions of the Xα energy levels for reacting molecules, as exemplified in Fig. 9 for the case of ethylene interacting with Ni, Pd, and Pt, are an approximate measure of the relative amounts of electron transfer between the reactants. Slater's transition-state procedure,[18] applied in conjunction with the SCF-Xα-SW method to the HOMO and LUMO of reacting molecules, leads to remarkably accurate ionization potentials, electron affinities, and electron excitation energies, including spin-orbital relaxation and the effects of averaging over many-electron multiplet states which are usually beyond the scope of conventional molecular-orbital theories.[20,21] The unique characteristics of the Xα orbitals described above greatly facilitate the application of orbital symmetry rules and the concept of electron flow between reacting molecules via the coordination chemical bonding of reactants to the catalyst. One may apply these concepts, where possible, in molecular-orbital studies of chemisorption and reactivity on metallic and bimetallic clusters.

As mentioned earlier, one can also carry out theoretical studies of the effects on chemisorption and reactivity of adding nonmetallic ligands to small transition-metal aggregates, in part, to model metal-support interactions in heterogeneous catalysis [see Fig. 8(a)] and, in part, to determine the electron-withdrawing effects of such ligands in homogeneous catalysis by isolated transition-metal complexes [see Fig. 8(b)]. As a simple prototype, one may consider the effects of adding planar electronegative ligands to a platinum-olefin complex. This is just the situation in the classic platinum-ethylene-chlorine complex, Zeise's anion $[Pt(C_2H_4)Cl_3]^-$, for which SCF-Xα-SW calculations have already been completed.[44] To illustrate the effects of the chlorine ligands on platinum-ethylene bonding, one may compare the molecular-orbital energies for $[Pt(C_2H_4)Cl_3]^-$ directly with those shown in Fig. 9 for $Pt-C_2H_4$. This comparison is made in Fig. 13. The σ orbitals (due to ethylene) of $Pt-C_2H_4$, i.e., the levels labeled $1a_1$ up to $1a_2$, remain relatively unperturbed by the addition of the Cl ligands. The localized Cl 3s levels are found at approximately -1.5 Ry in $[Pt(C_2H_4)Cl_3]^-$, as indicated in Fig. 13. A very striking effect occurs, however, for those levels that are derived primarily from the π and π* orbitals of ethylene. The levels in $[Pt(C_2H_4)Cl_3]^-$ which are derived from the ethylene π orbitals are labeled by "†" and those derived from the π* orbitals are labeled by "*". The broken line rectangle indicates the region in which the energies of predominantly nonbonding Cl 3p orbitals are found. The strongly antibonding orbitals (not shown) are found at approximately -0.3 Ry.

The highly electronegative chlorine ligands effectively withdraw electrons from the central Pt atom, giving the latter atom a formal oxidation number of +2, and promoting the bonding between ethylene and platinum by allowing the $4a_1$ and $5a_1$ orbitals, which correspond respectively to nonbonding Pt $d_{x^2-y^2}$ and slightly antibonding Pt d_{z^2} orbitals in $Pt-C_2H_4$, to become significantly bonding in $[Pt(C_2H_4)Cl_3]^-$. Moreover, the $3a_1$ and $2b_2$ orbitals, which are responsible for $Pt(d_{z^2})-C_2H_4(\pi)$ and $Pt(d_{yz})-C_2H_4(\pi^*)$ bonding in $Pt-C_2H_4$ are more strongly bonding in $[Pt(C_2H_4)Cl_3]^-$ due to the influence of the chlorine ligands.

A classic example in heterogeneous catalysis of the stereochemical effects of coordinating transition-metal atoms with electron-withdrawing ligands is the group of transition-metal halide crystals, such as $TiCl_3$, which catalyze the polymerization of olefins, as discussed by Ziegler and Natta[65,66] for which they received the Nobel Prize in Chemistry in 1963. The chemisorption and polymerization of the parent olefin (e.g., ethylene), as originally postulated by Cossee and Arlman,[67] is envisioned to occur in the vicinity of a transition-metal atom, specifically titanium in the +3 oxidation state, which is incompletely coordinated by chlorine ligands near the surface of an α-$TiCl_3$ crystal,

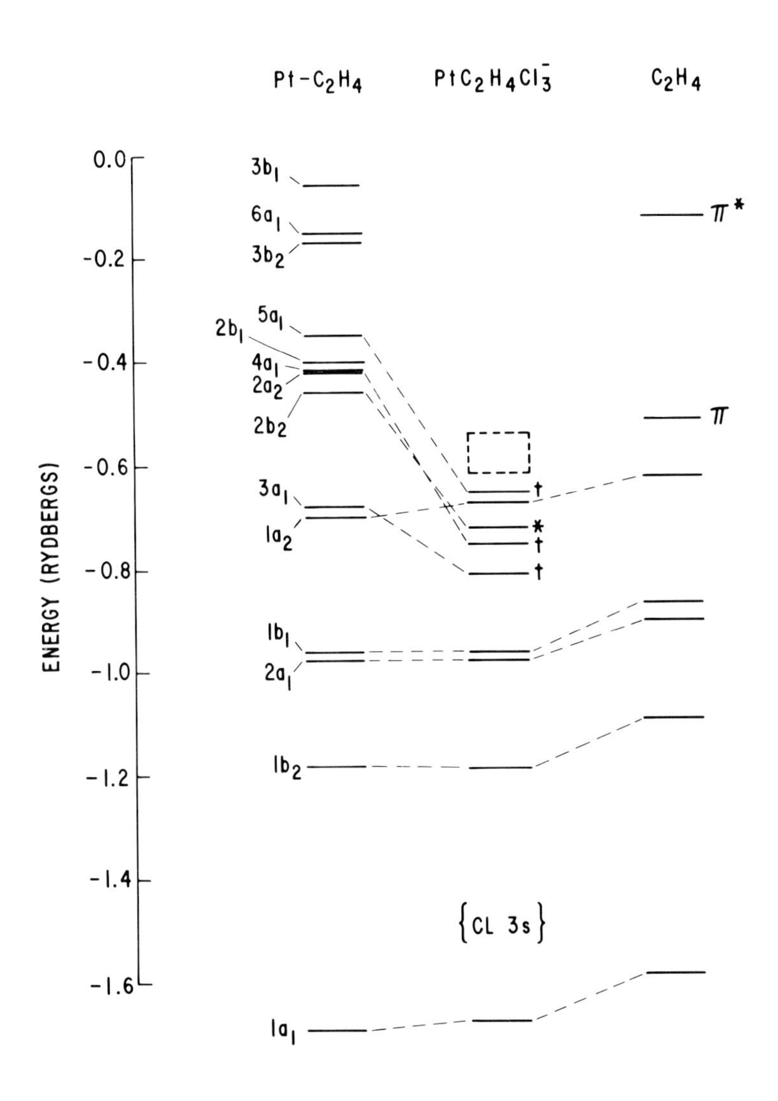

Figure 13. Comparison of calculated electronic energy levels for $Pt\text{-}C_2H_4$, $PtC_2H_4Cl_3^-$, and C_2H_4. The levels for $PtC_2H_4Cl_3^-$ are from Rösch, Messmer, and Johnson.[44]

as represented schematically in Fig. 14 (M = Ti, X = Cl, R = alkyl). Note that the geometry of the surface titanium-olefin complex is similar to that of the isolated transition-metal complex shown in Fig. 8(b) which homogeneously catalyzes olefin hydrogenation, although the Ziegler-Natta catalysis of olefin polymerization occurs heterogeneously. SCF-Xα-SW studies (described in Ref. 46) have recently been carried out for titanium-ethylene complexes which simulate the first step shown in Fig. 14. These results again suggest the crucial role of the electron-withdrawing ligands in determining the metal-olefin bonding. Such effects are also probably important in chemisorption and catalytic reactivity on the surfaces of transition-metal compounds such as NiO, TiO_2, and perovskites.[60]

As suggested earlier, the supporting environment of a metallic or bimetallic cluster may influence its electronic structure, interaction with adsorbates, and ultimately its catalytic activity. Most supporting materials are porous oxides and may effectively act to withdraw electrons from small metal clusters in contact with them, in much the same way as the chlorine ligands withdraw electrons from platinum and titanium in the complexes considered above. A further analogy with the above platinum- and titanium-olefin complexes suggests that electron depletion in a small supported metal cluster would allow it to interact with adsorbates in a different manner from an isolated metal cluster. For larger supported metal aggregates or particles, on the other hand, the electron withdrawing effects of the support may be largely screened out from the surfaces of these particles by the intervening metal atoms. Thus for these larger particles the principal effect on catalytic behavior might be determined by the surface area and morphology of the particles.

For example, platinum clusters containing fewer than six atoms and supported in the supercages of a Y-zeolite chemisorb less oxygen than larger platinum particles or crystallites.[33] If the effect of the zeolite framework is to withdraw electrons from the cluster, it would tend to increase the work function of the remaining electrons in the cluster, thereby making it more difficult to dissociate molecular oxygen through electron transfer from the supported metal cluster to oxygen molecular antibonding orbitals. One may test such naive arguments by combining theoretical studies of the effects of supporting environments, as described earlier, with theoretical studies of chemisorption. In other words, one may compare molecular-orbital results for adsorbate molecules interacting with isolated metal clusters with the results obtained for the same clusters embedded in molecular environments simulating typical supporting materials.

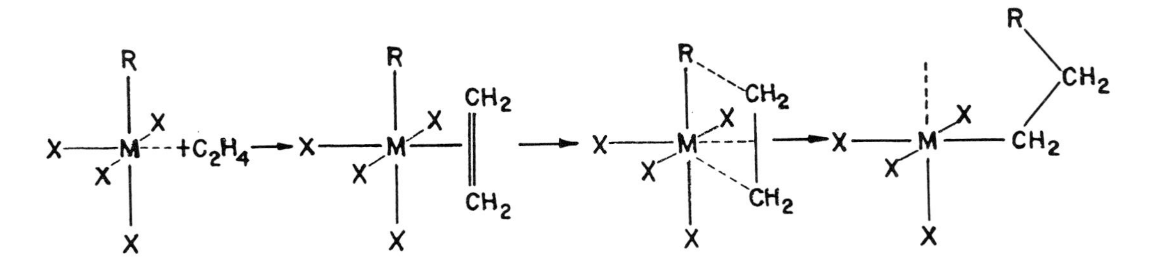

Figure 14. Proposed mechanism for the Ziegler-Natta catalytic polymerization of olefins (see Ref. 67). This mechanism is not universally accepted.

ACKNOWLEDGMENTS

I am very grateful to Dr. Richard P. Messmer of the General Electric Corporate Research and Development Center at Schenectady, New York, for his collaboration in many of the theoretical studies reported above and for his assistance in preparing this manuscript. I am also grateful to the Office of Naval Research, the National Science Foundation, Grants DMR74-15224 and GH33635, and the donors of the Petroleum Research Fund, administered by the American Chemical Society for supporting various phases of this research at M.I.T.

REFERENCES

1. M. Boudart, in Proceedings of the Robert A. Welch Foundation Conference on Chemical Research. XIV Solid State Chemistry, edited by W. O. Milligan (The Robert A. Welch Foundation, Houston, Texas, 1970), p. 299.

2. J. H. Sinfelt, in Annual Review of Materials Science, edited by R. A. Huggins, R. H. Bube, and R. W. Roberts (Annual Reviews, Palo Alto, California, 1972), V. 2, p. 641.

3. J. R. Anderson, Structure of Metallic Catalysts (Academic, New York, 1975).

4. Energy Bands in Metals and Alloys, edited by L. H. Bennett and J. T. Waber (Gordon and Breach, New York, 1968).

5. Electronic Density of States, edited by L. H. Bennett (National Bureau of Standards Spec. Pub. 323, Washington, D. C., 1971).

6. R. C. Baetzold, J. Chem. Phys. 55, 4363 (1971); J. Catal. 29, 129 (1973); R. C. Baetzold and R. E. Mack, J. Chem. Phys. 62, 1513 (1975).

7. M. Wolfsberg and L. Helmholz, J. Chem. Phys. 20, 837 (1953); R. Hoffmann, J. Chem. Phys. 39, 1397 (1963).

8. J. A. Pople and D. L. Beveridge, Approximate Molecular Orbital Theory (McGraw-Hill, New York, 1970).

9. G. Blyholder, Surface Sci. 42, 249 (1974).

10. A. J. Bennett, B. McCarroll, and R. P. Messmer, Surface Sci. 24, 191 (1971); Phys. Rev. B3, 1397 (1971).

11. D. J. M. Fassaert, H. Verbeek, and A. Van Der Avoird, Surface Sci. 29, 501 (1972).

12. L. W. Anders, R. S. Hansen, and L. S. Bartell, J. Chem. Phys. 59, 5277 (1973).

13. L. W. Anders, R. S. Hansen, and L. S. Bartell, J. Chem. Phys. 62, 1641 (1975).

14. J. C. Robertson and C. W. Wilmsen, J. Vac. Sci. Tech. 9, 901 (1972).

15. A. B. Anderson and R. Hoffmann, J. Chem. Phys. 61, 4545 (1974).

16. G. Blyholder, J. Chem. Phys. 62, 3193 (1975).

17. J. C. Slater and K. H. Johnson, Phys. Rev. B5, 844 (1972); K. H. Johnson and F. C. Smith, Jr., Phys. Rev. B5, 831 (1972).

18. J. C. Slater, in Advances in Quantum Chemistry, Vol. 6, edited by P.-O. Löwdin (Academic, New York, 1972), p. 1; J. C. Slater, The Self Consistent Field for Molecules and Solids, Vol. 4 of Quantum Theory of Molecules and Solids (McGraw-Hill, New York, 1974).

19. K. H. Johnson, J. Chem. Phys. 45, 3085 (1966); in Advances in Quantum Chemistry, Vol. 7, edited by P.-O. Löwdin (Academic, New York, 1973), p. 143.

20. K. H. Johnson, J. G. Norman, Jr., and J. W. D. Connolly, in Computational Methods for Large Molecules and Localized States in Solids, edited by F. Herman, A. D. McLean, and R. K. Nesbet (Plenum, New York, 1973), p. 161.

21. K. H. Johnson, in Annual Review of Physical Chemistry, Vol. 26, edited by H. Eyring, C. J. Christensen, and H. S. Johnston (Annual Reviews, Palo Alto, California, 1975), p. 39.

22. J. G. Fripiat, K. T. Chow, M. Boudart, J. B. Diamond, and K. H. Johnson, J. Molec. Catal. 1, 59 (1975); C. Y. Yang and K. H. Johnson (unpublished work).

23. R. P. Messmer, S. K. Knudson, K. H. Johnson, J. B. Diamond, and C. Y. Yang, Phys. Rev. B13, 1396 (1976).

24. R. P. Messmer, C. W. Tucker, Jr., and K. H. Johnson, Chem. Phys. Lett. 36, 423 (1975).

25. C. Y. Yang and K. H. Johnson, Paper Presented at March, 1976, Meeting of the American Physical Society, Atlanta, Georgia (to be published).

26. Cary Y. Yang, K. H. Johnson, and R. P. Messmer, Paper Presented at the March, 1976, Meeting of the American Physical Society, Atlanta, Georgia (to be published).

27. J. L. Carter and J. H. Sinfelt, J. Catal. 10, 134 (1968).

28. D. Vvedensky and K. H. Johnson, Paper Presented at the March, 1976, Meeting of the American Physical Society, Atlanta, Georgia (to be published).

29. J. Friedel, Can. J. Physics 34, 1190 (1956); J. Phys. Radium 19, 573 (1958).

30. D. H. Seib and W. E. Spicer, Phys. Rev. B2, 1676 (1970).

31. J. L. Carter and J. H. Sinfelt, J. Catal. 10, 134 (1968).

32. M. H. Dickens, C. G. Shull, W. C. Koehler, and R. M. Moon, Phys. Rev. Lett. 35, 595 (1975).

33. R. A. Dalla Betta and M. Boudart, in Proceedings of the Vth International Congress on Catalysis, J. W. Hightower, Ed. (North Holland Publishing Co., Amsterdam, 1973), p. 1329.

34. J. A. Tossell, D. J. Vaughan, and K. H. Johnson, Chem. Phys. Lett. 20, 329 (1973); J. A. Tossell, J. Am. Chem. Soc. 97, 4840 (1975); J. A. Tossell, J. Phys. Chem. Solids 36, 1273 (1975).

35. T. Herskovitz, B. A. Averill, R. H. Holm, J. A. Ibers, W. D. Phillips, and J. B. Weiher, Proc. Nat. Acad. Sci. U.S.A. 69, 2437 (1972).

36. C. Y. Yang, K. H. Johnson, R. H. Holm, J. G. Norman, Jr., J. Am. Chem. Soc. 97, 6596 (1975).

37. D. Case, Doctoral Thesis, Department of Chemistry, Harvard University (in progress); D. Case and M. Karplus, Chem. Phys. Lett. (in press).

38. J. G. Norman, Jr., and S. C. Jackels, J. Am. Chem. Soc. 97, 3833 (1975); J. G. Norman, Jr. (personal communication).

39. See, for example, Chem. and Eng. News, Sept. 24, 1973, p. 15; International Symposium on N_2 Fixation, Interdisciplinary Discussions, June 3-7, 1974, Pullman, Washington.

40. C. Y. Yang, Doctoral Thesis, Department of Materials Science and Engineering, M.I.T. (in progress); C. Y. Yang and K. H. Johnson (to be published).

41. F. A. Cotton, Quart. Rev. (London) 20, 389 (1966); Accounts Chem. Res. 2, 240 (1969).

42. M. R. Churchill, J. Wormwald, J. Knight, and M. J. Mays, J. Am. Chem. Soc. 93, 3073 (1971).

43. G. C. Bond, Catalysis by Metals (Academic, New York, 1962), p. 357.

44. N. Rösch, R. P. Messmer, and K. H. Johnson, J. Am. Chem. Soc. 96, 3855 (1974).

45. R. P. Messmer, in The Physical Basis of Heterogeneous Catalysis, edited by R. I. Jaffee and E. Drauglis (Plenum, New York, 1975), p. 261.

46. N. Rösch and K. H. Johnson, J. Molec. Catal. (in press).

47. G. C. Bond, in Homogeneous Catalysis, ACS Advances in Chemistry Series, edited by R. F. Gould, Vol. 70 (American Chemical Society, Washington, D. C., 1968), p. 25.

48. M. J. S. Dewar, Bull. Soc. Chim. France 18, C79 (1951); J. Chatt and L. A. Duncanson, J. Chem. Soc. 2939 (1953).

49. G. C. A. Schuit and L. L. van Reijen, Adv. Catal. 10, 242 (1958).

50. G. C. Bond and P. B. Wells, Adv. Catal. 15, 91 (1964).

51. N. Rösch and T. N. Rhodin, Phys. Rev. Letters 32, 1189 (1974).

52. F. Herman, A. R. Williams, and K. H. Johnson, J. Chem. Phys. 61, 3508 (1974); K. H. Johnson, F. Herman, and R. Kjellander, in Electronic Structure of Polymers and Molecular Crystals, edited by J. André, J. Ladik, and J. Delhalle (Plenum, New York, 1975), p. 601.

53. J. G. Norman, Jr., J. Chem. Phys. 61, 4630 (1974); Mol. Phys. (in press).

54. D. R. Salahub, R. P. Messmer, and K. H. Johnson, Mol. Phys. 31, 529 (1976).

55. V. Balzani and V. Carassiti, Photochemistry of Coordination Compounds (Academic, New York, 1970).

56. R. G. Woodward and R. Hoffmann, Accounts Chem. Res. 1, 17 (1968); The Conservation of Orbital Symmetry (Academic, New York, 1969).

57. K. Fukui, Bull. Chem. Soc. Japan 39, 498 (1966); K. Fukui and H. Fujimoto, Bull. Chem. Soc. Japan 39, 2116 (1966).

58. F. D. Mango and J. H. Schachtschneider, in Transition Metals in Homogeneous Catalysis, edited by G. N. Schrauzer (Marcel Dekker, New York, 1971), p. 223; F. D. Mango, Adv. Catal. 20, 291 (1969).

59. R. G. Pearson, Theoret. Chim. Acta (Berlin) 16, 107 (1970); Accts. Chem. Res. 4, 152 (1971).

60. T. E. Madey, J. T. Yates, Jr., D. R. Sandstrom, R. J. H. Voorhoeve, in Treatise on Solid State Chemistry, edited by N. B. Hannay (Plenum, New York, in press).

61. J. Halpern, in Proceedings of the 14th International Conference on Coordination Chemistry, Toronto, 1972, p. 698.

62. A. B. Kunz, M. P. Guse, and R. J. Blint, J. Phys. B8, L358 (1975).

63. W. G. Klemperer, J. Johnson, K. H. Johnson, and A. Balazs, Chem. Phys. Lett. (manuscript in preparation); W. G. Klemperer, J. Am. Chem. Soc. (manuscript in preparation).

64. J. Hinze, M. A. Whitehead, and H. H. Jaffe, J. Am. Chem. Soc. 85, 148 (1963).

65. K. Ziegler, E. Holzkamp, E. Breil, and H. Martin, Angew. Chem. 67, 541 (1955).

66. G. Natta and I. Pasquon, Adv. in Catal. 11, 1 (1959).

67. P. Cossee, J. Catal. 3, 80 (1964); E. J. Arlman, J. Catal. 3, 89 (1964); E. J. Arlman and P. Cossee, J. Catal. 3, 99 (1964). Note: The Cossee-Arlman mechanism is not universally accepted.